TK 9145 COL

£30 NETT

UNDERSTANDING NUCLEAR POWER

Understanding Nuclear Power

A technical guide to the industry and its processes

H. A. Cole

754

Gower Technical Press

Published by
Gower Technical Press Ltd,
Gower House,
Croft Road,
Aldershot,
Hants GU11 3HR,
England

Gower Publishing Company,
Old Post Road,
Brookfield,
Vermont 05036,
U.S.A.

British Library Cataloguing in Publication Data
Cole, H.A.
  Understanding nuclear power: a technical guide
  to the industry and its processes.
  1. Nuclear energy
  I. Title
  621.48   TK9145

Library of Congress Cataloging-in-Publication Data
Cole, H.A. (Harry A.)
  Understanding nuclear power.
  Includes index.
  1. Nuclear energy.   I. Title.
TK9146.C63   1987   621.48   87-7594

ISBN 0-291-39704-2

Printed in Great Britain at the
University Press, Cambridge

Contents

Illustrations

Figures

Tables

Preface

Part of my job involves travelling around the country presenting illustrated lectures to a wide variety of people on the subject of nuclear power. The audiences range from sixth formers, members of professional institutions and women's organizations, science teachers and members of Round Tables and Rotary clubs; in fact, a complete cross section of the general public, both young and old. It is during the inevitable question time following these lectures that I occasionally meet people who are apprehensive and sometimes downright worried about nuclear power. These are not emotionally unstable bigots or politically motivated people but nice, decent, sincere individuals who are apprehensive about certain – and sometimes all – aspects of nuclear power. In talking to these people, however, and asking them to tell me what it is that worries them, it is clear from their answers that in most instances their fears are based upon an unclear understanding of the nuclear industry and of the complex way in which energy is derived from nuclear fuel. Many who are worried about nuclear waste, for example, have no idea what it consists of, what it looks like or how it is formed, stored and administered. Others believe that this country imports raw nuclear waste from other parts of the world and having got it here then moves it around Britain as though it were domestic refuse. Some people still fervently believe that a nuclear reactor could explode like a bomb if it went out of control and that millions of people would be killed in the process.

There is, therefore, an obvious need for a book which can explain the complex subject of nuclear power in a way which can be readily understood by anyone, whether they be students, teachers, doctors, secretaries or musicians. The intention of this book is to do just that; its purpose is not to put the case for or against nuclear power but simply to say 'like it or not, this is nuclear power and this is how it works'. Having read the book you may or may not be more in favour of nuclear power but it is hoped you will be much better informed about it and therefore in a better position to participate in a meaningful debate with other rational people about its pros and cons.

H. A. Cole

Acknowledgements

In preparing this book the author gratefully acknowledges the generous and enthusiastic assistance received from the United Kingdom Atomic Energy Authority, the Central Electricity Generating Board, the National Radiological Protection Board, GEC Energy Systems, Atomic Energy of Canada, Westinghouse Electric Corporation, General Electric Company (USA), and from many colleagues at the Harwell Laboratory.

1 Britain's present and future energy requirements

In 1985 the amount of energy used in Britain was equivalent to the burning of 327 million tonnes of coal and it cost £40 000 million; an amount equal to an expenditure of £109 million a day, or £4.5 million every hour of every day! We didn't actually burn that amount of coal, in fact we burnt only 105 million tonnes of it, but if we take all the coal, oil, gas, hydro-electricity and nuclear fuels which were used and converted them into one common type of fuel – coal, for example – then in 1985 the energy used by the United Kingdom was equivalent to burning 327 million tonnes of coal (this is usually abbreviated mtce). This energy was used in lighting and heating homes and places of work and entertainment, running radios, televisions, video recorders, washing machines, deep freezers and refrigerators, preparing and transporting food, manufacturing cars, buses, trains, ships and aeroplanes, and providing the lubricants and fuel for them, manufacturing clothing and footwear, bed linen, carpets and furniture used in the home. Virtually everything we do involves the expenditure of energy in some way or other.

If the figure of 327 mtce is divided by 56 million – the approximate population of the UK – it works out to be almost 6 tonnes of coal equivalent for every man woman and child! On its own this figure is meaningless but when it is compared with similar figures for Bangladesh and the USA, for example, it is found that the figure for Bangladesh is less than 1 tonne of coal equivalent whereas it is almost 11.5 tonnes for the USA. There is therefore, an obvious link between energy consumption and

1

standard of living; *the higher the standard of living the greater is the energy required to sustain that standard.*

The expression 'standard of living' is not meant to imply bingo palaces, football stadiums and space invader machines but to convey such things as adequate shelter for one's family with adequate light, heat, food and clean water; proper sanitation and good medical attention; rudimentary schooling and a better expectation of life; all the things which any sensible person would identify as being the 'quality of life'; note that no luxuries like colour TV or central heating appear in this list! It is, therefore, impossible to make significant improvements in anyone's standard of living without, at the same time, bringing about an increase in the demand for energy.

If you have any doubts over the validity of this assertion, consider how you would go about building a badly needed hospital in some under-developed part of the world without using vast amounts of energy in manufacturing and transporting the bricks, cement, glass and steel used in its construction. Then, having 'built' it, try estimating the continuous expenditure of energy required to supply heating, lighting and food for its patients and staff, the energy required for the laundry, the sterilization plant and to operate energy intensive items such as the lifts, X-ray machines, etc. You will probably need a tanker of oil each day for the central heating alone! The list is near-endless and you will find that no matter how careful has been the implementation of energy conservation techniques the inescapable fact is that the *building and day-to-day running of the hospital has placed an additional and substantial load on the energy demands of the country in which it has been built*, even though its very existence has brought about an undeniable improvement in the standard of living of its inhabitants.

FORECASTING ENERGY REQUIREMENTS

Most people in Britain today enjoy a standard of living considerably better than that enjoyed by their parents when they were the same age. Young people are physically larger than their parents – a sure sign of plenty of food and good medical attention. People are living longer; they live in warmer and more comfortable surroundings and many own their own homes. Their jobs are less dangerous, they work fewer hours in cleaner and better conditions (coal mining is an excellent example of such progress) and they are looked after much better when they retire. Life is far from perfect but, for most of us, it was considerably less enjoyable 30 or 40 years ago.

Looking 20 to 30 years ahead it is reasonable to suppose that by then life will be much better than it is now and that generations of that period will look back on the 1980s as being a less pleasant period in which to have lived. This is as it should be; every generation has a moral responsibility to ensure that life is more enjoyable for each successive generation – always bearing in mind what is meant by standard of living and quality of life!

Should you be in any doubt about this then recall what living conditions were like in the 1920s when the 12-hour working day and 6-day week were the norm; coal miners died in their hundreds every year and worked in the most appalling conditions – no pithead baths and medical attention in those days; gas lighting and gas cookers were newly acquired luxuries; diphtheria, infantile paralysis and tuberculosis were horrifically commonplace; houses were cramped and draughty; food for many was poor and inadequate; only the rich owned cars and travelled abroad.

Compare this with life today. Almost everyone owns a car; virtually everyone owns a colour TV and more than one radio; many own their own homes and take holidays abroad. Houses are warmer and have piped hot water and central heating; indoor toilets and bathrooms are taken for granted. And yet, as everyone knows, there are still many thousands of people without homes and many thousands of homes which do not come up to a satisfactory standard. Medical facilities could be very much better and working conditions could be made safer and more pleasant. These are just the sort of things which future generations have the right to expect *but such improvements cannot be brought about without the expenditure of energy and it is the responsibility of the present generation to ensure that this energy will be available when it is needed.*

In the unhappy environment of a world recession and massive unemployment it seems crazy to be worrying about the energy demands of future generations; we surely have enough troubles of our own! It is also clear that if things stay as they are or actually get worse, there will be no need to worry about future energy demands because *there won't be any*! On the other hand, things were very much worse in 1926; there was terrible unemployment in Britain then and much poverty and illness through insufficient food; the outlook at the time was bleak indeed. And yet things did gradually improve and reached an all-time boom in the 1960s. Britain's Prime Minister of the day told us 'you have never had it so good'; and he was right! Planners of today *must* therefore assume that things are going to get better and they must ensure that future generations will have the energy resources to satisfy their higher energy demands.

It is, of course, impossible to predict with any accuracy what energy demands are likely to be 20 to 30 years ahead but we must at least try, even though we know we will be wrong in our estimates – past history shows this to be almost certain! All we can hope for is that, in being wrong, we shall have *over*-estimated the energy demand and not *under*-estimated it. In doing so we will, of course, be held responsible for bringing about a glut of energy resources and of wasting huge amounts of public money. People will write strongly worded letters to the national newspapers criticizing our incorrect forecasting but overlooking the fact that, because of our errors, they are able to do so from warm, comfortable and well-lit homes. They will also be able to criticize us in the comfort of heated buses, trains and cars which will take them to work as usual, and to read about our failures in their newspapers which will be delivered without difficulty each day; life will go on as usual with no inconveniences. If, on the other hand, we have *under*-estimated the energy demands then there will be power cuts and blackouts and fuel shortages – especially in the winter months when demand is greatest and there will be no newspapers delivered because the presses will have no electrical power. Homes will be without light and power; you will be unable to operate your TV, and the electrically operated pump in your central heating system will not operate and so your home will be cold as well as dark. And if there is a shortage of petrol then it will be rationed so that essential services can be maintained. Don't think this could not happen; it already has in the past.

Power cuts were regular occurrences during the winters of the 1950s because the power stations had neither the capacity nor the fuel (coal) to satisfy the demand for electricity, small though it was in those days of austerity. In fact, the demand on the National Grid system exceeded its capacity to supply electricity at normal frequency and voltage *every consecutive year* over the periods 1950 to 1956, 1961 to 1967 and 1969 to 1972, and once during 1973 to 1974 (*Handbook of Electricity Supply Statistics*, 1982, Table 21, page 33, Electricity Council). A few years later petrol rationing was imposed after closure of the Suez Canal in 1967, something which hadn't happened since the Second World War. The CEGB was criticized at the time for underestimating the expected demands of post-war Britain, just as it is being criticized today for having a massive 7 per cent surplus generating capacity (actually the surplus is 35 per cent but 28 per cent of this is a carefully planned surplus to take into account maintenance and unexpected breakdowns, and freak weather conditions; it is interesting to know that even this 35 per cent was insufficient to meet the demand which occurred during the televized marriage ceremony of Prince Charles and his bride and voltage reductions had to be made; no one criticized the CEGB on that day!). How inept of the CEGB not to have foreseen the nationalization of Iranian oil in 1951, or the Middle-East wars of 1948, 1967 and 1973, or the deposal of the Shah of Iran in 1979, or the

continuing war between Iran and Iraq. How naive of the CEGB to assume
that the standard of living and energy demands of the 1960s would not
continue to expand as they were doing at the time into the 1980s, and how
incompetent of its planners not to have foreseen the present worldwide
recession and massive unemployment and taken into account the reduction
it would have on its sales of electricity; after all it only takes ten years or so
to build a power station – surely they could plan that far ahead with greater
accuracy! *On the other hand, maybe forecasting future energy demands is
not as easy as it might at first appear*!

Future demands

In trying to estimate future energy requirements it is first necessary to
examine what has happened in the past and see if there are any trends or
fluctuations which might give an indication of what to expect in the future;
always bearing in mind that any information derived in this way should be
used with the utmost caution. Having studied past energy demands it is
customary to assume a range of scenarios for the future, based upon
existing energy demands and likely improvements in the economic state of
the country. Less obvious considerations include studies of how existing
energy is being used, and by whom, what conservation methods are likely
to be introduced during the period being considered, and what changes can
be expected in the style in which people live; for example, will there be
greater or less use of electrical products; is saturation of a type of product
likely thereby reducing its demand to replacement only; will more economical cars, lorries and aeroplanes be produced?

Figures 1.1 and 1.2 illustrate how the demands for electricity and
primary energy in Britain have varied over the past 35 years. (Primary
energy represents all forms of basic fuels such as coal, oil, gas, uranium etc.
Electricity is a secondary type of fuel because it is produced from a primary
fuel.) Note that despite the occasional fluctuation the trend, until recent
years, has been relentlessly upward. The downward trend in primary
energy demand in recent years is due, in part, to energy conservation but
mostly because of the recession and the closure of many energy intensive
industries such as steel plants, aluminium smelters, shipbuilding and car
assembly plants. Figures 1.3 and 1.4 illustrate the types of fuels used to
satisfy Britain's energy demands in 1985 and where they were used.
Figures 1.5 and 1.6 illustrate how the electricity was used by various sectors
of the community and also what types, and relative amounts, of fuels were
used in its generation; note the large contribution made by coal, 64.5 per
cent. The nuclear contribution of 17.7 per cent represents a fuel equivalent

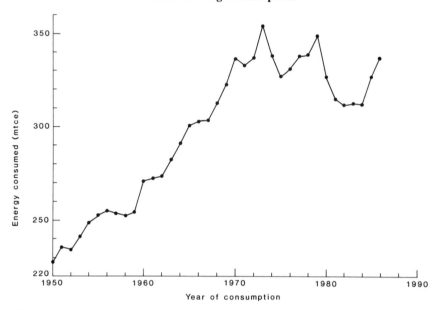

Figure 1.1   UK primary energy consumption 1950–1986 (temperature corrected)
Source: Digest of UK Energy Statistics, Dept of Energy

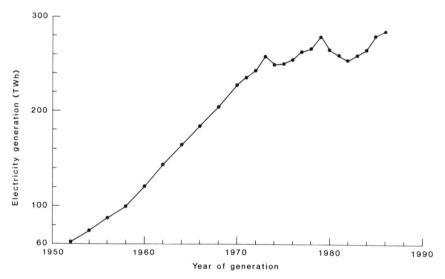

Figure 1.2   Electricity generated 1952–1986: UK public supply
Source: Digest of UK Energy Statistics, Dept of Energy

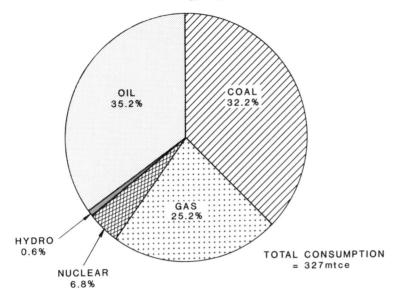

Figure 1.3   UK primary fuel consumption (1985)

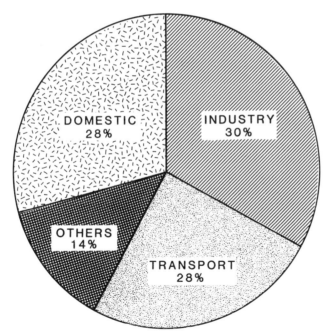

Figure 1.4   UK primary energy users (1985)

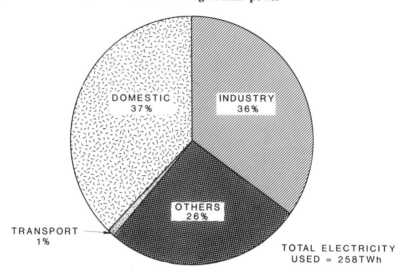

Figure 1.5   Electricity users (1985)

of over 20 million tonnes of coal; a very important contribution as far as
the generating boards are concerned because of the relatively high price of
coal. To Britain, however, this contribution is even more important from a
conservation point of view because coal is a much more valuable commodity than uranium. Uranium is an otherwise useless material, suitable only
for producing heat. Coal, on the other hand, can be used to produce heat,

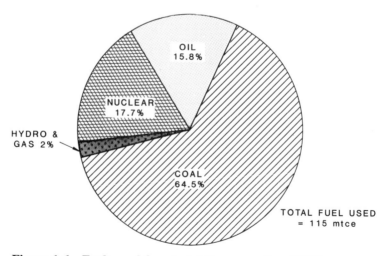

Figure 1.6   Fuels used for electricity generation (1985)

gas, oil, lubricants, fertilizers, etc., and be used as petrochemical feedstock. *Coal is the most valuable commodity Britain possesses and should never be used wastefully.*

The segments labelled 'others' in Figures 1.4 and 1.5 represent such users as schools, libraries, hospitals, government offices, theatres, insurance companies, etc.

Energy projections

The Department of Energy and a number of other organizations have in recent years published a variety of energy projections for the year 2000 and beyond and have produced widely varying figures (see Further Reading at the end of the book). Some projections are based on the assumption of a low growth rate in the country's economic wealth, that is its Gross Domestic Product (GDP), whilst others have optimistically assumed a much higher growth rate. One projection goes as far as to assume a modest growth rate without need for *any* additional energy, providing we all adopt a less energy-intensive life style and use existing energy in a more efficient way. It is not the intention of this book to praise or criticize the validity of such projections but only to explain why such projections have to be made and to comment on what has been published.

Early (1976) projections originating from the Department of Energy for the year 2000 have ranged from energy demands of 420 mtce for a low growth scenario (GDP falling to less than 1 per cent per annum), to as much as 760 mtce for a high growth scenario (GDP rising to 4.5 per cent per annum). More recent (1979) projections from the Department are for energy demands ranging from 445 mtce, for a 2 per cent per annum growth in GDP, to 510 mtce for a 3 per cent per annum growth. When compared with the figure of 327 mtce for 1985, these projections represent expected increases in energy demand ranging from 36 per cent to 56 per cent, depending upon the eventual growth in GDP. A very important constituent part of these projections is an allowance for the introduction of conservation methods which, it is hoped, will bring about a reduction in energy demand of 20 per cent; a very substantial amount. It will be interesting indeed for those of us still around in the year 2000 to see which, if any, of these projections turn out to be correct.

ENERGY RESOURCES

Estimating future energy demands is one thing; having sufficient resources available at the time to meet those demands is quite another. When compared with many other industrialized countries – France and Japan for example – Britain is in a very fortunate position with respect to energy. It is, at present, completely self sufficient in oil, it has an abundant supply of natural gas, a substantial nuclear power programme and at least 300 years' reserves of economically recoverable coal. With such good fortune it is natural to ask 'why then are we concerned about future energy demands?' The answer becomes obvious on a closer examination of the present situation; firstly, coal.

Coal

There were nearly 900 collieries in Britain in 1952; they employed 709 000 workers and produced about 228 million tonnes of coal. By the end of 1986 the number of collieries had fallen to 119, the number of workers had been reduced to 125 000 and the amount of coal they produced was only 94 million tonnes, that is 134 million tonnes *less* than that produced 30 years ago! (The figure has been affected by the tail end of the 1984/1985 miners' strike; coal production at the end of 1983 amounted to 119 million tonnes and the number of employees was 187 000.) When Sir Derek Ezra (now Lord Ezra) was the Chairman of the National Coal Board (NCB) he produced a corporate plan entitled 'Plan for Coal' (1974). In it he foresaw a progressive increase in the rate of coal extraction which he hoped would reach 170 million tonnes by the year 2000; an amount which, if achieved, is still 58 million tonnes less than that produced in 1952. Coal alone, therefore, could never satisfy the energy demands expected for the year 2000.

Oil and gas

Although impossible to predict with certainty, official estimates show that Britain's reserves of natural oil and gas will start to decline before the year 2000 and, at the current rates of consumption, will be fully depleted before the year 2010; then Britain will once more have to assume the role of a net oil importer. It is confidently predicted that most people alive today will witness the decline of Britain's oil and gas. Of course, gas could once again be produced from Britain's abundant coal reserves as it was for many decades before North Sea gas was discovered; this was the so-called 'town gas'. Also there is nothing unusual about producing oil from coal; South Africa derives most of its oil from coal liquefaction and considerable research is devoted into the process in Britain. The process is at present very expensive, however, and it is unlikely that sufficient coal would be available at the time of need. Home-produced oil and gas could not therefore possibly meet Britain's expected increase in energy demand for the year 2000.

Renewable energies

The total energy available in Britain from the so-called 'renewables' (sun, waves, wind, tides, geothermal heat, etc.) is enormous; far more than would ever be needed, no matter how wastefully it was used. Energy from the sun alone could, if converted into a more useful form, supply Britain with more than 80 times its present energy requirements. Unfortunately, it is a very low-grade diffuse form of energy and, quite apart from its unpredictability, is completely out of phase with maximum demand. Insolation in Britain (energy received from the sun) is typically 10 times weaker in mid-winter, when it is wanted most, than in mid-summer. Nevertheless, it is still a very useful form of energy and considerable progress has been made in both Britain and other parts of the world in the design and installation of solar panels – in which the sun's heat is used to supplement domestic hot water supplies – and photovoltaic solar cells in which the sun's light is converted directly into electricity; solar cell arrays are used to provide power in spacecraft and, for example, in unattended radio beacons.

Waves

Wave power, unlike solar power, is in phase with demand and much more predictable. On the other hand, although it is a relatively simple matter in the laboratory or on a lake to convert wave motion into mechanical motion and thence into electrical power, it is going to be very difficult to do the same thing at sea with huge structures weighing many hundreds of tonnes without them being destroyed or detached from their moorings during severe storms and mountainous seas. Preventing the ingress of corrosive salt water into the electrical generating plant of the structures is another problem being considered. Maintenance costs are likely to be high and there will almost certainly be environmental objections to the siting of huge, unattractive structures along many kilometers of the most attractive part of Britain's coastline where wave power is most prolific. Off the North West coast of Scotland, for example, waves of 3 metre height regularly appear at intervals of about 10 seconds and with a predictability of better than 70 per cent. Such waves have an electrical energy equivalence of about 50 kW per metre length and could, if enough wave-converting structures were installed, supply a substantial proportion of Britain's electricity demand. Unfortunately, this would involve the installation of structures which would not only spoil the appearance of the area but may also have a serious impact on the local ecology; waves reaching the shore are likely to be less than 20 per cent of their normal height and this reduction is expected to have a deleterious effect on the provision and mixing of nutrients for marine life. It is because of this that marine biologists are studying the effects of wave power devices on marine life.

The potential energy available from the waves around Britain's shores has never been in doubt. Designing and installing secure and reliable devices which will produce electricity at an economical price and make little adverse impact on the environment are more serious problems which, at present, appear to be a very long way from being resolved.

Wind

Wind power, like wave power, is in phase with demand but is much less predictable. On the other hand, the operating environment of an aerogenerator (wind-powered electrical generator) is likely to be much less hostile than comparable wave-powered devices and much easier to install and maintain. On the debit side there are very few suitable areas in Britain where the wind is strong enough and frequent enough to justify the expense of installing an array of aerogenerators. For 30 miles (48 km) or so around London, for example, there is 50 per cent annual calm and when the wind *does* blow its mean velocity is little more than 9 knots (16 km/ hour or 10 mph); so, quite apart from the environmental impact, there would be little point in installing aerogenerators in the London area since, for half the time, they would be completely idle.

Another disadvantage of wind energy is its low power density – typically only 100 W per square metre for a wind of 10 knots – which means that large areas of land are needed to accommodate the many aerogenerators required to produce a given power output. The disadvantage is made worse by the large separation distances required between adjacent aerogenerators so that one device does not 'steal' useful wind from its neighbour. For example, a 1000 MW wind-powered generating station would require 400 2.5 MW aerogenerators spaced over an area measuring 7.8 km × 7.8 km (12½ miles × 12½ miles); an area larger than the whole of the Isle of Wight! In comparison, a coal, oil, or nuclear-powered station of the same generating capacity would occupy an area little larger than that of Trafalgar Square.

In spite of these disadvantages wind power holds much promise for the future and considerable research is currently devoted to developing large aerogenerators which would be able to feed significant amounts of electrical power into the National Grid. This is no idle statement; a 200 kW aerogenerator was installed on land owned by the CEGB at Carmarthen Bay in December 1982, since when it has been feeding power into the

'Grid'. This generator will provide valuable operating experience and experimental data for the much larger (1 MW) aerogenerators which the CEGB is having designed.

In Scotland a 250 kW aerogenerator has been operating at Burgar Hill in Orkney since 1983 and providing data which was used in the design of a 3 MW (60 m diameter) aerogenerator which began operating in 1986 on the same site. The 3 MW machine was built by The Wind Energy Group (WEG) for the North of Scotland Hydro Electricity Board (NSHEB) who also own the site. Orkney represents an excellent case for the use of wind power; the wind is almost constant and strong enough to operate the most powerful aerogenerators yet devised. The 3 MW aerogenerator, although initially very expensive, will eventually pay for itself by the savings made from no longer having to make frequent and expensive deliveries of fuel oil to the island for the present gas-turbine generating set; the most expensive way of generating electricity!

Tides

Tidal power is a well-proven technology; the French have had a 240 MW tidal generating station at La Rance, Brittany, since 1966. In Britain there has been interest for more than 100 years in harnessing power from the river Severn by means of a barrage. However, several factors have prevented any commitment to constructing such a barrage: the enormous cost and long time-scale involved, the non-continuous nature of the power which it would produce, and, until recent years, the availability of much cheaper ways of generating electricity. All that is likely to change, however, and the publication of the Severn Barrage Report in 1986 gives hope that a structure may eventually be constructed close to the islands of Steep Holm and Flat Holm in the Severn estuary. The water retained by the proposed barrage, large though this would be, would still be insufficient to provide a continuous source of supply, and electrical power would remain dependent on the frequency of the tides. The report estimates that the proposed barrage could supply about 6 per cent of Britain's present electricity demand. This may not sound a lot but bear in mind that, once operating, the station would require no fuel deliveries and would therefore be independent of disruptive influences of Middle-East wars, strikes and adverse weather conditions. Remember also that, of all the renewable sources of energy discussed so far, tidal power is the only one which is absolutely predictable and guaranteed. A total of 17 potentially suitable sites in Britain have been assessed in a preliminary fashion for tidal power, some of which show good promise. The Mersey estuary, for example, appears to be a very good site for a tidal barrage. In addition to its good tidal range (nearly 8 m mean for spring tides) the river discharges into

Liverpool bay through a narrow deep channel which can be closed with a relatively short barrage less than 2 km long. A barrage located at this point would be about one-tenth the size of the Severn estuary scheme and could supply up to 1.2 terrawat-hours per year (TWh/y); equivalent to about 0.5 per cent of the combined electricity demand of England and Wales. Two sites are currently being examined, one between New Brighton and Bootle, the other between Birkenhead and a point close to the Garden Festival site. It is estimated that the proposed barrage would have an installed generating capacity of about 550 MW.

A detailed feasibility study of the Mersey site was initiated in October 1986 and was undertaken on behalf of the Mersey Barrage Company, a consortium of 17 British companies formed in February 1986 for the purpose of exploiting tidal power from the Mersey estuary as a commercial enterprise. Over half the cost of the £700 000 study was funded by the Department of Energy.

Geothermal heat

The existence of geothermal heat (heat from within the earth) is well known, as any coal miner will confirm; the deeper you go into the earth the warmer it becomes. Heat from the earth appears in the form of hot dry rocks or as aquifers which sometimes allow hot water (and sometimes steam) to escape to the surface. The warm water at Bath Spa is a good example of geothermal heat. Rain from the surrounding hills penetrates deep into the earth where it is heated before resurfacing at Bath. Research in recent years has shown the existence of vast amounts of subterranean hot water in various parts of Britain, all much greater than that in the Bath basin, and considerable research is being undertaken at the Cambourne School of Mines in Cornwall to see how heat from hot dry rocks may be extracted for useful purposes.

Geothermal heat is a potentially valuable source of energy and, in the long term, is likely to make a significant contribution to Britain's future energy demands. It is unaffected by storms, it is continuously available, its impact on the environment will be minimal and its associated apparatus is likely to require little maintenance. It is not, however, a truly renewable energy source because once the heat has been extracted it will take many decades for it to be replenished; this is because of the slow movement of heat due to the low thermal conductivity of the Earth's rocks.

Hydro-electricity

Natural-flow hydro-electricity (this is not the same as hydro-electricity generated from pumped storage) is one of the most well known forms of renewable energy sources, and one which has been successfully exploited in many parts of the world. Unfortunately, Britain does not have the sort of topography needed for hydro-electric generation and those areas which are most suitable have been almost completely exploited. At the time of writing, hydro-electricity contributes less than 2 per cent towards Britain's total supply and there is little hope of much improvement on this figure. This is a pity because hydro-electricity is cleaner and cheaper than *any other present form of generation* in Britain; it requires no deliveries of fuel or removal of waste products, it causes no pollution of the environment and its power output is continuous and guaranteed.

Conservation

Much has been said so far about the ability of coal, oil, gas, nuclear and the so-called 'renewables' to satisfy present and future energy demands in Britain but there is one thing which, theoretically, could comfortably satisfy all medium-term energy demands without the need for any additional power stations and without using any more energy than is being used at present; this is, of course, *conservation*, the most valuable commodity of all. Thankfully, Britain cannot claim to be the most wasteful user of energy in the world today although, sadly, it also cannot claim to be among the most efficient of users. The main reason for this is because the money Britain has to pay for its energy is much less than that paid by those less fortunate countries which do not have Britain's abundant reserves of coal, oil and gas. To give some idea of what could be achieved by energy conservation and, at the same time, to point out the enormous difficulties which either prevent or delay the implementation of an effective conservation policy, consider the generation of electricity.

In 1982 the National Coal Board extracted approximately 120 million tonnes of coal from Britain's collieries. Almost 80 per cent of this, that is 96 million tonnes, was sold to the power stations for the production of electricity. Some of the modern power stations have thermal efficiencies as high as 42 per cent whereas some of the much older stations have efficiencies as low as 28 per cent. The thermal, or steam-cycle efficiency of a power station is expressed by the ratio:

$$\text{thermal efficiency} = \frac{\text{amount of electricity generated}}{\text{corresponding amount of heat generated}} \times 100\%$$

The average efficiency figure, calculated over all stations, is approximately 33.3 per cent, that is one-third. This means that of the 96 million tonnes of coal burnt by the generating stations only one-third was actually used to generate electricity; the remainder – 64 million tonnes – was thrown away in the form of waste heat in the flue stack and cooling towers. In other words *a little over half of all the coal mined in Britain in 1982 was thrown away in the form of waste heat*. These 64 million wasted tonnes of coal represent about 20 per cent of Britain's total energy consumption for 1982 (317 mtce) and a substantial proportion of the additional energy demands expected towards the end of the century. Clearly, if this waste heat could be used, the energy problems anticipated for the future would be considerably reduced. However, things are seldom as simple as they at first appear; power station designs are optimized for the production of electricity, not hot water, and although two-thirds of the available fuel is lost as waste heat this heat is in the form of luke warm water and is therefore not recoverable. The temperature of the water at the foot of a cooling tower, for example, is little more than that of a heated swimming pool and well below that which could be used for domestic hot water. The enormity of the loss comes about because of the vast amount of cooling water used; many millions of gallons per hour for a large power station.

Power stations can, of course, be designed to reject heat at a much higher temperature and to operate less efficiently as an electricity producer; this was done on a small scale in Britain for many years (Battersea power station in London, for example), and on a much grander scale in Scandinavia. Under such operating conditions the rejected heat – usually in the form of steam – is transported through thermally-insulated pipes to homes and places of work, etc., by way of calorifier stations which transform the steam to water at temperatures of around 60°C. This arrangement ensures that someone living near to the power station receives water at the same temperature as someone living some distance away. Using a power station in this way is known as 'combined heat and power' (CHP) and with such a station its overall operating efficiency can be typically 70 per cent; a very worthwhile improvement and a much better utilization of its valuable fuel. CHP is not more widely used because when most of the existing 170 power stations were built, very few homes had a central heating facility and those which did were quite happy to make use of what was then *cheap* Middle-East oil. It is too late now to convert the large modern power stations to operate at much lower steam temperatures although the construction of new stations could be designed for CHP operation. Another important reason why CHP has not been more widely adopted is the enormous installation costs: if CHP was used in its fullest sense then every home would have to be supplied with a thermally insulated pipe (there are nearly 20 million dwellings in Britain!) and virtually every street in the entire country would have to be dug up to accommodate thermally-insulated buried pipes; an expensive undertaking by anyone's standard, bearing in mind that for obvious reasons power stations are located some distance from built-up areas. A more reasonable compromise might be to site smaller and less obtrusive power stations close to new towns, shopping precincts and factory complexes where they could be operated in the more efficient CHP mode.

Conservation makes good sense and is a worthwhile investment for the future. Unfortunately its implementation on the scale needed would be prohibitively expensive and could not be completed in the short time available.

Nuclear power

Nuclear power, being the main subject of this book, has been left to last in this short review of potential energy sources and will be examined in greater detail in the pages to follow. When nuclear power is compared with other sources of energy it is found to have the following important features:

1   *It is a well-established technology* both in Britain and in 32 other countries throughout the world. The first nuclear reactor in Britain was built nearly 40 years ago and is still working 24 hours a day.
2   *It is a safe technology.* During nearly 40 years of operating nuclear reactors in Britain *not one* member of the public has come to any proven harm through nuclear causes and only a few employees within the industry have lost their lives through what *might have been* nuclear causes. With an average mortality rate of about one death every eight years it is among the very safest of British high technology industries.
3   *It is inexpensive.* According to information published by the CEGB, electricity generated by the modern (AGR) nuclear stations is very much cheaper than that generated by modern coal and oil-fired stations and many millions of pounds are saved each year as a consequence.
4   *It saves coal and oil.* For every tonne of otherwise useless uranium burnt in a nuclear power station, 20 000 tonnes of much more valuable coal and oil are saved for future energy demands.

Despite these attractive features nuclear power stations are extremely complicated devices and therefore take much longer to build than do coal or oil-fired stations of the same size; they are also about 50 per cent more expensive to build. Once completed, however, the amount of fuel required by a nuclear station is only a small fraction of that required by a similarly-sized coal or oil station. A coal-fired station of 2000 MWe rating consumes about 20 000 tonnes of coal per day (5 bags every second!) and requires a train delivery every hour, 24 hours a day.

Nuclear power could never, in the short time available, provide all of the additional demand for electricity expected by the year 2000, bearing in mind that many of the early nuclear stations are already more than 30 years old and will need to be replaced within the next ten years or so. Projections by the Department of Energy expect nuclear power to contribute no more than about 50 per cent toward Britain's electricity demand by the year 2000 and that the remainder will have to be supplied mostly by coal, with some small contribution from oil and even smaller contributions from hydro-electricity and the less well established renewables.

Table 1.1 summarizes the main features contained in the publication *Energy projections for the year 2000*, published by the Department of Energy in 1979. The spread in the values given indicates the degree of uncertainty in the assumptions which had to be made regarding changes in economic growth and in the likely availability of the various energy sources. The demands take into account an allowance of 20 per cent for the implementation of conservation measures and also some small (2 per cent) contribution from the introduction of CHP. The lower figures assume an economic growth rate of about 2 per cent per annum, the higher figures representing about 3 per cent per annum. The most important figures to

Table 1.1
UK primary energy balance for the year 2000

Estimated demand		mtce
		445–510
Available supplies		
coal		137–155
gas		62– 65
oil		100
nuclear + hydro		88– 95
	Total	390–410
Energy gap		35–120

N.B. Figures do not add vertically

note from the table are the estimated short falls in energy, the so-called energy gap. This is the amount of energy which Britain would have to import in the way of oil, coal, liquefied gas, etc., from other countries in order to sustain the anticipated demand, at a time when there is likely to be a world shortage of energy; only the more prosperous countries will be able to support a high standard of living.

According to the projections the anticipated energy gap could be as low as 35 mtce, or as high as 120 mtce (a little more than Britain's total oil consumption in 1985); an average of 78 mtce. The anticipated demand could be as low as 445 mtce (a 30 per cent increase on the 1985 demand) or as high as 510 mtce (a 56 per cent increase on the 1985 demand; an average demand of 478 mtce (a 46 per cent increase on the 1985 demand)

REFERENCES

Baker, A. C., 'The Development of Functions Relating to Cost and Performance of Tidal Power Schemes and their Application to Small-scale Sites', Paper No. 16 to Institution of Civil Engineers Symposium, London, October 1986.

Tidal Power from the Severn, The Severn Tidal Power Group, Thomas Telford Ltd, 1986 (3 volumes).

2 History of nuclear power

No one individual can lay claim to having invented nuclear power. Like space flight, heart transplants and colour television, its manifestation was the culmination of the work of many dedicated scientists scattered throughout the world. It was the brilliant and, at times, seemingly outrageous theories and painstaking research of such people which pushed back the frontiers of scientific knowledge until, in 1942, the world's first nuclear chain reaction was demonstrated.

THE DISCOVERY OF RADIOACTIVITY

The story starts, if such a starting point exists, with the discovery of X-rays by Wilhelm Röntgen in 1895, quickly followed one year later by the accidental discovery of radioactivity by Henri Becquerel. As if providing the key to Pandora's box, there then followed in rapid succession some quite momentous discoveries and the creation of scientific theories which transformed the approach to the study of matter and of atomic physics in particular. In 1897 J. J. Thompson discovered the tiny particle known as the electron, its existence having been predicted by other scientists to account for some observed phenomenon. The following year, Marie Curie and her husband Pierre discovered the two radioactive elements polonium and radium. It was the work of Becquerel and the Curies which led the brilliant Ernest (later Lord) Rutherford to devote the rest of his life to studying all aspects of radioactivity and to explaining and naming what is

21

known as alpha and beta radiation. Rutherford was also responsible for proposing the nuclear structure of the atom and, in 1919, for being the first person to split the atom; in so doing he transformed one element into another (in this instance nitrogen into oxygen). During the 1920s Rutherford also postulated the existence of the neutron particle but it was 1932 before James Chadwick detected it in the world famous Cavendish laboratory at Cambridge.

Some years before the detection of the neutron, Frédéric Joliot (working with his wife Irène, the daughter of Marie Curie) mixed quantities of beryllium and radium together and, unknowingly at the time, created the world's first portable neutron source. What happened was the alpha particles emitted by the radioactive radium bombarded the non-radioactive beryllium atoms and caused them to emit what at the time was called 'beryllium radiation' but which, three years later, was identified by Chadwick as being neutron radiation. The discovery of the neutron is especially important since it is this particle which actually initiates and sustains the nuclear chain reaction so vital to nuclear power. It is also this particle which is found in one in every 5000 atoms of hydrogen and which causes the existence of heavy water, a substance which has played such an important role in the development of nuclear power. This particular form of 'heavy hydrogen' is known as deuterium and was discovered in 1932 by Harold C. Urey. When deuterium is mixed with oxygen it forms deuterium oxide (D_2O) otherwise known as 'heavy water'.

Other famous scientists of that era included Frederick Soddy (who introduced the idea of isotopes), C. T. R. Wilson (who built the first cloud chamber), the brilliant Niels Bohr (who perfected the atomic theory) and, of course, the incomparable Albert Einstein whose outstanding mind played so great a part in the work undertaken by every nuclear physicist.

Without intending to belittle the vital contributions made by many other scientists in these exciting times the next logical step in this potted history is the work undertaken by the Italian physicist Enrico Fermi. In 1934 he and his associates were experimenting with the use of neutrons as atomic projectiles instead of the alpha particles which were used by Rutherford to transmute nitrogen into oxygen and, in 1933, by Frédéric and Irène Joliot to produce artificial radioactivity (they were the first to produce radioisotopes). Three extremely important discoveries were made during these experiments, two of which made a vital contribution to the development of nuclear power.

The first discovery by Fermi was that a wide range of stable elements could be made radioactive (converted to radioisotopes) simply by subjecting them to neutron radiation, a discovery which many years later made Amersham International so successful! The second discovery was that the effectiveness of the neutrons in producing a radioisotope could be improved dramatically by slowing down, that is moderating, the speed of the neutrons before they reached the target material. This experiment demonstrated the importance of neutron moderation and later made possible the success of the world's first nuclear reactor.

The third and most important discovery was that uranium atoms, when irradiated with neutrons, could be made to split into two much lighter fragments of nearly equal size, each fragment being the nucleus of a completely different element; it was also found that each such fission event was accompanied by two, and sometimes three, neutrons. This, although Fermi didn't know it at the time – he didn't even realize that he had actually split the uranium atom – could be rated as the most important and most exciting discovery of all time in nuclear physics. Without it nuclear power would never have evolved; it would be difficult indeed to over-estimate its importance.

Some four years later, in 1938, the German scientists Otto Hahn and Fritz Strassman repeated Fermi's experiments with uranium; like him, they also found the unexpected presence of elements of a type completely unrelated to uranium but which always appeared in the target material after it had been irradiated with neutrons. Instead of dismissing this find, as Fermi had done four years earlier, they discussed it with the physicist Lise Meitner who, at that time, was working with Niels Bohr in Copenhagen. She discussed it with her nephew Otto Frisch and eventually put forward the theory that the bombarding neutrons must have caused some of the uranium atoms to split into two nearly equal fragments and that it was these 'fission fragments' which were being detected in the target material. The likelihood of such a phenomenon had never been considered before, but a few simple calculations showed that the total weight of the fragments (the new elements formed), plus that of the additional neutrons emitted during the experiments, was indeed almost identical to that of the uranium atoms which were assumed to have been destroyed in the process. The word 'almost' is important because the calculations of Meitner and Frisch had shown that the combined mass of the fission products was always slightly less than that of the uranium from which they were formed. They further proposed that this missing mass had been converted into energy exactly as Einstein had said was possible more than 30 years earlier. This was the first practical demonstration of Einstein's brilliant theory and it caused a flurry of excitement when Bohr visited the USA in 1939 and discussed it with the Master himself. Within a matter of days of the theory being announced, the work of Hahn and Strassman had been confirmed in four laboratories in the USA and later in France and Denmark.

The most important feature of the fission process as far as nuclear power is concerned is the fact that two, and sometimes three, neutrons are produced for every fission-producing neutron. The fact that more neutrons are produced than are used in the fission process had not gone unnoticed and scientists of the day were quick to realize that with a suitable neutron moderator and a large enough quantity of uranium it should be possible to initiate a fission chain reaction which, if properly controlled, would be self sustaining. A problem was that the presence of any impurities in the moderator, or in the uranium itself, might be sufficient to absorb those all-important fission neutrons which ought to have been captured by other uranium atoms, causing them to fission; the reaction, assuming it could be started, might just 'fizzle out'. Considerable research was therefore necessary before such an assembly could be built. A suitable material had to be found for the moderator and preliminary experiments were required to determine how much uranium would be needed to sustain the reaction; always assuming that such a quantity would be available! In parallel with this work, other experiments were necessary to determine how much impurity could be tolerated in the moderator and the uranium and, of vital importance, how the whole assembly could be controlled – especially if the reaction turned out to be much greater than had been thought possible! It transpired that more than 30 experimental assemblies had to be built before sufficient information became available but eventually a final design was produced and it was decided to construct the assembly at the University of Chicago; it was November 1942.

THE FIRST REACTOR

The world's first nuclear reactor – or atomic pile as it was then called – was successfully demonstrated in a squash court at the University of Chicago on 2 December 1942. It used graphite (a crystalline form of carbon) as the moderator and approximately 2 tonnes of uranium fuel. It could hardly be called a fully engineered device since it was little more than a pile of graphite blocks – hence its early name – interleaved with rods of uranium and with an emergency control rod suspended by a rope; a man standing by with an axe would, in an emergency, cut the rope and allow the control rod to drop into the core of the reactor. This rod, being made from cadmium (a high neutron absorber) would soak up all the surplus neutrons and would shut down the reactor; fortunately it was never needed. The operation of this reactor conclusively demonstrated the feasibility of producing a controlled chain reaction using uranium and confirmed the theoretical predictions of Enrico Fermi and his team who designed and constructed the reactor; a truly memorable occasion.

Figure 2.1   Calder Hall nuclear power station

25

The first fully engineered nuclear reactor to appear in western Europe was constructed at what is now the United Kingdom Atomic Energy Authority's Atomic Energy Research Establishment (AERE) at Harwell, near Oxford, in August 1947. Known as GLEEP (Graphite Low Energy Experimental Pile), this historic reactor has operated continuously, 24 hours a day, for over 40 years and is still in daily use for scientific experiments.

The first nuclear powered electricity generating station in Britain was built by the UKAEA in 1956 at Calder Hall, Cumbria, where it continues to supply electrical power to the National Grid after more than 30 years of operation. This station (Figure 2.1) is now owned and operated by British Nuclear Fuels plc (BNFL), a commercial subsidiary of the Department of Energy.

The first nuclear power station to be owned and operated by the electricity generating boards was built for the Central Electricity Generating Board (CEGB) at Berkeley, Gloucestershire, in 1962. In 1986 there were in Britain 16 nuclear power stations either in operation or undergoing commissioning trials, two more under construction and two research reactors which were also generating electricity (see Figure 2.2 and Table 2.1). The contribution made by these stations towards the total amount of electricity generated in Britain was about 21 per cent (1986) but this is expected to rise to about 25 per cent when all of the stations are operating.

WORLD NUCLEAR POWER (from the IAEA Bulletin, No. 1, 1987)

At the end of 1986 there were 396 nuclear power reactors, with a combined generating capacity of 272 315 MWe (Megawatts electrical), operating in 26 countries throughout the world. They made a 15 per cent contribution towards the world's electricity demand and between them have accumulated over 4200 reactor-years of operating experience. A further 137 reactors are at present under construction, the planned generating capacity of which is almost 122 000 MWe.

Table 2.2 lists the nuclear generating capacity of 12 countries arranged in descending order of installed capacity. Note that although France derives almost 65 per cent of its electricity by way of nuclear power (the highest in the world), its installed capacity is less than half that of the USA which derives less than 16 per cent of its electricity this way. Put another way, the installed nuclear generating capacity of the USA is greater than the combined nuclear generating capacity of France, Finland, Spain, Belgium, West Germany and the United Kingdom!

● **Nuclear Power Stations**
■ **Nuclear Fuel Sites**
▲ **R and D Engineering Establishments**

Figure 2.2   Britain's nuclear power industry

Table 2.1
Britain's nuclear reactors

MAGNOX stations	Date of commissioning	Nett capability
		MW sent out
Calder Hall*	1956	200
Chapelcross*	1958	200
Berkeley	1962	276
Bradwell	1962	250
Dungeness 'A'	1965	410
Hinkley Point 'A'	1965	430
Hunterston 'A'	1964	300
Oldbury on Severn	1967	416
Sizewell 'A'	1966	420
Trawsfynydd	1965	390
Wylfa	1971	840
AGR stations		Nominal capacity
		MW
Hinkley Point 'B'	1976	1320
Hunterston 'B'	1976	1320
Dungeness 'B'	1984	1200
Hartlepool	1981	1320
Heysham 'A'	1981	1320
Heysham 'B'	Late 1980s	1320
Torness	Late 1980s	1320
Other stations		Gross capacity
		MW
SGHWR, Winfrith*	1967	100
PFR, Dounreay*	1975	250

In this table, the reactors marked with an asterisk* were built as UKAEA experimental reactors. The others were built by British industry, on the basis of UKAEA designs, for the Generating Boards to supply electricity commercially to the National Grid.

Table 2.2
Nuclear power stations in operation at end of 1985

Country	Installed capacity (MWe)	% share of total electricity generated
1 USA	77 804	15.5
2 France	37 533	64.8
3 USSR	27 756	10.3
4 Japan	23 665	22.7
5 W. Germany	16 413	31.2
6 UK	10 120	19.3 (21% 1986)
7 Canada	9776	12.7
8 Sweden	9455	42.3
9 Spain	5577	24
10 Belgium	5486	59.8
11 Switzerland	2882	39.8
12 Finland	2310	38.2

CHRONOLOGICAL SUMMARY

1798   Klaproth discovers uranium
1895   Wilhelm Röntgen discovers X-rays
1896   Henri Becquerel discovers radioactivity
1897   J. J. Thompson identifies the electron
1898   Pierre and Marie Curie discover polonium and radium
1901   Max Planck introduces his quantum theory
1913   Niels Bohr introduces his model for the radioactive atom
1915   Albert Einstein introduces his theory of relativity
1919   Ernest Rutherford transmutes nitrogen into oxygen using alpha
         particles; the first person to split the atom
1920   Rutherford postulated the existence of the neutron
1923   A. H. Compton confirms the photon (particle) nature of X-rays by
         studying the 'Compton effect'
1928   Paul Dirac postulates the existence of the positron
1930   Ernest O. Lawrence invents the cyclotron
1932   J. D. Cockcroft and E. T. S. Walton use a particle accelerator (atom
         smasher) to bombard a lithium target with protons and so pro-
         duce the first case of nuclear disintegration by artificial means

1932   James Chadwick discovers the neutron

1932   Harold C. Urey discovers heavy hydrogen (deuterium)

1932   Carl D. Anderson detects the positron postulated by Dirac

1934   Frédéric and Irène Joliot use alpha particles to produce radio-isotopes

1934   Enrico Fermi uses neutrons to produce radioisotopes

1934   Enrico Fermi bombards uranium with neutrons and produces neptunium; he also becomes the first person to cause uranium fission but does not realize the fact

1938   Otto Hahn and Fritz Strassman repeat Fermi's experiments with uranium and also cause nuclear fission but do not recognize the fact; they discuss their results with Lise Meitner

1938   Lise Meitner and Otto Frisch put forward the theory of nuclear fission after studying the experimental results of Hahn and Strassman; they also assert that mass has been converted into energy, as Einstein said it could

1939   Niels Bohr visits Einstein in the USA to discuss nuclear fission

1939   Laboratories in the USA, Denmark and France independently confirm the phenomenon of nuclear fission

1939   Walter Zinn, Leo Szilard, Enrico Fermi, Herbert Anderson and H. B. Hanstein publish work on neutron yield from uranium fission

1940   Glen Seaborg discovers plutonium

1940   Uranium-235 isotope separated from uranium-238

1942   Enrico Fermi demonstrates the world's first atomic pile (nuclear reactor) in Chicago on 2 December

1945   Detonation of world's first atom bomb, code-named Trinity, at the Alamogordo bombing range, 150 miles North of Los Alamos, on 16 July

1945   First use of a nuclear weapon when atom bomb, code-named Little Boy, was dropped on Hiroshima on 6 August

1945   Second use of nuclear weapon when atom bomb, code-named Fat Man, was dropped on Nagasaki on 10 August

1946   Atomic Energy Research Establishment (AERE) set up at Harwell on 1 January, headed by Sir John Cockcroft

1947   Western Europe's first nuclear reactor constructed at AERE Harwell; known as GLEEP (Graphite Low Energy Experimental Pile), it was completed in August

1947   Springfields fuel manufacturing plant produces its first uranium

1951   Windscale (Sellafield) reprocessing facilities completed

1953   Capenhurst gaseous diffusion plant completed

1954 Control of atomic energy transferred from Ministry of Supply to Department of Atomic Energy. UKAEA comes into being in August; first chairman Sir Edwin Plowden

1956 Britain's first nuclear power station commissioned at Calder Hall, Cumbria; built for UKAEA

1957 The Windscale reactor fire

1959 The first Scottish nuclear power station officially opened at Chapelcross

1959 Dounreay Experimental Fast Reactor (DFR) begins operating

1962 Windscale Advanced Gas Cooled Reactor (WAGR) begins operating

1962 Britain's first commercial nuclear power station begins operating at Berkeley, Gloucestershire, for the CEGB

1968 Britain's only power-producing heavy water reactor operates at full power; known as SGHWR (steam generating heavy water reactor)

1970 Non-proliferation Treaty comes into force

1971 National Radiological Protection Board (NRPB) set up

1971 British Nuclear Fuels Limited (BNFL) set up

1972 2000-million year old Oklo reactor discovered in Gabon

1975 Prototype Fast Reactor (PFR) at Dounreay begins generating electricity

1975 The Browns Ferry reactor fire

1976 First commercial AGR reactor in Scotland commissioned at Hunterston

1977 DFR shut-down for good

1978 Joint European Torus (JET) fusion project begins at Culham, Oxfordshire

1979 Accident at Three-Mile Island nuclear power station

1981 Tamuz reactor in Iraq severely damaged by air attack

1982 Nuclear Industry Radioactive Waste Executive (NIREX) set up

1983 Sizewell PWR inquiry opens under leadership of Sir Frank Layfield, QC

1984 Sinking of the French ship Mont Louis carrying 350 tonnes of uranium hexafluoride (hex)

1985 Unit-2 reactor restarted at Three-Mile Island power station

1986 14-tonne container of hex ruptured at the Webbers Falls plant of the Sequoyah Fuel Corporation, Oklahoma

,1986 The Chernobyl reactor accident in USSR

1986 Sizewell inquiry gives go-ahead for construction of PWR

3 The meaning of radioactivity

Before examining the nuclear industry in depth it is first necessary to understand what is meant by the term 'radioactivity'. Consider the following simple analogy:

When a once-active volcano has given up all its energy it is said to be extinct, or inactive. Such a volcano is passive, absolutely stable and no longer a danger to anyone. An active volcano, on the other hand, is at times (which are unpredictable) highly unstable, during which it gives off vast amounts of energy; usually in the form of hot lava which can be very dangerous to anyone who gets too close to it. Some active volcanoes are more unstable than others and hence more dangerous. All active volcanoes become less active as time goes by and eventually become stable, that is inactive. This may take (in geological terms) a short time or a long time; no two active volcanoes are alike.

A non-radioactive substance is said to be inactive. Such a substance is passive, absolutely stable and presents no danger to anyone. A substance which is radioactive, on the other hand, is at times (which are unpredictable) highly unstable, during which it gives off vast amounts of energy, in the form of nuclear radiation which can be very dangerous to anyone who gets too close to it. Some radioactive substances are more unstable than others and hence more dangerous. All radioactive substances become less active as time goes by and eventually become stable – inactive.

From this much simplified analogy it can be deduced that radioactive substances are unstable substances which give off energy in the form of nuclear radiation, but which eventually become inactive and stable.

Having defined what is meant by the term 'radioactivity' it is now necessary to understand what is meant by nuclear radiation. Before this can be done, however, it is necessary to understand some simple chemistry.

THE STRUCTURE OF MATTER

Everything on land, in the sea, in the air, or in the sun and planets, is made up from individual elements, or a compound (mixture) of elements. Ordinary water, for example, is a substance made up of a compound of the elements hydrogen (chemical symbol H) and oxygen (chemical symbol O). Common table salt, on the other hand, is made up of a compound containing the two elements sodium (chemical symbol Na) and chlorine (chemical symbol Cl). Many elements exist entirely on their own; some examples are iron (Fe), lead (Pb), aluminium (Al), mercury (Hg), helium (He), copper (Cu) and gold (Au), although each of these may be mixed with others to form simple or complex compounds.

An element is something which cannot be destroyed or changed in any way (except by use of a nuclear reactor or 'atom smasher'), no matter what is done to it. It may be boiled, frozen, compressed, or immersed in an acid, and although it may change in form it will still be the same element as when the treatment started. Mercury, for example, is an element which appears as a solid piece of metal when frozen, or as a liquid at normal room temperature, or as a vapour when heated to a high temperature, but whatever form it may take it is *still mercury*.

The hydrogen which is contained in water can very easily be separated from the oxygen with which it is combined, to appear in its original form which, at normal temperatures, is a highly explosive gas! This simple example illustrates how it is possible for an element to take on a very different form when made to combine with another element to form a chemical compound.

There are 92 naturally-occurring elements found on earth, the lightest of which is hydrogen (H) and the heaviest uranium (U). The elements themselves usually appear in a form which is large enough to be seen – a gold ingot for example – but which itself is made up of billions upon billions of minute particles called *atoms*. An atom is the smallest quantity of any element which can exist and yet still possess all the chemical properties of that element. In a less accurate and non-scientific analogy, a single grain of sand represents the smallest quantity which can exist of the place known as the Sahara Desert.

A *molecule* is the smallest quantity of any chemical compound which can exist and yet still possess all the chemical properties of that compound.

Water, for example, is a compound which is made up of two atoms of hydrogen and one atom of oxygen and is given the chemical symbol H_2O. One single molecule of water therefore consists of two atoms of hydrogen and one atom of oxygen and represents the smallest possible quantity of water. If this molecule is split further the compound known as water is destroyed and the individual atoms of hydrogen and oxygen are separated.

One single molecule of common salt (sodium chloride) is composed of one atom of sodium and one atom of chlorine and is given the chemical symbol NaCl. If sodium chloride is combined with three atoms of oxygen a completely different compound is produced which is known as sodium chlorate ($NaClO_3$). Sodium chlorate is a deadly poison widely used as a weed killer. One single molecule of sodium chlorate comprises one atom of sodium, one atom of chlorine and three atoms of oxygen.

The atom

At one time it was thought that the atom was composed of a tiny solid particle which was totally impenetrable and electrically neutral, that is it possessed no electrical polarity. The work of Lord Rutherford and others, however, showed this theory to be incorrect and that the atom had an internal structure; they also showed that it was the differences in the composition of this structure which made the atom of one element different from that of another element. Their experiments showed that all atoms are composed of a very heavy, densely packed central region, known as the *nucleus*, surrounded by a very much lighter, loosely packed region formed from a diffuse cloud of tiny particles called *electrons*. The electrons in this cloud orbit the nucleus in the same way that the moon orbits the earth, each electron taking a different path as illustrated in Figure 3.1.

The nucleus itself is composed of two types of particles, each of similar mass and each about 1850 times heavier than that of an individual electron. The overall mass of an atom (nucleus plus electrons) is therefore due almost entirely to the mass of the nucleus. The particles contained within the nucleus are collectively known as the *nucleons* and are of two types: one is a positively-charged particle known as a *proton*, the other is a neutral particle known as a *neutron*. The positive charge possessed by one proton is identical to that possessed by one electron, and in a completely neutral atom (most atoms are normally neutral) the total positive charge possessed by the protons in the nucleus is exactly balanced by the total negative charge possessed by the orbital electrons. *In a neutral atom there are as many protons in the nucleus as there are orbital electrons.*

Just as the centrifugal force of the moon balances the attractive gravita-

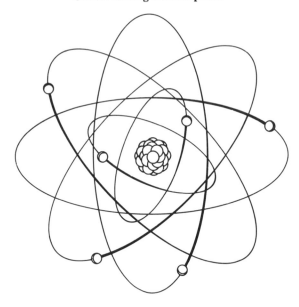

Figure 3.1   Simplified atom showing nucleus and orbital electrons

tional force of the earth and prevents it from crashing into the earth, so the centrifugal forces of the negative orbital electrons balance the attractive positive force of the nucleus and so prevent them being drawn into the nucleus.

Just as the moon is a very long way from the earth (about a quarter of a million miles), so the orbital electrons of the atom are a very long way away from the nucleus, even though the overall size of the atom itself is less than one millionth of one millimetre in diameter. This means that, in atomic dimensions, the atom consists mostly of space. To give some idea of the relative sizes and separation distances of the particles which make up the atom, consider the structure of the hydrogen atom; the lightest and simplest of all atoms. This consists of a nucleus containing a single proton (no neutrons), orbited by a single electron. Now imagine the electron to take on the size of a garden pea; if this were so the nucleus would be about 200 metres away and would be bigger than a double-decker bus!

The density (σ) of any element is equal to its mass (m) divided by its volume (v); the smaller the volume the greater the density. The atoms which make up an element are arranged so that the outermost of their orbital electrons are just touching each other; they cannot get any closer to each other than this. This means that the nuclei are, in atomic dimensions, very widely spaced. If all the orbital electrons could be removed from the atoms of an element and the nuclei allowed to come in contact with one another, the density of that element would be truly enormous. In fact, a heaped teaspoonful of nuclei would weigh more than one million tonnes! It is just this sort of situation which occurs in the so-called 'black holes' of outer space.

Ionization

Under normal circumstances the combined positive charge of all the protons within the nucleus of an atom is precisely balanced by the combined negative charge of all its orbital electrons and the atom is said to be electrically neutral. There are many instances, however, when this neutrality is destroyed; sometimes deliberately, by the use of scientific apparatus or by chemistry and sometimes through nature itself. Long distance radio communication, for example, is only possible because of thick layers of atoms lying above the earth's stratosphere which have had their neutrality destroyed by radiation from the sun. It is these layers, forming what is known as the ionosphere, which reflect radio waves back towards the earth and so enable them to travel over vast distances.

An electric arc discharge in a gas, for example, causes some or all of the orbital electrons of the gas to be torn away from their atomic orbits and, in the process, become free electrons. This leaves those atoms which have been affected in this way with a net positive charge because of the absence of the neutralizing action originally provided by the orbital electrons. Such atoms are said to have been *ionized* by the process and to have become *positive ions*. Ionization of this type can also be produced by lightning discharges in the atmosphere or by bombarding neutral atoms with atomic particles (protons, neutrons, electrons, etc.) used as high-speed projectiles or with nuclear radiation.

Ionization can also result in the formation of negatively-charged ions. This happens when an otherwise neutral atom is forced to accept one or more additional electrons into its orbital structure.

The phenomenon of ionization and the creation of positive and negative ions plays an important role in many aspects of nuclear physics, in particular the detection and measurement of nuclear radiation and the harmful effects of such radiation on living tissue.

Atomic number and mass

The number of protons in a nucleus defines the type of element represented by the atom containing that nucleus. The hydrogen atom, for example, has only one proton in its nucleus whereas oxygen has 8 and uranium has 92. This number is known as the *atomic number* of the element and is represented by the capital letter Z. Thus, when we say an element has a Z-number of 36 we know that its atoms each have 36 protons in their nuclei.

All known elements are tabulated in an internationally-recognized list known as the *Periodic Table*. The table starts with the element with Z-number 1, that is hydrogen and, at the time of writing, ends with the most recently-discovered element with a Z-number of 106, at present unnamed.

The *atomic mass* or *atomic weight* of an element represents the total number of nucleons (protons and neutrons) contained in each of its atoms and is represented by the capital letter A. The most common form of hydrogen, for example, has no neutrons in its nucleus and its mass number is therefore equal to its Z-number: $A = Z = 1$. The type of uranium atom known as uranium-235, on the other hand, has 143 neutrons in its nucleus as well as 92 protons and so its A-number is equal to $92 + 143 = 235$.

It is conventional when writing about the elements to use their chemical symbols, accompanied when necessary by indications of their Z- and A-numbers. The following examples illustrate the principle used throughout the world!

$$\text{mass number} \longrightarrow {}^{A}_{Z}X \longleftarrow \text{chemical symbol}$$
$$\text{atomic number} \longrightarrow$$

$$\quad {}^{16}_{8}O \qquad {}^{235}_{92}U \qquad {}^{28}_{14}Si \qquad {}^{204}_{80}Hg \qquad {}^{124}_{50}Sn$$

$$\quad oxygen \quad uranium \quad silicon \quad mercury \quad tin$$

Note that for all elements the number of neutrons in the nucleus is equal to $A - Z$.

When describing a particular type of element, either in writing or by word of mouth, it is common practice to use the name of the element, followed by its mass number; you will understand the reason for this shortly when 'isotopes' are discussed. For example, oxygen-16, uranium-235, silicon-28, etc.

Isotopes

Stated simply, isotopes are different types of the same basic thing. Sparrows, eagles, peacocks and swans are very different types of what is known as the 'bird' family. Similarly, roach, cod, merlin and eel are different types of the family known as fish. They are all *isotopes* of a particular family. So it is with the elements; very few appear in a unique form – most exist as a range of isotopes. Natural uranium, for example, when dug out of the ground and analysed, is found to consist (essentially) of two types of uranium. About 99.3 per cent is the most abundant uranium-238 isotope ($^{238}_{92}U$) and the remaining 0.7 per cent is the uranium-235 isotope ($^{235}_{92}U$). Similarly, naturally-occurring mercury is composed of a mixture of seven isotopes ranging from mass numbers 196 to 204 ($^{196}_{80}Hg$ to $^{204}_{80}Hg$). Note that it is the mass number (A) which varies from isotope to isotope within the same family identified by the constant Z-number. *The Z-number defines the family (element); the A-number defines the isotope of that family*. It follows then that different isotopes of the same element differ from one another solely because of the different number of neutrons contained in their nuclei. The isotope mercury-204, for example, has eight more neutrons in its nuclei than does the isotope mercury-196. Similarly, the uranium-238 isotope possesses three more neutrons than that possessed by uranium-235. The element which has the greatest number of naturally-

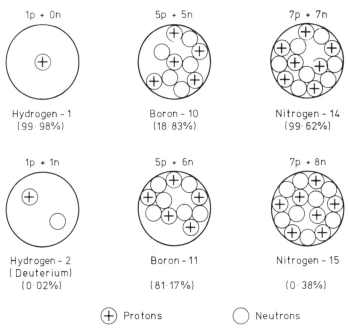

Figure 3.2   **Nuclear structure and relative abundances of some well-known isotopes**

occurring isotopes is tin (Sn). This has a range of ten isotopes extending from $^{112}_{50}$Sn to $^{124}_{50}$Sn. Of the 92 naturally-occurring elements only 21 exist as single species, for example beryllium ($^{9}_{4}$Be), aluminium ($^{27}_{13}$Al), cobalt ($^{59}_{27}$Co), gold ($^{197}_{79}$Au) and iodine ($^{127}_{53}$I).

The much simplified diagrams of Figure 3.2 illustrate the nuclear structure and relative abundances of a number of well known isotopes. The hydrogen-2 isotope ($^{2}_{1}$H) is also known as 'heavy hydrogen' (because its nucleus is twice as heavy as ordinary hydrogen) but more familiarly as deuterium ($^{2}_{1}$D). When two parts of deuterium are combined with one part of oxygen a special type of water is produced which is widely known as 'heavy water' (D_2O) but more scientifically as 'deuterium oxide'. In any sample of naturally-occurring water, molecules with two atoms of the H-1 isotope have an abundance of 99.986 per cent and molecules with two atoms of the H-2 (deuterium) isotope have an abundance of 0.014 per cent (one part in 7143). More is said about this very important substance later in the book.

Average mass

When an element exists in more than one isotopic form it is usual, when referring to its mass number, to specify a figure which corresponds to the

Table 3.1
Average mass numbers of selected elements

Element	Atomic number (Z)	Mass numbers (A) of isotopes	Average mass number (A)
Calcium	20	40, 42, 43, 46, 48	40.08
Zinc	30	64, 66, 67, 68, 70	65.38
Strontium	38	84, 86, 87, 88	87.63
Silver	47	107, 109	107.88
Tin	50	112, 114, 115, 116, 117, 118, 119, 120, 122, 124	118.7
Mercury	80	196, 198, 199, 200, 201, 202, 204	200.61
Lead	82	204, 206, 207, 208	207.21

average mass of all the isotopes which make up that element, making due allowance for their relative abundances. The element chlorine, for example, is known to exist in two isotopic forms, one having an atomic mass of 35 and an abundance of 75.529 per cent, the other having a mass of 37 and an abundance of 24.471 per cent. The weighted average of these two mass values is 34.453 and it is this value which is specified when referring to the natural form of chlorine: $^{34.453}_{17}Cl$.

Table 3.1 illustrates the principle of average mass for a few selected elements, while Table 3.2 gives Z- and mean A-numbers for all known elements.

Nuclides

Use of the word 'isotope' when referring to an element implies that the element exists in more than one form. In the case of beryllium and aluminium, however, this is clearly not the case because each of these elements, and others like them, exists in a single, unique, form. It was to avoid such misunderstandings and to provide for a more precise terminology, that the term *nuclide* was introduced in 1947.

The word nuclide is used when referring to the nuclear structure of an element without actually implying that it may or may not exist in many isotopic forms. We may, for example, state that there exist more than 280 naturally-occurring stable nuclides, or that uranium is a radioactive nuclide (radionuclide). Beryllium would be referred to as a 'single stable nuclide' rather than as a 'single stable isotope'. In spite of this you will find the words element, nuclide and isotope being used to mean the same thing.

Table 3.2
Alphabetical list of the elements

Name of element	Chem. symb.	Z- no.	Mean A-no.	Name of element	Chem. symb.	Z- no.	Mean A-no.
Actinium	Ac	89	227	Beryllium	Be	4	9.01
Aluminium	Al	13	26.98	Bismuth	Bi	83	208.98
Americium	Am	95	243	Boron	B	5	10.81
Antimony	Sb	51	121.75	Bromine	Br	35	79.90
Argon	Ar	18	39.95	Cadmium	Cd	48	112.4
Arsenic	As	33	74.92	Caesium	Cs	55	132.9
Astatine	At	85	210	Calcium	Ca	20	40.08
Barium	Ba	56	137.34	Californium	Cf	98	251
Berkelium	Bk	97	247	Carbon	C	6	12.01

Table 3.2   continued

Name of element	Chem. symb.	Z-no.	Mean A-no.	Name of element	Chem. symb.	Z-no.	Mean A-no.
Cerium	Ce	58	140.12	Osmium	Os	76	190.2
Chlorine	Cl	17	35.45	Oxygen	O	8	16
Chromium	Cr	24	51.99	Palladium	Pd	46	106.4
Cobalt	Co	27	58.93	Phosphorus	P	15	30.97
Copper	Cu	29	63.54	Platinum	Pt	78	195.09
Curium	Cm	96	247	Plutonium	Pu	94	242
Dysprosium	Dy	66	162.5	Polonium	Po	84	210
Einsteinium	Es	99	254	Potassium	K	19	39.10
Erbium	Er	68	167.26	Praseodymium	Pr	59	140.91
Europium	Eu	63	151.96	Promethium	Pm	61	147
Fermium	Fm	100	253	Protoactinium	Pa	91	231
Fluorine	F	9	18.99	Radium	Ra	88	226
Francium	Fr	87	223	Radon	Rn	86	222
Gadolinium	Gd	64	157.25	Rhenium	Re	75	186.2
Gallium	Ga	31	69.72	Rhodium	Rh	45	102.90
Germanium	Ge	32	72.59	Rubidium	Rb	37	85.47
Gold	Au	79	196.96	Ruthenium	Ru	44	101.07
Hafnium	Hf	72	178.49	Samarium	Sm	62	150.4
Helium	He	2	4.00	Scandium	Sc	21	44.95
Holmium	Ho	67	164.93	Selenium	Se	34	78.96
Hydrogen	H	1	1.008	Silicon	Si	14	28.08
Indium	In	49	114.82	Silver	Ag	47	107.87
Iodine	I	53	126.90	Sodium	Na	11	22.99
Iridium	Ir	77	192.22	Strontium	Sr	38	87.62
Iron	Fe	26	55.85	Sulphur	S	16	32.06
Krypton	Kr	36	83.80	Tantalum	Ta	73	180.95
Lanthanum	La	57	138.90	Technetium	Tc	43	99
Lawrencium	Lw	103	257	Tellurium	Te	52	127.60
Lead	Pb	82	207.2	Terbium	Tb	65	158.92
Lithium	Li	3	6.94	Thallium	Tl	81	204.37
Lutetium	Lu	71	174.97	Thorium	Th	90	232
Magnesium	Mg	12	24.30	Thulium	Tm	69	168.93
Manganese	Mn	25	54.94	Tin	Sn	50	118.69
Mendelevium	Md	101	256	Titanium	Ti	22	47.90
Mercury	Hg	80	200.59	Tungsten	W	74	183.85
Molibdenum	Mo	42	95.94	Uranium	U	92	238
Neodymium	Nd	60	144.24	Vanadium	V	23	50.94
Neon	Ne	10	20.18	Xenon	Xe	54	131.30
Neptunium	Np	93	237	Ytterbium	Yb	70	173.04
Nickel	Ni	28	58.71	Yttrium	Y	39	88.90
Niobium	Nb	41	92.90	Zinc	Zn	30	65.37
Nitrogen	N	7	14	Zirconium	Zr	40	91.22
Nobelium	No	102	254				

The point to remember is that the word element should be used when referring to the element as a whole, that is something composed of many billions of complete atoms, whereas the words nuclide and isotope should be used when referring to the nuclear structure of the individual atoms which make up the element.

THE CHEMISTRY OF RADIOACTIVITY

The 92 naturally-occurring elements exist in more than 320 isotopic forms, most of which (over 280) are stable, that is non-radioactive. A few, however, exhibit signs of radioactivity in some or all of their isotopes, the most well known of which are those of uranium and its daughter product radium ($^{226}_{88}$Ra); these particular elements are totally radioactive and do not exist in a stable form. Two other totally-radioactive elements which occur naturally are thorium ($^{232}_{90}$Th) and the gas radon ($^{222}_{86}$Rn) which is the daughter product of radium-226.

Some elements, although predominantly stable, exhibit very weak signs of radioactivity because one of their low-abundance isotopes is radioactive; the lower the abundance, the weaker is the radioactivity. Typical of this type of element is potassium (K). This exists in nature as a mixture of the three isotopes $^{39}_{19}$K, $^{40}_{19}$K, and $^{41}_{19}$K. The relative abundances of these are 99.10, 0.0118 and 6.88 per cent respectively, and it is the least-abundant K-40 isotope which is radioactive.

What makes an element radioactive or not is determined by the relative numbers of protons and neutrons in its nucleus, and their spacing from one another. The balance is quite critical and an element is just as likely to be radioactive for having too many protons in its nucleus as it is for having too many neutrons. The greater the number of nucleons in the nucleus, the greater is the likelihood of that element being radioactive. In fact, all of the nine naturally-occurring elements with atomic numbers (Z) above 83 (bismuth) exhibit some form of radioactivity; these all have mass numbers in excess of 209.

Artificial isotopes

So far only the isotopes of the 92 naturally-occurring elements have been considered, most of which are non-radioactive. With the advent of the particle-accelerator and the nuclear reactor, however, it became possible to produce a whole new range of additional, and previously unknown, isotopes from the naturally-occurring elements and also to create completely new elements which are no longer found on earth – each with its own range of isotopes.

Artificial (man-made) isotopes and elements are produced by bombarding certain elements with atomic projectiles. The projectiles may be electrons, protons, neutrons, deuterons (proton–neutron pairs from deuterium nuclei), or the nuclei of other elements, depending upon the machine being used and the type of isotope or element it is required to create. One of the simplest ways of creating a new element or isotope is by placing a small sample of a suitable target element in the core of a nuclear reactor and irradiating it with neutrons. The many billions of neutrons present in the core travel at very high speeds in all directions and many of them inevitably collide with atoms of the sample element being irradiated. When this happens an atom of the sample is forced to accept an additional neutron into its nucleus, an event which at once raises its atomic mass number by one unit and creates a new isotope of that element.

The absorption of an additional neutron by a nucleus is an event which is described by the term *neutron capture* and the susceptibility of an element or nuclide to such an event occurring is described by the term *neutron capture cross-section*. This term represents the ability of a nucleus to accept a neutron as a theoretical capture area; some nuclides will readily accept an additional neutron into their nuclei and are therefore said to possess a high neutron capture cross-section. Others, however, are more reluctant to accept additional neutrons and are therefore said to possess a very small capture cross-section. A simple analogy might be 'a wide goal mouth has a larger football capture cross-section than that of a narrow goal mouth'.

The process of isotope production by neutron capture is summarized by the following simplified expression:

$$_Z^A X + n \longrightarrow {}_Z^{A+1}X$$

where X is the chemical symbol of the target nuclide having an atomic number Z and atomic mass A, and 'n' is the absorbed neutron.

The result of the transition is an isotope of the element X having a mass number equal to A + 1; the change is known as *nuclear transformation*.

Another, less popular, method of isotope production is the process of *proton capture*. In this method a particle accelerator is used to bombard a target sample element with a powerful beam of protons. The process is similar to neutron capture but in this case the nuclei of the target element are forced to accept additional protons instead of neutrons. The result is very different from that of neutron capture because the process results in an increase, not only in the atomic mass number (A) but also in the atomic number (Z). The importance of this is that the alteration in Z-number means a complete change in the element itself since it is the number of protons in the nucleus which defines a particular element; see page 37. Changing one element into another in this way is known as *nuclear transmutation*. The process of proton capture is summarized by the following simplified expression:

$$_Z^A X + p \longrightarrow \ _{Z+1}^{A+1}Y$$

where Y is the chemical symbol for the new element which has been created and 'p' is the absorbed proton.

The following examples illustrate the difference between nuclear transformation by neutron capture, and nuclear transmutation by proton capture!

$$_{27}^{59}\text{Co} + n \longrightarrow \ _{27}^{60}\text{Co (unstable)}$$

$$_{13}^{27}\text{Al} + p \longrightarrow \ _{14}^{28}\text{Si (stable)}$$

Note that the stable cobalt-59 isotope is *transformed* into the unstable cobalt-60 isotope and that the stable aluminium-27 isotope is *transmuted* into the stable silicon-28 isotope.

All artificially-produced isotopes, except those which already exist in nature in a stable form, irrespective of their mode of production, are unstable and exhibit some form of radioactivity; when an otherwise stable nuclide is forced to accept an unwanted particle its delicately balanced stability is disturbed by the sudden acquisition of excess energy. Its reaction to this disturbance may be instantaneous or delayed over a long period of time, depending upon the degree of instability introduced. Irrespective of how long it takes, the nuclide eventually regains its original stability by shedding surplus energy in the form of nuclear radiation.

Nuclear radiation and decay

When an unstable nuclide gets rid of its excess energy it does so by emitting nuclear radiation; in doing so the nuclide is said to *decay*. This may take a few millionths of a second or many millions of years; no two nuclides are alike in this respect; each has its own 'radiation signature' and each is well known and documented. There are essentially seven types of radiation associated with radioactive decay; these are: alpha particles (α), beta

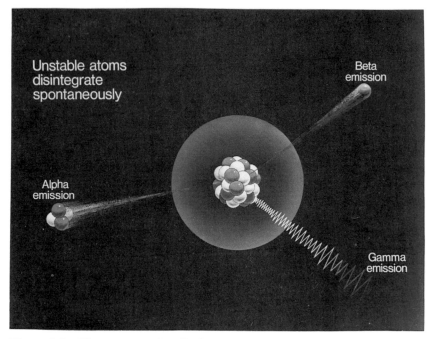

Figure 3.3   Three types of radiation

particles (β^-), X-rays (X), gamma rays (γ), neutrons (n), protons (p) and positrons (β^+). Three of these are shown in Figure 3.3

Alpha particles

An alpha particle is, in effect, the positively charged nucleus of a helium ($^4_2$He) atom and is composed of a tightly-bound cluster of two protons and two neutrons; its total mass is therefore equivalent to four atomic mass units, i.e. A = 4. In atomic terms, it is a massive particle and the largest to be emitted as a single entity in any mode of decay.

When a nuclide decays by alpha particle emission it loses two units of atomic number (two protons) and four units of atomic mass (two protons plus two neutrons). The nuclide is therefore *transmuted* into a completely different element whose Z-number is two units less and whose A-number is four units less than that of the so-called *parent nuclide*. The process is summarized by the simple expression:

$$^A_Z X \xrightarrow{\ \alpha\ } {}^{A-4}_{Z-2} Y$$

where Y is the newly created element known as the *daughter product*;

this itself may be unstable and decay by further α-emission – or some other decay mode – into another daughter product which may, or may not, be stable.

Some unstable elements, uranium-238, for example, go through 14 successive decay processes over a period of more than 4000 million years before reaching stability in the form of the element lead (Pb), whereas other elements may reach stability in a few minutes through a single decay process.

Alpha particle decay is usually, but not always, associated with the heavy unstable nuclides, for example uranium, thorium, plutonium, and is often accompanied by gamma radiation. Representative of alpha decay is the decay of the penultimate daughter product in the uranium-238 decay chain. Here the unstable polonium-210 nuclide ($^{210}_{84}$Po) decays by alpha emission to the stable nuclide lead-206 ($^{206}_{82}$Pb). The decay process for this reaction is given by the expression:

$$^{210}_{84}\text{Po} \xrightarrow{\alpha} {}^{206}_{82}\text{Pb (stable)}$$

Examples of other alpha-decay processes are (see Table 3.2):

$$^{238}_{92}\text{U} \xrightarrow{\alpha} {}^{234}_{90}\text{Th (unstable)}$$

$$^{146}_{62}\text{Sm} \xrightarrow{\alpha} {}^{142}_{60}\text{Nd (stable)}$$

$$^{212}_{83}\text{Bi} \xrightarrow{\alpha} {}^{208}_{81}\text{Tl (unstable)}$$

Beta particles

A beta particle is a negatively-charged electron. In fact the only difference between a beta particle and an ordinary electron is its name. A beta particle is an electron which originates from the nucleus of an atom. Beta particles are emitted when an unstable nuclide has too many neutrons. It comes about when one of the neutrons changes into a proton and an electron and the electron is ejected from the nucleus as a beta particle. In the process the nucleus gains one extra proton, hence its atomic number increases by one unit and a nuclear transmutation occurs. No loss of mass occurs, however, and so the mass number remains unchanged. The process is known as beta decay and is summarized by the simple expression:

$$^{A}_{Z}\text{X} \xrightarrow{\beta^-} {}^{A}_{Z+1}\text{Y}$$

Beta decay is not confined to a particular group of elements and is just as likely to occur with the lighter elements as with the heavier ones; it is also usually accompanied by X- or γ-radiation. The following reactions are representative of the beta decay process:

$$^{116}_{49}\text{In} \xrightarrow{\beta^-} {}^{116}_{50}\text{Sn (unstable)}$$

$$^{210}_{83}\text{Bi} \xrightarrow{\beta^-} {}^{210}_{84}\text{Po (unstable)}$$

$$^{14}_{6}\text{C} \xrightarrow{\beta^-} {}^{14}_{7}\text{N (stable)}$$

Positrons

A positron (β^+) is the positive equivalent of the negative electron, or beta particle (β^-), and like the beta particle it originates from the nucleus.

Positrons are emitted when an unstable nuclide has too many protons. It comes about when one of the protons changes into a neutron and a positive electron, and the electron is ejected from the nucleus as a positron. In the process, the nucleus loses one proton, so its atomic number decreases by one unit and a nuclear transmutation occurs. No loss of mass occurs, however, and so the mass number remains unchanged. The process is known as *positron decay* and is summarized by the simple expression:

$$^{A}_{Z}\text{X} \xrightarrow{\beta^+} {}^{A}_{Z-1}\text{Y}$$

Positron emission only occurs with artificially-produced nuclides and it is very rare among the heavier elements; it is almost always accompanied by X- or γ-radiation. The following reactions are representative of the positron decay process:

$$^{164}_{70}\text{Yb} \xrightarrow{\beta^+} {}^{164}_{69}\text{Tm (unstable)}$$

$$^{110}_{49}\text{In} \xrightarrow{\beta^+} {}^{110}_{48}\text{Cd (stable)}$$

$$^{64}_{29}\text{Cu} \xrightarrow{\beta^+} {}^{64}_{28}\text{Ni (stable)}$$

Beta decay

Because of the close similarity of the beta and positron decay processes (they both affect only the Z-number of an unstable nuclide), and the fact that the two particles differ only in the polarity of their charges, the term 'beta decay' is sometimes used to describe both processes, the difference being implied in the written word by use of the symbols β^- or β^+.

Although not widely used, the word *negatron* is sometimes used to distinguish the negative electron (beta particle) from the positive electron (positron).

Neutrinos

Neutrinos and anti-neutrinos are nuclear particles having no electrical charge and virtually no mass. They are created whenever a nuclide decays by the β^- or β^+ processes, but their description is outside the scope of this book.

Electron capture

A decay process which is very similar to positron decay, and whose reaction product is identical, involves the capture of one of the orbital electrons which surround the nucleus. The process is known as *electron capture* (EC), or simply K-capture since the electron which is captured is usually one of those occupying the so-called K-shell orbit; this is the electron orbit closest to the nucleus (see Figure 3.1).

The captured electron upsets the normally positive charge of the nucleus and in so doing neutralizes one proton and causes it to convert to a neutron. This loss of a proton causes a reduction in the Z-number of the nuclide by one unit and hence a nuclear transmutation to occur; no loss of mass occurs in the process, however, and so the mass number of the new nuclide remains the same as the nuclide from which it was formed. The process of electron capture is summmarized by the simple expression:

$$\ce{^{A}_{Z}X} \xrightarrow{\text{EC}} \ce{^{A}_{Z-1}Y} \quad \text{or} \quad \ce{^{A}_{Z}X} \xrightarrow{\text{K}} \ce{^{A}_{Z-1}Y}$$

which is seen to result in a transmutation identical to that associated with positron decay.

The important difference between the electron capture and positron decay processes is that no positrons are emitted from electron capture. There is, however, considerable X-radiation associated with electron capture as electrons from other orbits (shells) move in to fill the vacancy left by the captured electron and rearrange their positions. The situation is very similar to an aircraft holding pattern in which aircraft are 'stacked' at various carefully controlled heights. As soon as one aircraft is allowed to land, all the other aircraft move down to occupy a position in the next lower orbit. They do this by shedding surplus energy, that is they throttle back slightly so that they 'drop' to the required height of the lower orbit. The same sort of thing happens when an electron from one of the outer orbits (shells) drops into an orbit which is closer to the nucleus. It gets rid of its surplus energy in the form of an X-ray, the energy of the X-ray being determined by the amount of energy which has to be got rid of.

Electron capture decay, like positron decay, occurs in artificial nuclides which have too many protons. Unlike positron decay, however, electron capture occurs in a wide range of nuclides, including those with a high atomic number where it is the most dominant of the two processes. The following reactions are representative of the electron capture decay process:

$$^{206}_{83}\text{Bi} \xrightarrow{\text{K}} {}^{206}_{82}\text{Pb (stable)}$$

$$^{186}_{79}\text{Au} \xrightarrow{\text{K}} {}^{186}_{78}\text{Pt (unstable)}$$

$$^{37}_{18}\text{Ar} \xrightarrow{\text{K}} {}^{37}_{17}\text{Cl (stable)}$$

Many radionuclides decay by a combination of both electron capture and positron emission, a good example of which is the decay of cobalt-58 into iron-58. In this nuclide 85 per cent of all transitions are by way of electron capture and the remaining 15 per cent via positron emission. This process is summarized by the expression:

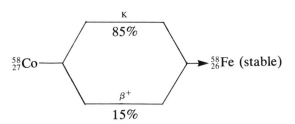

Some radionuclides decay in a very complex manner involving many different processes. Iodine-126, for example, decays by beta particle emission, positron emission and electron capture, all of which is accompanied by up to seven separate gamma rays. In the decay of this particular nuclide, 55 per cent of the transitions are by way of electron capture to tellurium-126; 1.3 per cent by way of positron emission, also to tellurium-126, and 44 per cent by beta particle emission to xenon-126. The decay scheme is summarized by the following expression:

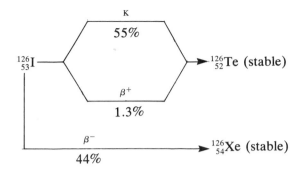

Isomeric transition

A detailed examination of this mode of decay is outside the scope of this book. Put simply, however, it comes about because it is possible for two otherwise identical nuclides to exist side by side and yet for only one of them to show signs of instability. The instability exists not because the nucleus has too many or too few protons but simply because following a nuclear transition certain nuclides need to rearrange their nucleons and orbital electrons into a more stable configuration. A nuclide does this by shedding surplus energy in the form of gamma rays. A nuclide which shows this form of instability is said to be in an *excited* or *metastable* state and is indicated by the lower-case letter 'm' following its mass number, for example tin-119m, or $^{119m}_{50}\text{Sn}$.

The important thing to remember about the isomeric transition mode of decay is that it involves no changes in atomic mass or atomic number and hence no nuclear change occurs. The nuclide is simply 'shaking off' unwanted energy so that it can exist in what is known as its *ground state* and which may, or may not be a stable state. The nuclide may do this nearly instantaneously or after some measurable delay.

Some representative examples of metastable nuclides are:

Antimony-122m   ($^{122m}_{51}\text{Sb}$)

Technetium-97m   ($^{97m}_{43}\text{Tc}$)

Protoactinium-234m   ($^{234m}_{91}\text{Pa}$)

X- and γ-radiation

X-rays, γ-rays, radio waves, infrared waves, ultraviolet waves, sunlight and all other forms of visible light are all different forms of electromagnetic waves existing within what is known as the *electromagnetic spectrum*. All such waves travel at the speed of light (in a vacuum) and differ from one another only in the magnitude of their wavelengths, and hence frequencies. The mathematical relationship between wavelength (λ) expressed in metres (m), and frequency (f) expressed in Hertz (Hz), is given by the expression:

$$\lambda = \frac{c}{f}$$

where 'c' is the velocity of light = 3×10^8 metres per sec.

A red-coloured light, for example, appears different from one coloured green simply because the wavelength of its emitted light is different. Radio waves used for CB radio are limited to a narrow band of electromagnetic waves which is centred on 27 metres. Radio waves used for TV reception, on the other hand, use much shorter wavelengths in the region of one-third of a metre.

Within a very narrow band of wavelengths electromagnetic waves become visible to the human eye. This part of the spectrum is known as the *visible spectrum* and extends from infrared at the long wavelength end of the band up to the ultraviolet short-wavelength end of the band; within this band lie all the colours found in the rainbow.

Radio waves have wavelengths which lie well above (longer) those found in the visible spectrum whereas X-rays and γ-rays have wavelengths which lie well below it. Figure 3.4 shows the relationship between various occupants of the electromagnetic spectrum.

X-rays and γ-rays, sometimes referred to as *photons*, differ from one another in name only. X-rays originate from outside the nucleus of the atom whereas γ-rays originate from within the nucleus; they are otherwise

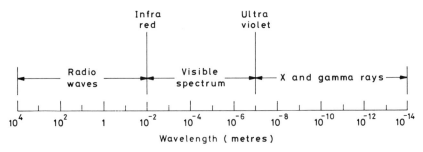

Figure 3.4   The electromagnetic spectrum

identical in form. They both extend over a broad band of the electromagnetic spectrum and their wavelengths often overlap. In general however, it can be assumed that γ-rays are of much shorter wavelength than X-rays.

X-rays are emitted by an unstable atom whenever one or more of its orbital electrons moves from a high-energy shell to a lower-energy shell closer to the nucleus, the surplus energy of the electrons being carried away by the X-rays. X-rays can also be produced by other means unconnected with unstable nuclei, for example X-ray machines.

γ-rays are produced whenever an unstable nucleus re-adjusts its energy level to regain stability. It may do this by the emission of a single γ-ray or by a shower of γ-rays of different energy occurring at different times.

X-rays and γ-rays are described by their energy or 'hardness'. A very low-energy ray, for example, would be described as being 'soft' and could be completely absorbed by a thin sheet of aluminium foil. A very high-energy ray, on the other hand, would be described as being 'hard' and could quite easily penetrate a thick sheet of lead. A close analogy is the way in which electromagnetic rays from the sun are described as being 'more powerful' at noon than at dawn on the same day; a fact which any sunbather will verify on a cloudless mid-summer day! The noon-time rays from the sun may be looked upon as being 'harder' than those which occur at dawn.

X-rays and γ-rays should be looked upon as radar-like pulses containing short-duration 'packets' of electromagnetic energy; in fact the only difference between a radar pulse and an X- or γ-ray is the magnitude of the wavelength contained within the pulse. Since they all travel at the same fixed speed (the speed of light), the difference in energy between two similar pulses can only be in the duration of the pulse and the amplitude and wavelength of the electromagnetic waves within the pulse.

The energy content of an X-ray or γ-ray is expressed in terms of a unit of measurement known as the *electron-volt*, usually abbreviated 'eV'. This unit is also used to express the kinetic energy of moving particles such as alpha particles, protons, neutrons, etc., and is equivalent to the kinetic energy acquired by a single electron in being accelerated through a potential difference of one volt. An electron in being accelerated through one thousand volts would acquire a kinetic energy of 1000 eV usually abbreviated to 1 keV.

X-rays, γ-rays, alpha and beta particles, etc., possess energies which typically range from a few keV to a few MeV (millions of eV). The iron-55 radionuclide, for example, emits a very soft K-shell X-ray during its decay process, the energy of which is only 6 keV. The sodium-24 nuclide, on the other hand, emits a very powerful gamma ray of 2.65 MeV. Table 3.3

Table 3.3
Energy of radiation of selected nuclides

Nuclide	Energy of radiation		
	alpha	beta	gamma
Uranium-235	4.5 MeV	none	95–185 keV
Iodine-131	none	250–810 keV	80–720 keV
Strontium-90	none	540 keV	none
Radium-226	4.7 MeV	none	188 keV

summarizes the energies of the various types of radiation emitted by a few well-known nuclides.

The neutron

The neutron is a nuclear particle which possesses no electrical charge and whose mass is nearly identical to that of the proton. It is ejected spontaneously by a very small number of unstable nuclides but more often it is ejected at high speed by a nuclide which has been made unstable by a bombarding particle. Once ejected from the nucleus the neutron is itself unstable and seldom lasts more than about 15 minutes, after which it decays into a proton and an electron.

There are many ways in which a nuclide can be made to eject a neutron from its nucleus; one of these is the so-called *alpha-n* reaction, usually abbreviated (α,n). In this reaction a target nucleus is bombarded by alpha particles causing it to eject a neutron from its nucleus and also to transmute it into another element. Bearing in mind that an alpha particle is a helium-4 nucleus ($_2^4$He) and that a neutron has a mass number of one but zero charge, then the α,n reaction can be summarized by the following simplified expression:

$$_Z^A X + _2^4 \alpha \longrightarrow _{Z+2}^{A+3} Y + _0^1 n$$

Note that on both sides of the expression the sum of the mass numbers is equal to $A + 4$ and that the sum of the atomic numbers is equal to $Z + 2$.

The following list is representative of the α,n reaction on certain nuclides; on the right-hand side of each reaction is the recommended 'shorthand' method of writing the expressions:

$$_4^9 Be + \alpha \longrightarrow _6^{12} C + n \qquad _4^9 Be\ (\alpha,n)\ _6^{12} C$$

$${}_3^7\text{Li} + \alpha \longrightarrow {}_5^{10}\text{B} + \text{n} \qquad {}_3^7\text{Li}\,(\alpha,\text{n})\,{}_5^{10}\text{B}$$

$${}_9^{19}\text{F} + \alpha \longrightarrow {}_{11}^{22}\text{Na} + \text{n} \qquad {}_9^{19}\text{F}\,(\alpha,\text{n})\,{}_{11}^{22}\text{Na}$$

$${}_{13}^{27}\text{Al} + \alpha \longrightarrow {}_{15}^{30}\text{P} + \text{n} \qquad {}_{13}^{27}\text{Al}\,(\alpha,\text{n})\,{}_{15}^{30}\text{P}$$

Speed of decay

Much emphasis has so far been placed on what is meant by radioactivity and on the many ways in which an unstable nuclide might decay. It is now necessary to examine what is meant by such terms as *half-life* and *decay constant*.

Imagine a room recently fitted with 1000 brand-new identically-made light bulbs and that these are arranged to be switched on all the time. We know that they are not going to last forever and that eventually they will all fail – of that we can be certain. We also know that some will last very much longer than others. What we cannot possibly know is just how long it will take for all of them to fail, nor the precise time at which any particular bulb will fail. The failure process is therefore random and unpredictable. What we can do, however, is to take a series of measurements on a large installation of bulbs over a long period of time and determine an average lifetime or failure rate; this could take a very long time. Another way of expressing the life expectancy of the bulbs would be to determine how long it takes for half of them to fail and to call this figure the *half-life* of the installation. Suppose, for example, the half-life figure for a 1000-bulb installation is quoted as being 600 hours. This means that although it is impossible to know which individual bulbs are going to fail, we can be fairly certain that 500 of them will fail after 600 hours of use. We also know that if these are not replaced then a further 250 (half the remainder) will fail after a further 600 hours, and so on.

This simplified example is meant only to illustrate the close similarity of bulb failure rate with radioactive decay. Even a tiny quantity of radioactive material will be composed of billions upon billions of individual identical atoms, each one of which will eventually decay by one or more of the processes recently described. No one could possibly predict with any certainty which atom is going to decay next, nor how long it will be before that happens. An unstable atom may take many years before finally decaying, whereas another atom of the same type will decay within a fraction of a second after being made unstable. It is precisely for this reason that the decay rate of a radionuclide is specified in terms of its half-life, a period of time expressed in seconds, minutes, hours, days or years.

The half-life figure quoted for a radionuclide means that during this period of time, one-half of its atoms will have decayed, leaving only half the number which previously existed at the start of the half-life period. It does not matter when a particular measurement might be made; the fact is that there will always be half the number of unstable atoms left at the end of any half-life period than existed at the start of that period.

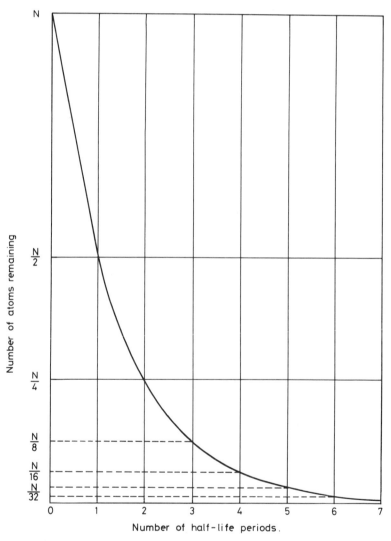

Figure 3.5   The meaning of half-life

The mathematical relationship between half-life and the number of remaining atoms is said to follow an exponential law and has the shape of the curve shown in Figure 3.5. The letter 'N' represents the number of atoms which existed at the beginning of the first measurement commencing at time 't = 0'.

The mathematical expression which is used to calculate half-life values is:

$$N = N_0 e^{-\lambda t}$$

where N is the number of atoms remaining after a time interval equal to 't' and λ is what is known as the *decay constant* for that particular nuclide. N_0 represents the number of atoms which existed at the start of the measurement.

The decay constant is described by the expression:

$$\lambda = \frac{0.693}{t_{\frac{1}{2}}}$$

where $t_{\frac{1}{2}}$ is the half-life value for the radionuclide being considered.

To understand how these expressions are used, consider the following example.

You have in your possession a strontium-90 radioactive source which you know was manufactured 10 years ago; how much of its initial radioactivity remains? The half-life of strontium-90 is 28 years.

To answer this question it is first necessary to determine the decay constant; this is done as follows:

$$\lambda = \frac{0.693}{t_{\frac{1}{2}}} = \frac{0.693}{28} = 0.0247$$

Re-arranging the expression for half-life and substituting for t = 10 years and $\lambda = 0.0247$

$$\frac{N}{N_0} = e^{-0.247 \times 10}$$

$$= e^{\frac{1}{2.47}}$$

$$= 0.78 = 78\%$$

Table 3.4
Half-lives of selected radionuclides

Nuclide	Half-life
Radium-226	1620 years
Iodine-131	8.04 days
Iodine-123	13 hours
Plutonium-239	24 000 years
Plutonium-238	86 years
Silver-110	24 seconds
Cobalt-60	5.26 years
Copper-66	5.2 minutes
Lead-204	140 000 million million years
Caesium-137	30 years
Germanium-77	11 hours
Uranium-238	4500 million years

Thus, after a period of ten years a strontium-90 source will have lost 22 per cent of its initial radioactivity.

Table 3.4 shows how half-life values vary between various radionuclides

Safety of long half-life

Imagine identical amounts of two different radionuclides. One amount is composed of plutonium-239 with a half-life of 24 000 years, the other of iodine-131 with a half-life of 8 days. Imagine further that there are two million atoms in each amount and that the decay of every atom from either amount is accompanied by the emission of a γ-ray of constant energy.

At the end of 24 000 years the Pu-239 sample will have lost one million atoms through nuclear decay and will have emitted one million γ-rays. This represents an average γ-ray emission rate of 41.7 per year – *less than one γ-ray per week!*

Now consider the I-131 sample. This also loses one million atoms during the period of one half-life but because the half-life is only 8 days instead of 24 000 years, the corresponding average γ-ray emission rate is 125 000 per day, or 1.45 per second!

Clearly, then, the sample with the very short half-life is considerably more dangerous to human life (as far as γ-radiation is concerned) than is the sample with the very long half-life. This may not necessarily be true for a radionuclide which is taken into the body since many other very important factors such as toxicity, type of radiation, energy and abundance of radiation have to be taken into account (this is explained in Chapter 18). Nevertheless, as a general rule it is reasonable to assume that, for external sources of radiation, *radionuclides with very long half-lives are likely to be less harmful than those with very short half-lives*.

The decay chain

Now that the various decay processes have been explained, it is possible to illustrate a complex decay scheme which involves many of them. The subject of the illustration is the uranium decay chain shown in Figure 3.6. Note from the diagram that alpha decays are indicated by vertical falls in which four units of atomic mass (A) are lost and that beta decays are indicated by horizontal shifts in which one unit of atomic number (Z) is gained; this form of presentation was chosen for clarity. Most of the transitions are accompanied by X- or γ-radiation.

Note that the Pa-234m nuclide decays to Pa-234 by isomeric transition (IT) and that the final element in the decay chain is stable lead-206. Note also that the decay of $^{238}_{92}U$ to $^{206}_{82}Pb$ involves a loss of 32 units of atomic mass $(238 - 206)$, 10 units of atomic number $(92 - 82)$ and involves 10 transmutations and 19 different nuclides. The diagram also shows that polonium (Po), astatine (At) and bismuth (Bi) decay by both α and β processes and that some elements (uranium, bismuth and polonium for example) appear in more than one isotopic form. The half-lives of the nuclides involved in the decay chain have been omitted from the diagram for clarity and are listed in Table 3.5 (y = years, d = days, h = hours, m = minutes, s = seconds).

Radiation penetrability

When alpha and beta particles travel through a thick sheet of material they give up some or all of their kinetic energies as the result of multiple collisions with the atoms which make up the absorbing material. It follows that the larger the size of the moving particle, the greater is the likelihood of it colliding with nearby atoms and the more shallow is its depth of penetration likely to be in any material. Alpha particles, for example, are relatively massive and are unable to travel very far in any material without being completely absorbed. In fact, alpha particles are completely absorbed by human skin, by a thin sheet of paper, or by about 50 mm (2 inches) of air. Beta particles, on the other hand, consist of individual electrons which are seven thousand times smaller in size than an alpha

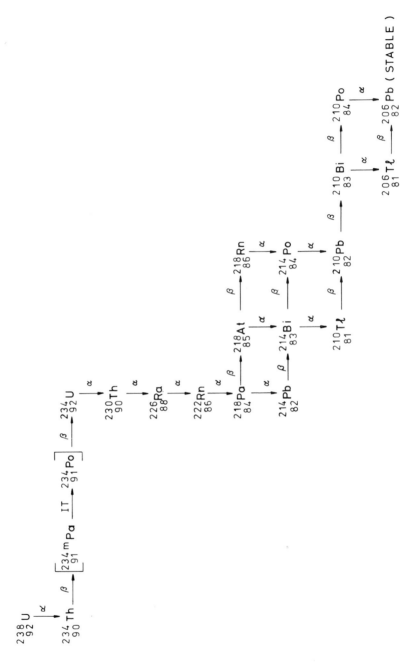

Figure 3.6   Decay scheme for uranium-238

Table 3.5
Half-lives of nuclides in uranium decay chain

U-238	4.5×10^9 y	Th-234	24.1 d
Pa-234m	1.18 m	Pa-234	6.66 h
U-234	2.5×10^5 y	Th-230	8×10^4 y
Ra-226	1620 y	Rn-222	3.825 d
Po-218	3.05 m	At-218	1.3 s
Rn-218	0.019 s	Pb-214	26.8 m
Bi-214	19.9 m	Po-214	1.6×10^{-4} s
Tl-210	1.3 m	Pb-210	22 y
Bi-210	5.01 d	Po-210	138.4 d
Tl-206	4.2 m	Pb-206	*stable*

particle and are therefore able to penetrate much further into material before being absorbed.

Beta particles also differ from alpha particles in that they are emitted by atoms, or generated by particle accelerators, with an enormous spread of energies, some of which may be classified as being very 'soft' (a few keV) whilst others may be classified as very 'hard' (a few hundred MeV). It is because of this that it is necessary to know the energy of a particular beta particle before its penetrability can be estimated.

Being negatively charged, beta particles are strongly influenced by the repulsive fields produced by the orbital electrons which surround the atoms making up the material through which the beta particle is travelling. The greater the Z-number of the material the greater is the number of its orbital electrons and the greater is the strength of the repulsive fields produced by its atoms. The type of material through which the beta particle is travelling therefore influences the degree of absorption.

Although dense materials with a high Z-number are very effective in absorbing beta particles, they also give rise to a type of electromagnetic radiation known as *Bremsstrahlung* – a German word meaning 'braking radiation', or 'slowing-down radiation'. The energy of the Bremsstrahlung produced in this way is often more penetrating than the beta radiation which produced it and for this reason it is common practice to use materials of low Z-number e.g. perspex when shielding against beta radiation.

The absorption of beta radiation in matter is a very complex process and depends upon many factors. In general, however, it may be assumed that virtually all beta radiation is absorbed by a 6 mm ($\frac{1}{4}$ inch) thickness of brass, or 24 mm of perspex, or 10 metres of air.

X-rays and γ-rays, being uncharged, have much greater penetrating powers than those of alpha and beta particles and considerable thicknesses of lead, steel or concrete are often required for satisfactory shielding from such radiation. The way in which X-rays and γ-rays are absorbed in material is an extremely complex one and involves three different mechanisms, each of which depends upon the energy of the incident radiation and the nature of the material through which it is passing. Other factors which have to be taken into consideration when calculating shielding thickness include such things as whether the radiation is collimated in the form of a pencil-like beam or arriving from all directions as a broad beam.

Some idea of the penetrability of gamma radiation may be obtained from a study of the cobalt-60 radionuclide. This simultaneously emits two gamma rays each time one of its atoms decays, the average energy of which is 1.25 MeV. The thickness of lead required to reduce the initial intensity of this radiation by a factor of ten is 46 mm (1.8 inches); an additional 35 mm (1.4 inches) is required to reduce it by a further factor of ten. In other words, about 81 mm (3.2 inches) of lead is required to reduce the intensity of 1.25 MeV gamma rays to 1 per cent of their initial value. Had the energy of the gamma radiation been 2.5 MeV instead of 1.25 MeV, the thickness of lead required to reduce the initial intensity to 1 per cent would

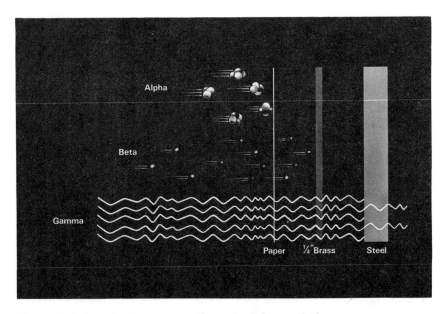

Figure 3.7   Penetrating power of α-, β- and γ-radiations

have had to be 117 mm (4.6 inches); the corresponding thickness of concrete would be about 720 mm (28 inches). Lead is clearly much more effective at absorbing gamma radiation than is concrete but it is also much more expensive!

Figure 3.7 demonstrates the relative penetrabilities of alpha, beta and gamma radiation.

Nuclear fission

The most important mode of nuclide decay, as far as nuclear power is concerned, is the phenomenon known as *nuclear fission*; in fact if it were not for nuclear fission there would be no nuclear power! Nuclear fission is a term used to describe a very rare phenomenon in which a nucleus splits (fissions) into two nearly equal parts: the so-called *fission products*. Only one of the 92 naturally-occurring elements has the ability to do this: the rare 235-isotope of uranium (its relative abundance in naturally-occurring uranium is only 0.7 per cent). Although U-235 is spontaneously fissile (fissions without external influences) it does so infrequently and in most instances it can be totally disregarded. On the other hand U-235 can be made to fission quite readily by forcing its nucleus to accept an additional neutron. Doing this makes the nucleus violently unstable and it quickly breaks up into two nearly equal fragments. The fragments produced by one atom are usually different from those produced by another atom so that the fissioning of, say, 10 atoms will produce 20 different products. Figure 3.8 demonstrates the process. The U-235 nucleus shown in Figure 3.8 absorbs a neutron and at once breaks up into two nearly equal fragments to form two completely different elements: in this case strontium-93 ($^{93}_{38}$Sr) and xenon-140 ($^{140}_{54}$Xe); the event is also accompanied by the emission of three neutrons. The notation used to describe this event is given by:

$$^{235}_{92}U + {}^{1}_{0}n \longrightarrow {}^{236}_{92}U \longrightarrow {}^{93}_{38}Sr + {}^{140}_{54}Xe + {}^{1}_{0}n + {}^{1}_{0}n + {}^{1}_{0}n$$

or, more briefly, as:

$$^{235}_{92}U \, (n,f) \longrightarrow {}^{93}_{38}Sr + {}^{140}_{54}Xe + 3n$$

Note that there are no alterations in the total atomic mass, or in the total atomic number, and that both sides of the expressions balance: total A-units = 236 and total Z-units = 92.

If a second U-235 atom is made to fission, this also will break up into two nearly equal fragments but these are very unlikely to be exactly the same

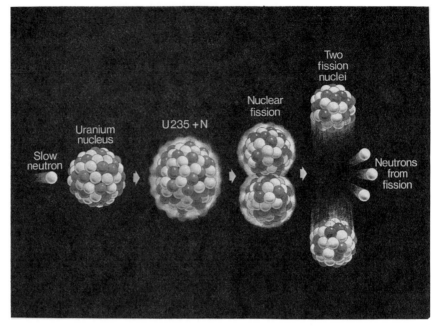

Figure 3.8   Fission of the uranium-235 atom

size, and hence type, as those produced from the preceeding fission event
and the number of neutrons produced is also likely to vary. The same thing
happens during other successive fissions and, in fact, a whole range of
different fission products are produced as more and more fissions occur.
The following examples illustrate just three of the 37 or so different events
which occur during the fissioning of U-235. Note that in each example the
total A- and Z-numbers balance on each side of the expressions.

$$^{235}_{92}U(n,f) \longrightarrow {}^{137}_{55}Cs + {}^{97}_{37}Rb + 2n$$

$$^{235}_{92}U(n,f) \longrightarrow {}^{140}_{56}Ba + {}^{93}_{36}Kr + 3n$$

$$^{235}_{92}U(n,f) \longrightarrow {}^{147}_{57}La + {}^{87}_{35}Br + 2n$$

Fission yield

The likelihood of a U-235 nucleus splitting into two identical fragments is
very small indeed; in fact it is typically only 0.01 per cent that is 1 in
10 000. In practice the splitting is very asymmetric and the products

produced from a large number of fission events fall into two broad mass-number groups, as illustrated in Figure 3.9. The so-called light-mass group extends from about mass numbers 85 to 104 and is centred on mass number 95 corresponding to zirconium-95 ($^{95}_{40}$Zr); this has a fission yield of about 6.3 per cent. The heavy-mass group extends from about mass numbers 130 to 149 and is centred on mass number 140 corresponding to barium-140 ($^{140}_{56}$Ba); this also has a fission yield of about 6.3 per cent. The valley between the two peaks in the graph represents the likelihood of the U-235 nucleus splitting into two virtually identical fragments, one having a mass of 117 and corresponding to rhodium-117 ($^{117}_{45}$Rh), the other having a mass of 118 and corresponding to silver-118 ($^{118}_{47}$Ag).

Experiments have shown that the range of fission product mass numbers extends from about A = 72, corresponding to zinc-72 ($^{72}_{30}$Zn), to A = 162, corresponding to dysprosium-162 ($^{162}_{66}$Dy), and that the average number of neutrons emitted is approximately 2.5 per fission event. These figures apply specifically to the fissioning of U-235 nuclei using bombarding neutrons which have been deliberately slowed down (moderated) to what are called *thermal energies*; such energies correspond to velocities of about 2200 metres per second (5000 mph). The importance of neutron moderators is discussed in more detail in Chapter 15.

Fission product energies

Under normal circumstances the 92 protons and 143 neutrons which make up the nucleus of a U-235 atom are tightly bound together by a very powerful nuclear force known as the *binding energy*. When this force is overcome by a bombarding neutron the U-235 nucleus splits in two and the resulting fission products fly apart in opposite directions at very high speed; it is as though the nucleus has exploded. These relatively enormous particles eventually come to rest after multiple collisions with the atoms of the material through which they travel, during which they give up their kinetic energies ($\frac{1}{2}$ mv^2) in the form of frictionally-generated heat. *It is this nuclear-created heat which is used by power stations to boil water, produce steam and drive turbines to generate electricity. It is the fundamental principle of nuclear power.*

Additional heat, albeit not nearly as great as that generated by the fission products, comes about from the kinetic energies of the fission neutrons and also from the nuclear radiation emitted by the fission products, *all of which are radioactive*.

The type of nuclear radiation produced by the fission products is predominantly β- and γ-activity because most of the products have an excess of neutrons in their nuclei. Many such nuclei decay by way of a

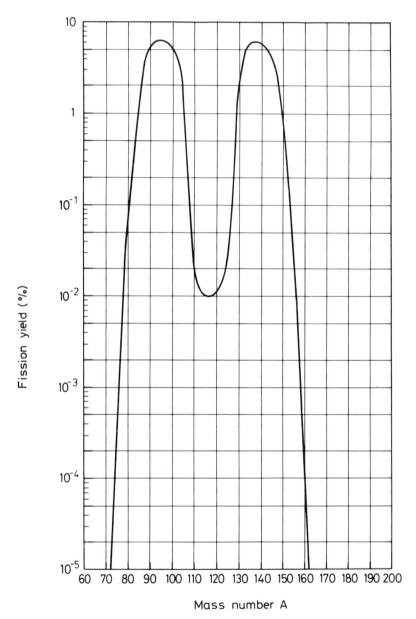

Figure 3.9   Fission product distribution and yield from U-235

multi-stage decay chain before achieving stability, as demonstrated by the decay process of the barium-141 fission product:

$$^{141}_{56}\text{Ba} \xrightarrow[18\,\text{m}]{\beta} {}^{141}_{57}\text{La} \xrightarrow[3.9\,\text{h}]{\beta} {}^{141}_{58}\text{Ce} \xrightarrow[32.5\,\text{d}]{\beta} {}^{141}_{59}\text{Pr (stable)}$$

The transition to the stable praseodymium-141 nuclide is by way of three beta decays with half-lives of 18 minutes, 3.9 hours and 32.5 days respectively.

The total energy released by the fissioning of a single U-235 atom is approximately 205 MeV and is distributed as shown in Table 3.6. About 87 per cent of the total energy is released at the instant of fissioning and is due to the fast-moving fission products and fission neutrons, plus some small amount from the accompanying gamma radiation. The remainder of the energy is released over a relatively long period of time as the fission products decay and emit beta and gamma radiation. Virtually all the energy possessed by the neutrinos is lost by their escape from the reactor assembly.

It is the slow release of energy by fission product decay which demands a continued source of cooling for the core long after the fission process of the reactor has been terminated, and it was this which presented so many problems during the accident at the Three Mile Island nuclear power station in 1979 (see Chapter 19).

For various reasons the amount of energy usefully extracted from the fissioning of a single U-235 atom is about 200 MeV. To put this in more readily-understood terms, the fissioning of 1 g of U-235 in a nuclear

Table 3.6
Energy released by fission of a uranium-235 atom

Source of energy	Energy released	
	MeV	% of total
Fission products	167	81.5
Fission neutrons	5	2.4
All gamma rays	13	6.3
Beta particles	8	4.0
Neutrinos	12	5.8
Total	205	100.00

reactor would yield as much heat as would the burning of 3 million grammes (3 tonnes) of coal!

Chain reaction

It has been shown that when a U-235 nucleus is made to fission it emits two, and sometimes three neutrons which fly off at very high speed. If it could be arranged for these neutrons to be captured by other U-235 nuclei then these nuclei would also fission and even more neutrons would be emitted. Under such circumstances, once the fission process had been started the neutron population in a sample of U-235 would double for each successive fission event and an 'avalanche' situation would quickly result and vast amounts of energy would be released in the form of heat and nuclear radiation; this, in fact, is what happens in a nuclear reactor.

Nuclear 'avalanches' of the type just described are more accurately known as *chain reactions*. They form the basis of all nuclear reactors but can only come about if certain very important criteria are fulfilled. In the first place the neutrons emitted by the fissioned nuclei travel at speeds which are much too fast for them to be captured by other U-235 nuclei and so they must first be slowed down (moderated). Secondly, there must be a minimum quantity of fissile material available for such a chain reaction to be sustained, otherwise it will just 'flare up' and 'fizzle out'; remember, the U-235 content of natural uranium is only 0.7 per cent, the remainder being the non-fissile U-238 isotope. Thirdly, the physical arrangement of the uranium must be such that a *critical mass* exists, which is the minimum quantity of uranium of a given physical arrangement which will support a chain reaction. The best possible physical arrangement is a sphere since this demands the smallest quantity of uranium to create a critical mass. On the other hand, it would be impossible to create a critical mass, no matter how much uranium was used, if it were distributed in the form of a thin film over a large surface area.

Fourthly, the uranium itself and the material used for moderating the neutrons must be of exceptionally high purity or the impurity atoms will themselves capture those neutrons which would otherwise contribute towards sustaining the chain reaction. Additional considerations will be discussed when the nuclear reactor materials are described (see Chapters 14 and 15).

It is at this point interesting to note that the U-238 content of natural uranium is itself a contaminant since its presence dilutes the much less abundant U-235 isotope by a factor of about 140:1. This means that 140 times more uranium is required to create a critical mass of U-235 than would otherwise be necessary if the uranium was in the pure U-235 form.

It is also the reason why a nuclear reactor could never explode like an atom bomb; atom bombs are constructed from pure U-235 or pure plutonium-239; nuclear fuel is too heavily diluted by U-238.

Figure 3.10 illustrates the principles of a nuclear chain reaction using uranium-235. A single neutron (n), after being slowed down by a moderator, causes a U-235 nucleus to fission, an event which results in the production of two fission products (FP) and two fission neutrons. These neutrons, after being moderated, go on to cause the fission of two more U-235 nuclei and the creation of four more fission products and four more neutrons.

It is important to realize that the diagram shown in Figure 3.10 is a neatly simplified version of what happens in reality. It is impossible to segregate and moderate individual neutrons, since many billions of them

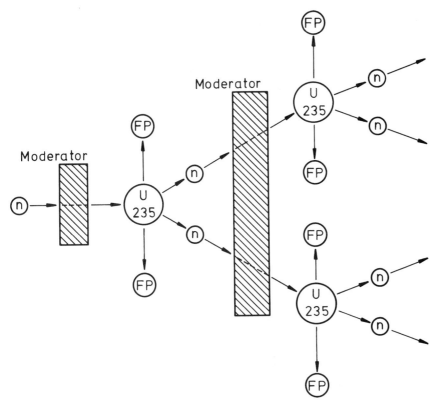

Figure 3.10   The creation of a nuclear chain reaction in U-235

are created in every cubic centimetre of uranium and they have to 'battle' their way out of the uranium before they ever come in contact with the moderator. Nevertheless the diagram is valuable in demonstrating the principles of a chain reaction.

Transuranic elements

Transuranic elements are so called because their atomic numbers extend beyond that of uranium (Z = 92) and, at the time of writing, range from Z = 93 (neptunium) to Z = 106 (as yet unnamed); they are all man-made and all radioactive (some are found in nature but at concentrations so small as to be negligible).

The first transuranic element to be discovered had an atomic number of 93 and an atomic mass of 238. Because of its numerical proximity to uranium it was given the name neptunium (Np), the name being derived from the planet Neptune – the next outermost planet in the solar system after Uranus, from which uranium had been named. This was in the winter of 1940–41; within a year another transuranic element had been discovered. This was found to be the daughter product of neptunium. It had an atomic number of 94 and an atomic mass of 239. It was given the name plutonium (Pu) after the planet Pluto, the next outermost planet beyond Neptune.

Plutonium is an extremely important element as far as nuclear power is concerned because, like U-235, it can be fissioned by neutrons and used as fuel in a nuclear reactor. There is, however, one very important difference between these two elements; *Pu-239 fissions better with fast (unmoderated) neutrons whereas U-235 fissions better with slow (moderated) neutrons*. You will appreciate the importance of this later when different types of nuclear reactors are discussed.

Plutonium is a silvery-white metal which has a density of 19.8 g/cm^3 and a melting point of about 640°C. For comparison, uranium has a density of 19.1 g/cm^3 and a melting point of 1133°C. Plutonium is known to exist in at least 15 isotopic forms ranging from $^{232}_{94}Pu$ to $^{246}_{94}Pu$, of which only Pu-239 and Pu-241 are fissile.

By far the most abundant of the plutonium isotopes is Pu-239. This is produced in a nuclear reactor by neutron capture in the U-238 content of the fuel. Unlike U-235, U-238 does not fission when it captures a thermal neutron; instead it is transformed into uranium-239. This is an unstable nuclide with a half-life of 23.5 minutes and decays by beta emission into neptunium-239 ($^{239}_{93}Np$). This nuclide is also unstable and with a half-life of 2.35 days decays by beta emission into plutonium-239 ($^{239}_{94}Pu$). The notation which describes the process is given by:

$$^{238}_{92}U + n \longrightarrow {}^{239}_{92}U \xrightarrow{\beta^-} {}^{239}_{93}Np \xrightarrow{\beta^-} {}^{239}_{94}Pu$$

Plutonium-239 is also unstable and decays by alpha emission into U-235. However, because of the very long half-life of Pu-239 (24 000 years), the build-up of U-235 is so slow that it may be completely disregarded. Once formed, therefore, Pu-239 can be looked upon as being stable and will build up at the same rate as U-239 is formed from neutron capture in U-238.

More will be said about the formation of other plutonium isotopes when nuclear reactors are described in Chapters 8 and 9.

Unified atomic mass unit

As has been explained earlier, the atomic mass is the total number of nucleons (protons and neutrons) contained in a particular nucleus. The U-238 nucleus, for example, is known to contain 238 nucleons (92 protons + 146 neutrons) and is therefore approximately 238 times heavier than the nucleus of elementary hydrogen which contains only a single proton and has an atomic mass of one. Although it is convenient to use atomic mass in this way for comparison purposes it does not give an indication of the actual weight of the nucleus in more familiar 'grams'. To do this it is necessary to make use of what is called the *unified atomic mass unit* (u). This is an internationally-agreed unit of mass defined as $\frac{1}{12}$th of the mass of the carbon-12 atom ($^{12}_{6}C$), the nucleus of which contains 6 protons and 6 neutrons. Measurements have shown that the weight of this atom is equal to 20.04×10^{-24} grams, from which it is calculated that $1\ u = 1.67 \times 10^{-24}$ grams.

If the masses of the proton, the neutron and the electron are each expressed in terms of the unified atomic mass unit it is found that the weight of the proton is 1.0072763 u whilst that of the neutron is slightly greater at 1.0086654 u; the weight of the electron on the other hand at 0.00055 u is almost 1850 times less than either of these particles.

THE MEASUREMENT OF RADIOACTIVITY

Radioactivity is measured in curies or, more recently, in becquerels. The curie (abbreviated Ci) was named after the 19th century French physicist Madame Curie and is used to indicate how many atoms in a given mass of radioactive material are disintegrating per second by way of radioactive decay. The unit was originally based on 1 gram of radium-226 whose atoms were known to disintegrate at the rate of 3.7×10^{10} per second. The curie was later more generally defined as being any mass of any radioactive material whose atoms are disintegrating at the rate of 3.7×10^{10} per second.

For example, the radioactivity of any mass of any radioactive substance – whether composed of a single type of element or of a mixture of elements – is said to be 10 Ci if measurements indicate that its atoms are disintegrating at a rate equal to 37×10^{10} disintegrations per second, a figure ten times greater than that which defines one curie; such a substance could be said to have a radioactivity equivalent to ten grams of radium.

As with other units, multiples and subdivision of the curie are indicated by prefixes: Megacuries (MCi, 10^6 Ci); kilocuries (kCi, 10^3 Ci); millicuries (mCi, 10^{-3} Ci); microcuries (μCi, 10^{-6} Ci); nanocuries (nCi), 10^{-9} Ci); picocuries (pCi, 10^{-12} Ci).

The curie is a unit which is used to define the amount of radioactivity from a given mass of material; it is not used to define the concentration. It is analogous to expressing the total population of a country rather than its population density. One millicurie of radioactivity, for example, may be measured in 1 million tonnes of seawater or in a piece of iodine-131 not much bigger than a pin head.

Another point which should be noted when studying the curie is that it gives no indication of the type of radiation which is emitted by the decaying atoms, nor the energy of this radiation; other units are required to do this as you will see.

The becquerel

The curie, along with many other units used to describe various aspects of radioactivity, has now been replaced by a new set of SI (Système International) units. The replacement for the curie is the becquerel, abbreviated Bq and named after the French physicist Henri Becquerel who discovered the phenomenon of radioactivity in 1896.

The becquerel is a much smaller unit than the curie and is defined quite simply as being the disintegration of one atom per second in any quantity of any radioactive material. The same prefixes as were used with curies are used for multiples and subdivision of becquerels, but with Giga- (10^9) and Tera-(10^{12}) as well.

Conversion factors

Since 1 becquerel is equivalent to 1 disintegration per second (dps) of radioactive atoms and 1 curie is equivalent to 3.7×10^{10} dps, then:

$$1 \text{ Bq} = \frac{1}{3.7 \times 10^{10}} \text{ Ci}$$

$$= \frac{10^{12}}{3.7 \times 10^{10}} \text{ pCi}$$

$$= 27 \text{ pCi}.$$

Similarly,

$$1 \text{ Ci} = 3.7 \times 10^{10} \text{ Bq}$$

$$= \frac{3.7 \times 10^{10}}{10^{9}} \text{ GBq}$$

$$= 37 \text{ GBq}.$$

Table 3.7 summarizes some equivalents between the two sets of units.

RADIATION DOSE

The sun's rays at mid-day on a cloudless summer's day in Spain, if distributed evenly over someone sunbathing, will cause that person to receive an evenly distributed dose of solar radiation.

If the sunbather restricts his exposure time to, say, 15 minutes each day for 8 consecutive days then, at the end of that time, he will have subjected himself to an effective exposure time of $8 \times 15 = 120$ minutes $= 2$ hours of mid-day sunshine and will have received a harmless total radiation dose resulting in, it is hoped, a healthy-looking golden tan.

Table 3.7
Conversion table for curies and becquerels

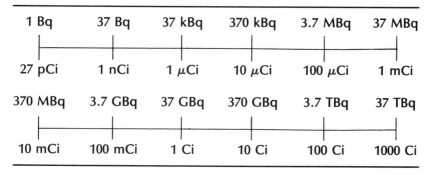

1 Bq	37 Bq	37 kBq	370 kBq	3.7 MBq	37 MBq
27 pCi	1 nCi	1 μCi	10 μCi	100 μCi	1 mCi

370 MBq	3.7 GBq	37 GBq	370 GBq	3.7 TBq	37 TBq
10 mCi	100 mCi	1 Ci	10 Ci	100 Ci	1000 Ci

In this example the sunbather has wisely restricted his exposure time so that the total absorbed dose is harmless.

It takes little imagination to guess what would happen if the sunbather were to absorb the same total solar radiation dose in one continuous exposure time of two hours from, say, 11 am to 1 pm. The solar radiation would, in this case, cause damage to the person's skin at a rate greater than the body's natural repair mechanism and would result in severe sunburn and (maybe) lasting harm to the body. The following conclusions may be drawn from these examples:

1 The total absorbed dose is dependent upon the intensity of the solar radiation and the total exposure time; very different results would have been obtained if the experiments had been conducted at the same spot in less intense mid-winter sunshine, or in more northerly climes.

2 The likelihood of harm being caused by solar radiation is determined not by the total absorbed dose but by the rate at which it is received. Within reason, virtually any total dose could be absorbed without harm, provided it was received in small enough doses spread over a long period of time.

The same could be said about the drinking of whisky. Drinking five bottles one after the other without a break would result in almost certain death whereas spread over a year or so in small doses would arguably be beneficial to the recipient!

Everything which has been said so far about solar radiation and its effects applies equally well to nuclear radiation, which is not surprising when one remembers that γ-rays, X-rays and light-rays are all part of the same electromagnetic spectrum and differ only in the magnitudes of their wavelengths.

The unit used for many years to specify the absorbed dose of nuclear radiation was the rad (an acronym derived from *r*adiation *a*bsorbed *d*ose) but this has now been replaced by the SI equivalent known as the gray (abbreviated Gy), from the name of Louis Harold Gray, a British scientist whose researches into radiation dosimetry contributed much towards a better understanding of the subject.

Absorbed dose is, in effect, the amount of energy dissipated (usually in the form of heat) when radiation is absorbed in a mass of material. This is why it is defined scientifically in terms of the units of energy (the erg or joule) and mass (the gram or kilogram). In purely scientific terms the rad is defined as an energy deposition of 100 erg per gram (erg/g), or 0.01 joules per kilogram (J/kg):

$$1 \text{ rad} = 100 \text{ erg/g}$$
$$= 0.01 \text{ J/kg.}$$

The relationship between the rad and the gray is:

$$1 \text{ rad} = 0.01 \text{ Gy}$$
$$1 \text{ Gy} = 100 \text{ rad} = 10^4 \text{ erg/g} = 1 \text{ J/kg.}$$

Multiples and sub-multiples used for the gray and the rad are the same as those used for the curie and the becquerel.

To give some idea of the hazards associated with absorbed dose, it is generally assumed that a whole-body radiation dose of about 400 rads (4 Gy) will result in a 50 per cent chance of death within a few days. Doses below 100 rads, however, although causing radiation burns, sickness and loss of hair, are virtually non-lethal.

Rems and sieverts

When studying the effects of radiation on living tissue it is found that the rad (or gray) is an accurate and satisfactory unit to use, provided its use is limited to radiation derived from X-rays, γ-rays and electrons (for example β-particles). Under such circumstances the biological damage caused by these types of radiation is very closely related to the energy which they deposit in living tissue; such a relationship fails to hold, however, for radiation derived from the much heavier neutrons, protons, alpha-particles and fission fragments and it is necessary to introduce a correction factor when studying the biological effects of these types of radiation. The correction factor so used is known as the quality factor, or more simply the 'Q'-factor, and has the following values for the type of radiation described:

$Q = 1$ for electrons (β-particles), X-rays and γ-rays

$Q = 10$ for protons and fast neutrons

$Q = 20$ for alpha particles and fission fragments.

When the Q-factor is incorporated into the rad as a multiplying factor the absorbed dose is said to have been converted into a biological dose equivalent and is expressed in another unit known as the rem (derived from *r*adiation *e*quivalent *m*an).

$$1 \text{ rem} = 1 \text{ rad} \times Q$$

At relatively low and intermediate dose levels the rem is a most valuable unit in the field of radiological protection because it indicates the biological effectiveness of all types of nuclear radiation, including delayed effects such as cancer and leukaemia. At high dose levels, however, experiments have shown that application of the Q-factor is inappropriate and that the rad is the more accurate unit to use.

The sievert (abbreviated Sv and derived from the name of Rolph Sievert, a Swedish scientist and at one time Chairman of the International Committee for Radiological Protection – ICRP) is the SI equivalent and replacement for the rem and is related to it, the rad and the gray as follows:

$$1 \text{ Sv} = 100 \text{ rem}$$
$$= 100 \text{ rad} \times Q$$
$$= 1 \text{ Gy} \times Q$$
$$1 \text{ rem} = 0.01 \text{ Sv}$$

The same multiples and sub-multiples are used for all four units.

Table 3.8 summarizes the units which have been described and illustrates the relationship which exists between them.

Table 3.8
Relationship between units of radioactivity and dosage

Unit	Used for measurement of	Definition	Equivalence
Curie (Ci)	Total activity	3.7×10^{10} dps	37 GBq
Becquerel (Bq)	ditto	1 dps	27 pCi
Rad (rad)	Absorbed dose	100 erg/g (0.01 J/kg)	0.01 Gy
Gray (Gy)	ditto	10^4 erg/g (1 J/kg)	100 rad
Rem (rem)	Biological effective dose	rad $\times$ Q	0.01 Sv
Sievert (Sv)	ditto	Gy $\times$ Q	100 rem

GLOSSARY OF TERMS

Alpha particle (α)   The positively-charged nucleus of a helium-4 atom ($^4_2$He). It comprises 2 protons and 2 neutrons. Many unstable nuclides decay by the emission of alpha particles.

Atom   The smallest quantity of any element which can exist and yet still possess all the chemical properties of that element.

Atomic mass (A)   The number of nucleons (protons and neutrons) contained in the nucleus of a particular atom.

Atomic number (Z)   The number of protons contained in the nucleus of a particular nuclide.

Becquerel (Bq)   The SI unit used to express the total radioactivity of any mass of material; it is equivalent to one atomic disintegration per second. The becquerel has now replaced the previously used unit known as the curie.

Beta particle (β⁻)   A negatively-charged electron which originates from the nucleus of an unstable atom. It is created when a neutron changes into a proton and is ejected from the nucleus at high speed.

Compound   A mixture of elements which may, or may not be, of the same type.

Electron   A negatively-charged particle having a mass equivalent to 0.00055 u (unified atomic mass units). The magnitude of the negative charge is equal to the positive charge of the proton.

Electron capture (EC); also known as K-capture   A radioactive decay process in which the nucleus of an unstable atom captures an electron from the K-shell orbit.

Gamma rays (γ)   Packets of electromagnetic radiation emanating from the nucleus of an unstable atom.

Gray (Gy)   The SI unit used to express the absorbed dose of nuclear radiation; it is equivalent to 10^4erg/g of dissipated energy. The gray has now replaced the previously used unit known as the rad.

Half-life   The time taken for half the number of atoms in a radionuclide to decay.

Isotope   A different type of the same element. Isotopes of the same element differ only in the number of neutrons contained in their nuclei.

Molecule   The smallest quantity of any compound which can exist and yet still retain all the chemical properties of that compound.

Neutron   A neutral particle having a mass equivalent to 1.0086656 u (unified atomic mass units).

Nuclear transformation or *nuclear conversion*   This is brought about when a nuclide is forced to accept an additional neutron into its nucleus. The process results in the creation of an isotope whose mass number (A) is one unit higher than that of the original nuclide. Most radioisotopes are produced in this way. Nuclear transformation does not involve any change in Z-number and hence does not bring about any change in the element represented by the target nuclide.

Nuclear transmutation   The changing of one element into another. Nuclear transmutation comes about through the process of radioactive decay, or from the impact of atomic particles emitted by a radioactive nuclide or generated by a nuclear reactor or particle accelerator.

Nucleons   The particles contained in the nucleus of an atom, i.e. protons and neutrons, used when referring to a particle in the nucleus without specifying its type.

Nucleus   The central core of an atom containing protons and neutrons.

Nuclide   A word used when referring to the nuclear structure of an element without implying that it may or may not exist in more than one isotopic form.

Positron (β^+)   A positively-charged electron which is otherwise identical to the electron. It is formed in the nucleus of an unstable atom when a proton changes into a neutron and is ejected from the nucleus at high speed.

Proton   A positively-charged particle having a mass equivalent to 1.0072763 u (unified atomic mass units).

Q-factor   A multiplying factor used in conjunction with the sievert to take into account the differing biological effects of various types of nuclear radiation.

Sievert (Sv)   The SI unit used to express the biological equivalent of absorbed dose of nuclear radiation; it is equivalent to the gray, multiplied by the relevant Q-factor. The sievert has now replaced the previously-used unit known as the rem.

Unified atomic mass unit (u)   The international unit of mass equal to $\frac{1}{12}$th of the mass of the carbon-12 atom. Its weight is equal to 1.67×10^{-24} grams.

X-rays   Packets of electromagnetic radiation emanating from the regions occupied by the orbital electrons of an unstable atom.

4 A summary of the nuclear fuel cycle

Figure 4.1 illustrates the operating principles of what is known as the *nuclear fuel cycle* and, in effect, summarizes what is to appear in the following chapters.

The operation starts with the exploration and mining of uranium ore in many countries as far apart as Canada, Australia, South Africa and the Soviet Union. After grading and some initial chemical treatment the ore is reduced to a concentrated form which, because of its bright yellow colour, is known as *Yellow Cake*. This is exported all over the world to countries such as Britain, France and Germany which possess the technology required for manufacturing nuclear fuel elements for their own – and other countries' – nuclear reactors.

On arriving at the fuel element manufacturing plant the ore is dissolved in acid and the uranium separated from the residue. At this stage about 0.7 per cent of the uranium is U-235. The uranium is then converted either into solid rods of pure uranium metal or mixed with fluorine to form a compound known as uranium hexafluoride – more commonly known simply as hex.

The rods of natural uranium are then clad in metal containers made from a magnesium–aluminium alloy known as Magnox and sold as fuel elements to the first generation of British nuclear power stations which, because of the type of fuel they use, are called Magnox stations; there are eleven of these in Britain and two overseas.

The uranium in the form of hex is sent to the uranium enrichment plant

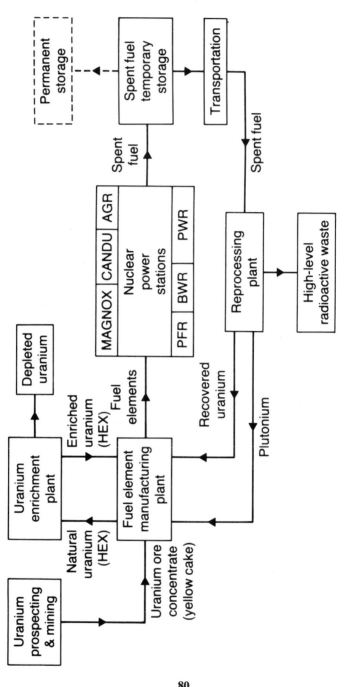

Figure 4.1   The nuclear fuel cycle

where its U-235 content is artificially raised from its normal concentration of 0.7 per cent up to a level which is typically between 2 per cent and 3 per cent. After this has been done the uranium – still in the form of hex – is returned to the fuel element manufacturing plant where it is used to make small ceramic pellets from enriched uranium dioxide (UO_2) powder. The residue from the enrichment process is a type of uranium whose U-235 content has been severely reduced; this is known as depleted uranium and large quantities of it are stockpiled at the British enrichment plant at Capenhurst in Cheshire. More will be said about depleted uranium when the so-called fast breeder reactor (FBR) is described.

The enriched UO_2 pellets fabricated at the fuel element manufacturing plant are packed into metal containers made from either stainless steel, or from a zirconium alloy known as *Zircaloy*, and sold to nuclear power stations which make use of this type of fuel. The stainless-steel-clad fuel is used in Britain's second generation of nuclear power stations which are based on the AGR type of reactor (advanced gas cooled reactor); the Zircaloy-clad fuel is used in power stations throughout the world which are based on the BWR (boiling water reactor) and PWR (pressurized water reactor) types of reactor. There are five AGR stations in operation in Britain and two more under construction. At the time of writing there are no BWR or PWR reactors operating in Britain.

The so-called CANDU reactor (the name is an acronym derived from CANadian Deuterium Uranium) also uses uranium dioxide fuel but its uranium is not enriched.

After remaining in the reactor for typically five years or so, the fuel elements have to be removed from the core and replaced with fresh ones. This is not because all of the fissile material in the fuel, that is the U-235, has been 'burnt up' but simply because the fissioning of other U-235 atoms has resulted in an unacceptable build-up of fission products (see page 63). If these are allowed to remain in the reactor core they will 'soak up' more and more valuable neutrons, which would otherwise be used to maintain the nuclear chain reaction, until they eventually stop it working altogether. There are other important reasons why the build-up of the fission product impurities makes it necessary to remove the fuel from the reactor but these will be discussed in greater detail in Chapter 10.

Some of the fission products (iodine-131 for example) have half-lives of only a few days whereas others (caesium-137 for example) have half-lives of many years. It is the intense radioactivity of the short-lived fission products and the associated heat which they produce, which makes it necessary to store the spent fuel elements for 100 days or so after it has been removed from the reactor before it can be transported elsewhere.

This temporary storage takes place on the site of the nuclear power station and usually in water-filled 'cooling ponds'. What happens now depends upon the policy of the power station operator. In Britain, France and many other countries, the spent fuel is transported to what is known as the *reprocessing plant* (which may be in another country) where the unburnt uranium, and some plutonium, is separated for further use and the fission product residues are stored as high-level radioactive waste. In the USA and Canada, on the other hand, the economic advantages of recovering the unburnt uranium are quite small since both have substantial reserves of uranium and are, in fact, uranium exporters. In these countries, therefore, the spent fuel is transported to a permanent repository where it is safely stored without further treatment. This method of operation is known as the once-through cycle.

The uranium which is recovered from the spent fuel at the reprocessing plant is sent back to the fuel manufacturing plant and after suitable treatment is sent to the enrichment plant to have its original U-235 content restored. On return to the fuel manufacturing plant the uranium is used to make more fuel elements.

The plutonium which is recovered from the spent fuel originates in the reactor when a small amount of the U-238 content of the fuel is transmuted to Pu-239 by way of neutron capture (see p. 70). The plutonium recovered in this way is used to make fuel elements for the fast breeder reactor and will be described more fully in Chapter 9.

5 Uranium mining and exploration

URANIUM RESOURCES

Uranium is a relatively abundant element widely distributed throughout the world in concentrations ranging from about 0.001 to 4 per cent. In fact, uranium is more abundant than silver and mercury and it has been estimated that more than one million million (10^{12}) tonnes of it exists within the first kilometre or so of the earth's crust, albeit at concentrations which, at present, make it uneconomic to recover. In addition to the uranium found in the earth's crust, about 4000 million tonnes are estimated to be present in the world's oceans, although at extremely weak concentrations: in the region of three parts per billion, or three atoms of uranium for every one thousand million molecules of water!

Excluding the communist bloc there are eight countries with huge uranium deposits: Australia, Brazil, Canada, Namibia, Niger, South Africa, the USA and Zaire. In addition to these countries significant deposits of uranium are found in Argentina, France, Gabon, Greenland, India and Sweden. Britain has no economically-recoverable reserves of uranium and imports all its requirements, mostly from Canada.

Mining of uranium is at present considered economic when the uranium concentration of the ore exceeds about 0.05 per cent, that is 500 parts per million (ppm). A report published by the Organization for Economic Cooperation and Development (OECD) in 1978 indicated that within the non-communist world there are about 2.3 million tonnes of known uranium reserves of economically-recoverable concentrations and a further 2.7 million tonnes of so-called estimated additional resources, making 5 million tonnes overall. However, since large areas of the world have yet to be explored for uranium these figures are far from definitive and are of value only for general guidance.

The world's largest operating mine, Key Lake, began production in October 1983 in Saskatchewan, Canada, and is now producing uranium oxide (U_3O_8) at the rate of nearly 5500 tonnes per year. The production from the Key Lake mine, combined with that from Canada's other uranium mines, has placed Canada in first place among the free world's uranium producers. Key Lake is a shallow, open-pit mine with estimated reserves of nearly 89 000 tonnes of U_3O_8 at an average grade of 2.35 per cent; one of the highest in the world.

URANIUM REQUIREMENTS

The total amount of natural uranium required to sustain a nuclear reactor throughout its operational life depends upon many factors including, for example, the type of reactor, its thermal efficiency and operating load factor; whether or not it uses enriched fuel; whether or not the spent fuel is reprocessed and, if reprocessed, the value assigned to the 'tails' at the uranium enrichment plant, and the burn-up rating of the fuel.

To give some idea of lifetime fuel consumption, a 1000 MWe AGR reactor operated for 30 years at 65 per cent load factor, and with a fuel burn-up rating of 18 000 MWd/t, would consume approximately 3100 tonnes of natural uranium; this is the amount of uranium which must be extracted from the ore supplied to the fuel manufacturing plant and assumes that the spent fuel is recycled through the reprocessing plant. If the burn-up rating of the fuel is increased to the design target of 24 000 MWd/t then the lifetime fuel consumption falls to 2450 tonnes. If the spent fuel is not reprocessed then the lifetime consumption figures rise to 4200 tonnes for 18 000 MWd/t burn-up and to 3250 tonnes for 24 000 MWd/t burn-up. Comparable figures for a PWR reactor of the same power rating, and 33 000 MWd/t target burn-up, are 3350 tonnes *with* spent fuel reprocessing, and 4300 tonnes *without* reprocessing. These figures are summarized in Table 5.1. The important points to observe from the table are, firstly, that the act of reprocessing reduces considerably the lifetime uranium requirements for both types of reactors (about 25 per cent for the AGR and 22 per cent for the PWR). Secondly, for the 18 000 MWd/t burn-up rating of the AGR there is very little difference in the lifetime uranium requirements of the two types of reactors. However,

Table 5.1
Lifetime uranium requirements for the AGR and PWR reactors

Reactor type	Natural uranium requirements (tonnes)*	
	Without reprocessing	With reprocessing
AGR (18 000 MWd/t)	4200	3100
AGR (24 000 MWd/t)	3250	2450
PWR (33 000 MWd/t)	4300	3350

*These figures assume a 30-year life at 65% load factor and a power rating of 1000 MWe.

when the AGR is eventually operated at its target burn-up rating of 24 000 MWd/t the difference between the two types of reactor will be quite substantial, the AGR requiring approximately 25 per cent less fuel than the PWR.

URANIUM PROCUREMENT

The responsibility of ensuring Britain's supplies of uranium rests with the British Civil Uranium Procurement Directorate (BCUPD) and its executive unit the British Civil Uranium Procurement Organization (BCUPO). These organizations were set up in 1979 by the Central Electricity Generating Board (CEGB, which supplies the staff for BCUPO), the South of Scotland Electricity Board (SSEB) and British Nuclear Fuels plc (BNFL). The functions of BCUPO are to monitor developments in world policies relating to energy, and nuclear energy in particular, and to estimate what effects, if any, these policies may have on the security and cost of Britain's uranium supplies. The organization maintains close working contacts with member organizations on such matters as international nuclear and non-proliferation policies and UK nuclear fuel cycle and reactor policies. In addition it provides an advisory service to UK government departments and representation at international level. BCUPO also participates in the activities of the Uranium Institute, an international association of many large companies engaged in uranium mining, processing and consumption.

An interesting and very important feature of uranium is that its contribution to the overall cost of nuclear-generated electricity is considerably less than that of coal and oil in the fossil-fuelled stations. The result is that nuclear generating costs are relatively insensitive to the price of uranium. It is competition within the widely dispersed uranium market itself, coupled with enrichment and reprocessing costs, which have the greatest influence on the price of uranium. Uranium is also easy to transport and because of its high energy density requires very little storage space. The consequence of all this is that, unlike many other types of fuels, the amassing of large strategic and commercial stocks of uranium is both feasible and sensible.

An unfortunate factor which has an important influence on the availability of uranium is its undeniable link with atomic weapons, bearing in mind that virtually all nuclear explosives are derived from the separation of U-235 from naturally-occurring uranium, or from the production of Pu-239 from the transmutation of naturally-occurring U-238. It is the fear of nuclear proliferation which makes uranium a politically-influenced commodity and which has, in some countries, resulted in severe restrictions on its export.

URANIUM MINING

In 1789 a German apothecary by the name of Martin Klaproth was given a sample of a blackish ore which later became known as pitchblende. In analysing the sample in his laboratory he identified a hitherto unknown substance which he christened uranium. He chose this name after the planet Uranus which itself had been discovered earlier in the same decade (1781) and which derived its name from the Greek god of the heavens. Although unaware of it at the time, what Klaproth had discovered was not the pure element uranium but its oxide. It was not until 1841 that Eugene Peligot managed to isolate pure uranium at the Conservatoire des Arts et Metiers in Paris. Further work on this newly-discovered material led to the production of various uranium salts which were found to be unusually bright in colour and which were quickly made use of in the commercial world for imparting a greenish-yellow tint to leather, pottery, glass and wood.

Pitchblende is a massive and non-crystalline variety of the primary mineral uraninite, a form of uranium oxide containing also the oxides, sulphides and arsenides of lead, iron, thorium and other elements. Uraninite is brownish-black in colour and occurs in granite rocks and metallic veins. Pitchblende is found in veins in Norway, Bohemia, Zaire, Canada and, in small quantities, in Cornwall.

Uranium is found in many geological formations; in phosphate deposits, in shales, in granite and sandstone deposits, often as a secondary mineral. A typical uranium ore is *carnotite*. This is found as a yellow impregnation in sandstone, particularly in Colorado, USA, and contains a hydrated mixture of uranium, vanadium and potassium. It is frequently mined primarily for its vanadium content; the uranium, if extracted, is a valuable by-product. *Autunite* is another valuable ore of uranium. It, too, is yellow in colour and is a hydrous mixture of uranium, calcium and phosphorus. Often found in close proximity to autunite is the ore known as *torbernite*. Rich-green in colour, this ore contains uranium, copper and phosphorus and is sometimes known as copper (or cupro-) uranite. Valuable uranium of about 0.02 per cent concentration is extracted from the gold ore residues from South African gold mines. Furthermore, the pyrites (the geological term for sulphide of iron, commonly known as fools gold because of its brassy-yellow colour) also present in the residues can be processed and used to produce sulphuric acid which is used in processing the uranium.

Yellow cake

To reduce the amount of material which must be transported between the uranium mine and the nuclear fuel manufacturing plant the uranium ore is first concentrated by a chemical leaching process using a powerful acid or alkali, depending on the nature of the ore. Ores containing large amounts of calcium carbonate, for example, are leached with sodium carbonate whereas others may be leached with sulphuric acid. The resulting ore concentrate is the brightly coloured 'Yellow Cake' (shown in Figure 5.1) consisting mostly of a mixture of various sorts of uranium oxide ranging from UO_2 to U_3O_8. The concentration of uranium oxide in the yellow cake is typically within the range 70 to 90 per cent.

URANIUM EXPLORATION

Many of the rich uranium deposits which have been identified throughout the world were discovered by inexperienced, and often part-time, prospectors armed only with a primitive Geiger counter which responded to gamma, and sometimes beta, radiation. To be more accurate, what the prospectors discovered at the time were surface rocks or areas of terrain which showed signs of being radioactive. What the prospectors did not

Figure 5.1 The Mary Kathleen uranium mine in Queensland, Australia

know was whether the radioactivity originated from uranium (or its decay products), or from thorium or potassium; the three most common naturally-occurring radioactive materials. This is where the specialized knowledge and experience of the nuclear geologist comes into play. By using more sophisticated instrumentation and special chemical techniques he can determine from a sample supplied by the prospector not only its general chemical composition but also how much of the radiation being emitted is from alpha, beta and gamma radiation. He can further deduce, by carrying out a spectral analysis of the radiation, the isotopic composition of the radioactive content of the sample and, in so doing, determine if the parent atom is present with its decay products, and in the correct proportions, that is in *equilibrium*. This is very important information because the geologist knows that gamma radiation from uranium alone is very weak and that the gamma radiation detected by the prospector's Geiger counter in the field, if not from thorium or potassium, is almost certainly from the uranium decay products – principally bismuth and radium. If some or all of the decay products have been leached out of the uranium-bearing ore by rainfall, for example, then the prospector's radioactive sample is virtually useless because it contains little or none of the uranium he has been seeking. On the other hand, what the prospector *has* done is to bring to the attention of the geologist areas of abnormally high radioactivity which might be worthy of further investigation. The geologist is well aware of the effects of leaching and by visiting the site from where the sample was obtained he may be able to apply his knowledge of the local geology to pinpoint the location of the uranium from which the decay products were leached. This may involve the in-situ examination of a few samples taken from just below the surface, or even drilling a metre or so into the ground using a portable percussion drill and measuring the radon gas content of the hole. If the results of a preliminary survey are encouraging then this will usually be followed up by a more detailed car-borne (and sometimes air-borne) survey lasting many weeks.

Geological instrumentation

Electronic instruments used for uranium exploration and in-situ analysis fall into two main categories. The types used for initial large-area surveys of terrain are usually mounted in aircraft and powered from 'mains-type' power supplies. Similar instrumentation is used in vehicles for follow-up surveys of promising areas, or on ships for sea-bed surveys. Instruments of this type are generally heavy, bulky, not weatherproof, and are often used in the laboratory for routine work. The second category of instrumentation is the portable, battery-operated types designed specifically for the field geologist or prospector. Instruments of this type are small, lightweight, completely weatherproof and able to operate reliably for long periods over a wide temperature range. Instruments of this type enable the geologist to roam freely over rough terrain in all types of weather ranging from the high temperatures of the southern Sahara in mid-summer to the cold of the Northern Territories of Canada in mid-winter.

The accuracy and reliability of portable instruments varies considerably with price but the professional geologist uses types which are extremely rugged and fully sealed against the ingress of dust, moist air and water. They are able to operate from sub-zero to temperatures in excess of 50°C and to survive the many bumps and accidental immersions which always happen during field surveys. A good example of a modern field instrument is the four-channel gamma spectrometer illustrated in Figure 5.2. This particular instrument is fully portable, battery operated, and was designed by the author at The Harwell Laboratory to meet the needs of the field geologist. It is hermetically sealed against moisture and able to operate with guaranteed accuracy over the temperature range −10°C to 60°C. It may be carried by a shoulder strap or mounted within a vehicle or aircraft, or on a ship, and connected to a wide variety of radiation detectors which are also hermetically sealed and designed to suit the application. The type of detector shown in Figure 5.2 may be held in the hand or connected to an extension rod so that it may be used to examine an inaccessible rock face or crevice. Much larger and more sensitive detectors may be mounted on the floor of an aircraft, or on the top of a pole fitted to the roof of a land survey vehicle, and connected to the instrument by a long length of ordinary coaxial cable. Detectors of this type are often fitted inside a shock-proof stainless steel container and towed along the seabed behind a ship to survey the radioactive content of the seabed. In this particular application the coaxial cable may be up to 1000 m in length and protected with an outside layer of galvanized steel strands. Much smaller detector probes of about 25 mm diameter are often used to log the radioactive content of deep boreholes to determine the presence, depth and thickness of uranium-bearing strata.

The instrument is able to measure the intensity of the gamma radiation it detects and to sort it into four energy bands or channels. One band is devoted to total radiation detected, irrespective of origin; the others are devoted separately to detecting radiation originating from uranium, thorium and potassium. This ability to carry out an 'on the spot' analysis is extremely valuable to the geologist and is a facility which was denied the early prospectors. Naturally, it is essential that the pre-set energy bands

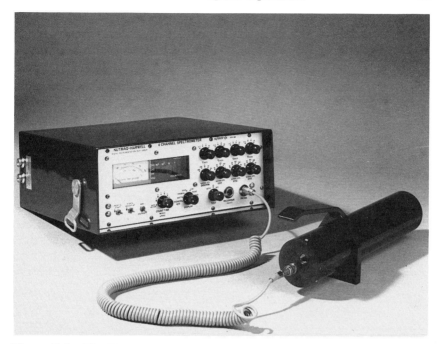

Figure 5.2   The Harwell 4-channel portable spectrometer

remain absolutely stable, irrespective of wide variations in temperature, otherwise the information presented to the geologist and his subsequent analyses will be of little value. In this instrument the stability is guaranteed by an internal self-checking circuit which immediately senses any drift due to temperature or ageing of critical components and automatically adjusts the relative positions of the energy bands to compensate for the initial drift. It is this facility which enables the instrument to be used over such a wide temperature range and which makes it such a useful tool.

REFERENCE

World Uranium Potential: an International Evaluation, Joint Report by the OECD Nuclear Energy Agency and the IAEA, December 1978.

6 Fuel element manufacture

One of the world's biggest fuel element manufacturing plants is that operated by British Nuclear Fuels plc (BNFL) at Springfields in Lancashire. The plant, established in 1946, processes many thousands of tonnes of uranium ore each year and from it manufactures nuclear fuel elements for British and overseas reactors.

The uranium arrives at the plant as yellow cake contained in 200 litre drums (Figure 6.1). It is first sampled and assayed and then dissolved in hot nitric acid. This produces a slurry comprising impure uranyl nitrate (liquefied uranium in nitric acid) and some insoluble impurities. The solid impurities are removed by filters and the uranyl nitrate is purified by mixer–settler solvent extractors using tributyl phosphate in odourless kerosene (TBP/OK) as the extracting agent. This stage in the process is a very important one in the production of nuclear fuel because it removes undesirable impurities such as boron and cadmium, both of which have very high neutron capture cross-sections. It also removes impurities which may contaminate the uranium hexafluoride (UF_6) – hex – produced at a later stage.

The dilute solution of purified uranyl nitrate which emerges from the solvent extractor stages is concentrated by a series of evaporator units, rotary kilns and fluidized beds, and in the process is converted into uranium trioxide (UO_3). The next stage in the process is the conversion of UO_3 into uranium tetrafluoride (UF_4). This is a two-stage process involving the hydrogen reduction of UO_3 into uranium dioxide (UO_2) and the

Figure 6.1   Arrival of uranium ore concentrates at the fuel manufacturing plant

hydrofluorination of UO_2 into UF_4. The resultant UF_4 is then conveyed to intermediate storage hoppers prior to its subsequent conversion to uranium metal for Magnox fuel elements, or to uranium hexafluoride (hex) for the uranium enrichment process.

URANIUM METAL PRODUCTION

The solid rods of pure uranium metal used in Magnox-type fuel elements are produced from the UF_4 by first mixing it with magnesium metal raspings and then compressing the mixture to form pellets weighing about 3 kg. These are then stacked inside a graphite-lined stainless steel vessel and loaded into a heating furnace which is held at a temperature of between 600°C and 900°C. This initiates a reaction between the UF_4 and the magnesium and the melting of the uranium metal content of the mixture. The molten uranium falls into a base catchpot and, after cooling, forms a metal billet weighing about 350 kg. It is from billets like this that uranium rods of various shapes and sizes are later produced to suit different types of reactors.

A typical Magnox fuel element produced by the Springfields plant is shown in Figure 6.2; it is of the type used in the Calder Hall reactors. It consists of a solid rod of natural uranium metal weighing about 11.5 kg and measuring 28 mm diameter and 1 m in length. It is clad in a tight-fitting Magnox can and has a heat-energy equivalence of nearly 150 tonnes of coal. The outer surfaces of the can take the form of a spiral arrangement of herring-bone cooling fins through which the cooling gas flows when the element is in the reactor core. Each of the four reactors at the Calder Hall power station has more than 10 000 fuel elements of this type and a total uranium inventory of over 100 tonnes.

HEX PRODUCTION

The production of uranium hexafluoride (hex) for the uranium enrichment plant is based upon the chemical reaction of uranium tetrafluoride (UF_4) with elemental fluorine (F). The reaction takes place in a fluidized bed chemical reactor containing calcium fluoride granules maintained at a temperature of about 460°C.

The UF_4 is injected into the reactor at a carefully controlled rate by a screw-type feeder and results in the generation of gaseous uranium hexafluoride (UF_6). The gaseous hex is then passed through a series of filters to remove any entrained solids and then to a condenser which is held at a temperature of -30°C. The low temperature of the condenser causes the

Figure 6.2   A typical Magnox fuel element

hex to solidify. When this has taken place the condenser is removed from the fluidized bed reactor and its contents raised to a temperature of 90°C, causing the hex to liquefy. Finally, the liquefied hex is poured into containers like those shown in Figure 6.3 and allowed to cool to normal ambient temperatures, whereupon it reverts once more to the solid phase. The hex is now in a form suitable for shipment to the enrichment plant.

The nature of hex

At temperatures below about 60°C, hex is a transparent crystalline solid which has a whitish appearance not unlike that of a frosted light bulb. Hex is a very dense material, because of its uranium content – which also accounts for it being mildly radioactive.

Solid hex is a highly corrosive material which reacts vigorously with water, alcohol and ether. On the other hand it is relatively inert in atmospheres of oxygen, nitrogen or dry air and it is not excessively corrosive when in contact with copper, nickel, aluminium and teflon. It will,

Figure 6.3 Containers of uranium hexafluoride (hex) at the Springfields fuel element manufacturing plant. Each container weighs 12 tonnes

however, attack glass rapidly unless the glass is absolutely free from water. If allowed to come in contact with moist air it will react vigorously with the water content of the air and form a highly toxic liquid called hydrogen fluoride (HF). This is one of the world's most dangerous substances. It fumes strongly in air, that is it forms a cloud of tiny particles, and if the liquid is brought in contact with human skin it will penetrate the skin and rot the bone marrow of the body's skeleton.

When hex at normal atmospheric pressure is slowly heated in a closed container it sublimes, that is it changes directly into a gas without going through the liquid phase, as soon as the temperature exceeds about 60°C. When the gaseous hex so formed is slowly cooled to below 60°C the hex de-sublimes, that is it changes back into a solid, and reforms as an accumulation of fluffy snow-like material. In the gaseous form, hex is a very unpleasant choking poison.

If the heating is carried out whilst the hex is under pressure, it will not sublime but will, instead, slowly change into a clear, colourless, mobile liquid of high density. It is in this form that hex is conveyed from point to point around the enrichment plant and fed into empty transport containers.

There was much concern expressed in the world's media when, in the autumn of 1984, a small freighter called the *Mont Louis*, carrying containers of hex from a French port for enrichment in the Soviet Union and eventual return to France, was in collision with a cross-channel steamer and sank in the eastern reaches of the English Channel. 'What would happen', it was loudly asked, 'if one of the containers of liquid hex lying either in the sunken hold of the *Mont Louis* or on the seabed itself should fracture or burst?' The answer is that little danger of any kind would have been caused, even if one of the containers had burst – none of them did and all were eventually recovered intact. There would, of course, have been a violent reaction between the escaping hex and the sea but the massive dilution and turbulence afforded by the sea, coupled with the offshore wind, would have quickly dispersed the hex and its dangerous reaction products such as HF and reduced them to harmless levels.

It is perhaps of interest, in passing, to ask why a French freighter should have been carrying hex for enrichment in the Soviet Union in the first place, when France has fuel-enrichment capacity of its own in abundance. The reason shows up in one of the more unexpected effects of the United States McMahon Act of 1946. France was one of the first European countries after the War to decide on a national policy of producing electricity by the use of nuclear power. Opting for the Westinghouse design of the boiling water reactor (BWR – see Chapter 8), it needed enriched uranium for its fuel. Denied this by the USA, France entered into a long-term commercial contract for certain tonnages of hex to be enriched for it every year in the Soviet Union. The contract has not yet expired.

OXIDE FUEL

After having its U-235 content increased at the enrichment plant the hex is returned to the fuel element manufacturing plant where it is used to make oxide fuel for British AGR and American-type BWR and PWR reactors. Although the fuel elements used in these reactors differ considerably in physical size and shape and use different types of cladding material they all contain small ceramic pellets which have been formed from uranium dioxide (UO_2) powder containing varying amounts of U-235 to suit each type of reactor.

The first stage in the conversion of enriched hex into ceramic oxide fuel takes place in a rotary kiln in which hex vapour is made to react with hydrogen and steam. This results in the formation of ceramic grade

Figure 6.4   Enriched UO_2 ceramic pellets for the British AGR being examined at BNFL's Springfields plant

uranium dioxide powder which is fed into storage hoppers. After grinding to the correct texture and mixing with a binding agent the UO_2 powder is poured into moulds where 60-tonne multi-stage hydraulic presses form them into pellets of the required shape and size. These so-called 'green' pellets are then heated in furnaces to a temperature of 800°C, a process which removes the binding agent by evaporation.

After the debonding process the pellets are fed to sintering furnaces where they are heated in a hydrogen atmosphere to a temperature of 1650°C. On leaving the furnaces the pellets appear in a very hard ceramic form like those used in the British AGR reactor and shown in Figure 6.4. The pellets shown in the picture are approximately 14.5 mm in diameter and 14.5 mm in height and weigh about 20 grammes. The heat energy equivalence of each pellet is about 1.5 tonnes of coal.

Oxide fuel elements

After careful inspection the oxide pellets are automatically loaded into tubular-shaped cans made from stainless steel – for British AGRs – or from Zircaloy (an alloy of zirconium) for BWRs and PWRs. The cans are then

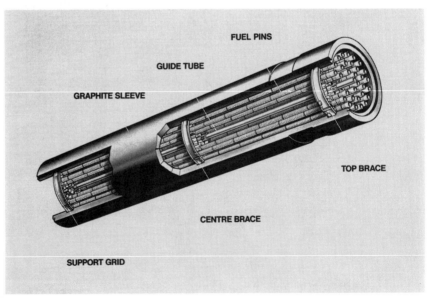

Figure 6.5   An oxide fuel element of the type used in the British AGR

filled with helium gas to aid heat transfer and the ends are sealed with leak-tight caps. The cans when filled in this way are known as *nuclear fuel pins*. Each AGR fuel pin contains 68 pellets and possesses a heat energy equivalence of about 83 tonnes of coal.

A *nuclear fuel element* is formed from a cluster of fuel pins and takes many forms, depending upon the type of reactor in which it is to be used. Some clusters are of circular shape whereas others may be of square or hexagonal shape. Figure 6.5 illustrates the physical arrangement of an AGR fuel element cluster formed from a circular grid arrangement of 36 pins. The pins are supported and separated from each other by support grids at the centre and at each end of the cluster and are enclosed in a double layer graphite sleeve. The sleeve functions as both a physical containment and as part of the overall moderator which will be discussed more when the nuclear reactor is examined. Figure 6.6 shows AGR fuel elements being assembled at the manufacturing plant. Each element has the heat energy equivalence of about 3000 tonnes of coal.

Figure 6.6 Fuel elements for the British AGR being inspected at the manufacturing plant

FAST-REACTOR FUEL

The fuel used in the core of the Prototype Fast Reactor (PFR) takes the form of an assembly of cylindrically-shaped ceramic pellets like those shown in Figure 6.7; each pellet measures 5 mm in diameter and 7 mm in length. Its chemical composition is very different from that of the fuels used in the AGR, BWR and PWR reactors and consists of a homogeneous mixture of (typically) 25 per cent plutonium dioxide (PuO_2) and 75 per cent UO_2. The main fissile constituent of the fuel is the Pu-239 content of the PuO_2 which functions in the same way as does the U-235 content in other types of fuel.

The photograph shown in Figure 6.8 illustrates the physical appearance of a PFR fuel element. The element, also known as a fuel 'sub-assembly', is formed from a hexagonally-shaped cluster of 325 stainless steel fuel pins containing the pellets, the cluster being held rigidly in position by a tight fitting outer wrapper. The pins are each of 2.74 m in length and 5.8 mm in outside diameter and have a wall thickness of 0.4 mm. Each pin contains

Figure 6.7   Mixed oxide fuel pellets for the PFR

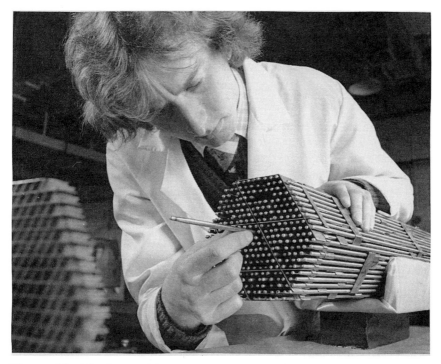

Figure 6.8 Final assembly of a PFR fuel element

910 mm of PuO_2/UO_2 pellets symmetrically distributed about its centre and 227.5 mm of UO_2 pellets at each end; making an overall pellet length of 1.365 m. The uranium in the UO_2 pellets is of the depleted type (uranium with very little U-235) and plays only a small part in maintaining the chain reaction when present in the core. The uranium content of the PuO_2/UO_2 pellets is of natural composition and contains 0.7 per cent U-235.

The centre region of the pin containing the length of PuO_2/UO_2 pellets forms the fuel content of the pins and the lengths of depleted UO_2 pellets at each end form what is known as the *axial breeder*; this will be explained in greater detail in Chapter 9. The total number of pellets of all types in each fuel pin is about 270, which means that each completed sub-assembly contains more than 74 000 pellets.

BNFL's Springfields plant manufactures all the uranium components used in the core of the PFR, for example, UO_2 breeder pellets and UO_2 breeder assemblies described later; it also produces UO_2 in the form of powder which is eventually blended with PuO_2 at the company's Sellafield plant to produce the PuO_2/UO_2 core pellets. The Sellafield plant uses the UO_2 powder and UO_2 pellets it receives from Springfields to produce the PuO_2/UO_2 core pellets and to assemble the individual fuel pins and completed sub-assemblies ready for use in the PFR.

PWR FUEL

Fuel used in the pressurized water reactor (PWR) is in the form of small pellets of enriched uranium dioxide which are similar in appearance to the pellets used in the AGR fuel elements. The U-235 enrichment, however, is slightly higher for the PWR and varies typically between 2.1 and 3.1 per cent, depending upon its intended position in the reactor core.

PWR pellets are approximately 8.2 mm in diameter and 13.5 mm in

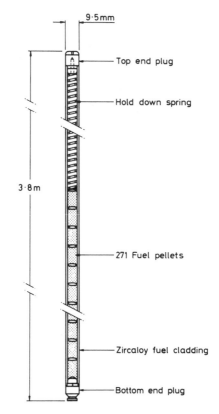

Figure 6.9   A typical PWR fuel rod

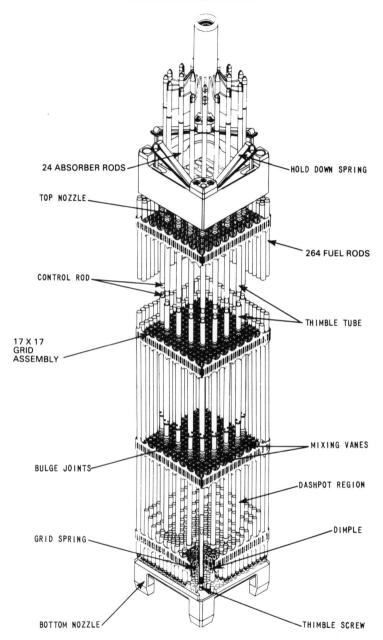

24 ABSORBER RODS

HOLD DOWN SPRING

TOP NOZZLE

264 FUEL RODS

CONTROL ROD

THIMBLE TUBE

17 X 17
GRID
ASSEMBLY

MIXING VANES

BULGE JOINTS

DASHPOT REGION

DIMPLE

GRID SPRING

BOTTOM NOZZLE

THIMBLE SCREW

Figure 6.10   A fuel and control rod assembly of a Westinghouse PWR (courtesy of Westinghouse Electric Corporation)

length and 271 of them are contained in a helium-filled Zircaloy tube to form a fuel pin, or fuel rod, like that shown in Figure 6.9. The pellets are slightly dished at each end to allow for themal expansion and are held in tight contact with one another by a hold-down spring at the top of the rod. The rods are 3.8 m in length, 9.5 mm outside diameter, and 264 of them are arranged in a square shaped cluster to form a PWR fuel assembly like that shown in Figure 6.10. The rods are spaced 12.6 mm (centre to centre) from each other and held in position along their lengths by eight spacer grids containing 289 holes arranged in the form of a 17×17 matrix measuring 214 mm $\times$ 214 mm.

The 25 holes in the spacer grids not occupied by the 264 fuel rods are fitted with hollow tubes called thimbles. These function as guide tubes for a cluster of 24 control rods (neutron absorbers) which move up and down inside the thimbles to control the reactor. The 25th hole is located in the centre of the assembly and is used to house monitoring instrumentation, for example a neutron detector.

More will be said about this type of fuel assembly when the PWR is described.

7 Uranium enrichment

99.3 per cent of all naturally occurring uranium (U_{nat}) consists of the non-fissile isotope U-238. The remaining 0.7 per cent, that is one atom in every 140 atoms of U_{nat}, consists of the fissile isotope U-235. It is from the U-235 content of natural uranium that virtually all of the world's nuclear power stations derive their energy to produce electricity.

Some nuclear reactors – the British Magnox and the Canadian CANDU for example – are designed to operate with natural uranium as fuel, whereas others – the British AGR and the American BWR and PWR reactors for example – require the U-235 content of the fuel to be artificially enriched from its normal abundance of 0.7 per cent up to a level which is typically between 2 and 3 per cent.

The artificial enrichment of natural uranium takes place in what is known as an enrichment plant, an example of which is the Capenhurst plant in Cheshire operated by British Nuclear Fuels plc (BNFL). Since U-235 does not exist in nature as a separate isotope – if it did there would be no need for enrichment – it cannot simply be added to natural uranium in order to increase the U-235 abundance. Instead, all enrichment processes are based upon techniques which reduce the U-238 content of U_{nat}, thereby leaving the residue with a higher than normal content of U-235. By way of analogy, the salt content of sea water can be enriched simply by evaporating some of the water, a process which leaves behind a liquid which has a greater concentration of sodium chloride (salt) than is normal; this is what happens in the Dead Sea!

Because isotopes of the same element have virtually identical chemical properties, it is not possible to separate them economically by purely chemical means. It is, however, possible to exploit the physical differences between them, even though such differences may be very small. In the case of natural uranium, for example, the physical difference between the two isotopes is only three units of atomic mass; a difference of only 1.26 per cent. Nevertheless it is this small difference which has been successfully exploited for many years by enrichment plants in various parts of the world.

ENRICHMENT PROCESSES

There are many ways in which the U-238 and U-235 isotopes can be separated from one another but only three have been used on a large scale and these all make use of natural uranium in the form of the gas uranium hexafluoride (UF_6), hex, a product widely used during the fuel element manufacturing process.

Since one atom of natural uranium combines with six atoms of fluorine to form one molecule of UF_6, it follows that hex is really a mixture of two isotopic forms, one formed from U-238 and the other from U-235, i.e. $^{nat}UF_6 = {}^{238}UF_6 + {}^{235}UF_6$. Bearing in mind that fluorine exists in a single isotopic form and has an atomic mass of 19, the total mass of the $^{238}UF_6$ molecule is equal to $(1 \times 238) + (6 \times 19) = 352$. Similarly, the mass of the $^{235}UF_6$ molecule is equal to $(1 \times 235) + (6 \times 19) = 349$. Note that although the difference in mass between the two molecules is still the same as that between the two uranium isotopes, that is three units of mass, this difference is only about 0.86 per cent of the total molecular mass.

Electromagnetic enrichment

The first of the three main separation processes is the electromagnetic method, widely used in the early days of nuclear research. The process is best understood by referring to the much simplified diagram shown in Figure 7.1.

Hex gas molecules are ionized in the ion source so that each acquires an overall positive charge. The resultant positive ions are then released into the evacuated flight tube where they come under the attracting influence of the negative potential of the accelerating electrode; a thin disk with a hole at its centre. By the time the ions reach the disk they are travelling at very high speeds and some of them are captured by the disk itself. Some, however, pass through the centre hole of the disk and emerge as a parallel beam of fast moving positive ions.

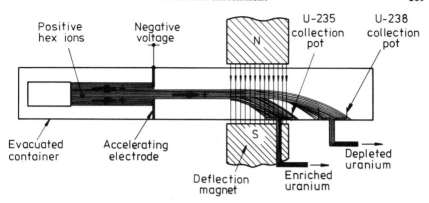

Figure 7.1   Principles of the electromagnetic method of uranium enrichment

Shortly after leaving the accelerating electrode the ions come under the influence of the magnetic field produced by the deflection magnet. The degree of deflection introduced by the magnetic field depends upon the mass of the particles moving through its area of influence, which means that the lighter U-235 gas molecules will be deflected to a greater extent than will the heavier molecules containing U-238. The result is a partial separation of the beam into two unequal parts, one slightly rich in U-235 molecules, the other slightly deficient in U-235 molecules. The separation is, of course, far from perfect and both beams contain substantial quantities of U-238 molecules. Nevertheless, if collection pots are placed as shown in Figure 7.1 it is possible to collect quantities of hex which are abnormally rich in U-235 content. The efficiency of the enrichment can be improved substantially by repeatedly passing the enriched gas through the same process so that the enrichment improves with each traverse of the flight tube.

The hex gas which is collected from the U-238 collection pot has an abnormally low U-235 content – typically 0.2 per cent – and is said to contain *depleted uranium*. More will be said about depleted uranium when the Fast Reactor is described.

Gaseous diffusion

Although the electromagnetic method of isotope separation is extremely efficient, and is still widely used in nuclear research laboratories for the extraction of ultra-pure isotopes from multi-isotope materials, its throughput is very low and hence unsuitable for commercial-size uranium enrichment plants.

A more suitable method of uranium enrichment and one which has, until recent years, been used in all enrichment plants throughout the world is based upon the technique known as *gaseous diffusion*. The technique is based upon 'Graham's Law of Diffusion' which states that lighter gases diffuse more readily through a porous barrier than do those which are heavier. Expressed in a more scientific way the law states that *the rate at which a gas diffuses through a porous barrier is inversely proportional to the square root of the density of the gas molecules*. For a gas such as hex, therefore, the heavier gas molecules containing U-238 atoms will diffuse less readily through a given barrier than will the lighter gas molecules containing U-235 atoms. The reason for the phenomenon is simply that, at the same temperature, the molecules of a light gas move, on average, with a higher speed than do those of a heavier gas.

The simplified diagram shown in Figure 7.2 illustrates the principles of U-235 enrichment by the process of gaseous diffusion. Hex gas at about 60°C and low pressure is fed into the cylinder containing a porous barrier, the holes in which are about one-millionth of a centimetre (10^{-6} cm) in diameter. The surface area of the barrier is chosen so that the impedance it offers to the flow of the gas is approximately equal to the impedance offered by the small bleed pipe at the top of the container. Thus, about half the gas flows through the barrier and half through the bleed pipe.

For the reasons given above, the gas which flows through the barrier will contain a higher concentration of U-235 molecules than that which flows through the bleed pipe. The output from the barrier is therefore composed of enriched hex gas and that from the bleed pipe is composed of depleted hex gas, usually referred to as the *tails*. Unfortunately, the degree of enrichment from such an arrangement is very small indeed and a great many stages are required in sequence before enrichment levels of 2–3 per cent are achievable.

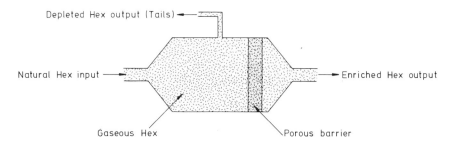

Figure 7.2   Simplified diagram of a single gaseous diffusion stage

The theoretical maximum separation of the two hex molecules is given by the expression:

$$s = \sqrt{\frac{^{238}UF_6}{^{235}UF_6}} = \sqrt{\frac{352}{349}} = 1.0043$$

where 's' is the so-called *separation factor*.

Bearing in mind that only half the gas flow actually passes through the barrier and that the practical realization of 's' will always be less than perfect, separation factors of about 1.0014 are the best that can be expected from a single stage. For 'n' successive stages connected in a series cascade (that is a sequence), the separation factor is given by the expression: $s = (1.0014)^n$. Thus, for one diffusion stage ($n = 1$), the U-235 abundance is increased from its normal value of 0.7 per cent to $0.7 \times 1.0014 = 0.70098$ per cent. Similarly, for 1000 stages ($n = 1000$), the normal U-235 abundance is increased to a level equal to $0.7 \times (1.0014)^{1000} = 0.7 \times 4.05 = 2.83$ per cent. This is about the enrichment levels used in British AGRs and American-type PWRs.

Diffusion cascade

Figure 7.3 illustrates the way in which a series cascade of diffusion stages are interconnected for the enrichment of U-235. The arrangement was first introduced in the USA in 1945 and is based upon a design used in Germany in 1932 for the separation of the isotopes of hydrogen, carbon, nitrogen and other elements. The individual pumps are controlled in such a way that the depleted hex from one stage is fed back to the preceeding stage where it is mixed with the less-depleted gas being fed to that stage.

Gas centrifuge

The gaseous diffusion method of uranium enrichment, although reliable and still widely used in the USA and other parts of the world, suffers from the disadvantages of high operating costs due to the vast amounts of electrical energy it uses and long construction times because of the enormous size and complexity of the plant making it difficult to match production with demand.

In the UK, research into gas centrifuge technology has been going on for many years. In the late 1960s it was selected in preference to diffusion technology for the next increment of UK enrichment plant capacity

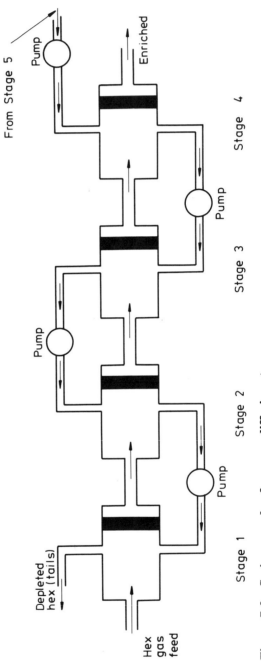

Figure 7.3 Series cascade of gaseous diffusion stages

112

because of its better economics, greater flexibility in construction and its ability to compete at plant power consumption capacities 10 per cent of those of enrichment plants.

The operating principle of the gas centrifuge is based upon the well-known fact that when a gas composed of mixed isotopes is injected into a rapidly rotating cylinder, the centrifugal force acting on the gas molecules causes the heavier ones to drift towards the outside wall of the cylinder whilst the lighter ones tend to remain close to the spin axis of the cylinder. The method is particularly suited to the separation of U-238 and U-235 molecules in gaseous hex because the separation factor depends upon the

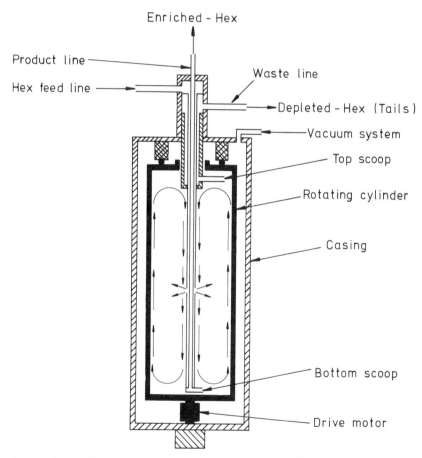

Figure 7.4   Schematic arrangement of a gas centrifuge

Figure 7.5 A Capenhurst gas centrifuge cascade hall

absolute *difference* of the molecular masses rather than on their *ratio* as in the diffusion method, thereby yielding separation factors up to 20 times better. You will recall from page 108 that the molecular mass difference between $^{238}UF_6$ and $^{235}UF_6$ is equal to three units of mass, i.e. 352 − 349.

The operating principles of the gas centrifuge are illustrated in Figure 7.4. Gaseous hex is injected into the rotating cylinder at a point about half-way down the length of the spin axis. The very high speed of the cylinder – typically more than 50 000 rpm – causes the heavier $^{238}UF_6$ molecules to drift towards the outer wall and the lighter $^{235}UF_6$ molecules to remain close to the spin axis. The separation efficiency is enhanced by forcing the gas to circulate within the cylinder, from top to bottom, so that it follows the flow lines indicated in the diagram. Circulation of the gas causes the heavier (depleted hex) molecules to flow towards the top of the cylinder, close to the wall, where they are collected by the top 'scoop', and the lighter (enriched hex) molecules to flow towards the bottom, close to the spin axis, where they are collected by the bottom scoop.

The rotating cylinder is rigidly mounted within an evacuated environment to remove the effects of drag which would otherwise be introduced by the surrounding air. Individual centrifuge assemblies are relatively small and in a modern enrichment plant, such as that at Capenhurst, many thousands of them are interconnected in a series-parallel cascade arrangement like that shown in Figure 7.5.

SEPARATIVE WORK UNITS

The operating costs of an enrichment plant and the prices it charges to its customers depends upon the U-235 content of the hex supplied by the customer as feed material and how much of it must be pumped through the cascades to achieve the required amount of output product, at the required enrichment level, for a specified depletion level of the hex 'tails'. It is because of these many inter-dependent variables that enrichment costs are based upon what are called *separative work units* (SWUs). These represent the amount of work needed to bring about the required separation and are expressed in kilogrammes of separative work.

For example, a customer may supply the enrichment plant with a quantity of hex containing natural uranium and request that 100 tonnes of enriched output product be produced from it having a U-235 content of 3 per cent. The customer further states that the hex should be recirculated until the U-235 content of the 'tails' has fallen to 0.2 per cent. How much feed material must be supplied to meet this requirement?

Calculations, made using formulae which are beyond the scope of this book, show that 548 tonnes of natural uranium are required as feed material and that the total effort required by the plant would be equivalent to 431 000 kg of separative work. It is interesting to note that had the customer allowed the 'tails' to be discarded at 0.3 per cent U-235 instead of the original 0.2 per cent, then less recirculation of the hex would have been necessary – hence less SWUs would have been accumulated – but more feed material would have had to be supplied to make up for the greater quantity of uranium wasted in the 'tails'. In fact, the total amount of feed material would have risen to 657 tonnes but the separative work would have fallen to 342 000 kg. The choice open to the customer on whether to supply more feed material or to incur more separative work charges is based upon relative costs, contractural terms and, perhaps, on political decisions.

URENCO

Development work on the gas centrifuge method of U-235 enrichment began independently in the United Kingdom, the Federal Republic of Germany and the Netherlands. In 1970, however, these three countries came together at the *Treaty of Almelo* and formed the jointly-owned *Ur*anium *En*richment *Co*mpany known as URENCO Limited. The purpose of the treaty was to promote the industrial development and commercial exploitation of the gas centrifuge process as a collaborative venture and to co-ordinate production and marketing and sales activities from a central point to the benefit of the three national shareholders. URENCO operates three enrichment sites, one at Capenhurst in England, one at Almelo in The Netherlands and one at Gronau in West Germany. The headquarters of URENCO is based at Marlow, Buckinghamshire (UK). An associate company dealing with the centrifuge technology, and known as CENTEC GmbH, is based in Germany. The British operating company is known as URENCO (UK) which is largely owned by British Nuclear Fuels plc (BNFL).

The total operating capacity of the enrichment plants operated by URENCO was 1600 tonnes of separative work per annum in 1986 but is planned to rise to between 3000 and 5000 tonnes per annum by the end of the century. Britain's first gas centrifuge enrichment plant, with an initial capacity of 200 SWUs per annum, began operating at Capenhurst in 1977 and uses less than one-tenth of the electrical power consumed by an equivalent gaseous diffusion plant. Present capacity is 400 tonnes of separative work and there is further capacity of 300 tonnes under construction.

While it is anticipated that centrifuge technology will provide the most economical enrichment service through to the next century, extensive research into laser technology is being undertaken by BNFL and its partners in URENCO, as well as the UKAEA.

8 The nuclear power station: thermal reactors

A nuclear power station is, in many ways, very similar to a coal-fired power station (see Figures 8.1, 8.2). It produces heat to boil water to produce high-pressure steam and it uses identical types of turbo-generator sets to generate electricity from the steam. The differences lie in the way in which the heat is produced. In the coal-fired station heat is produced by burning coal in a furnace; in the nuclear station the heat is produced by 'burning' uranium (or plutonium) in a nuclear reactor. The word 'burning' when referring to fuel in a nuclear reactor is really a misnomer because there is no combustion, ash, flames or smoke associated with the burning of nuclear fuel, unlike that associated with the burning of coal. Nevertheless, the terms 'burning' and 'burn-ups' have become widely accepted throughout the world when referring to nuclear fuel and will therefore be used in this book in the same way; burn-up is explained in Chapter 9, page 184.

Heat from a nuclear reactor is generated from the fissioning (splitting) of atoms in the critical mass of nuclear fuel contained in its core. The chain reaction which keeps the fissioning process going is maintained by the presence of the moderator which surrounds each fuel element. This slows down the fast-moving neutrons which are generated by each fission event and in so doing ensures that enough of them are captured by other fissile atoms in the core.

The considerable heat which is created in the fuel by the fissioning process is removed by pumping a coolant through the core and using it to transfer the heat to a heat exchanger containing water; the heat exchanger is usually referred to as the boiler, or steam generator.

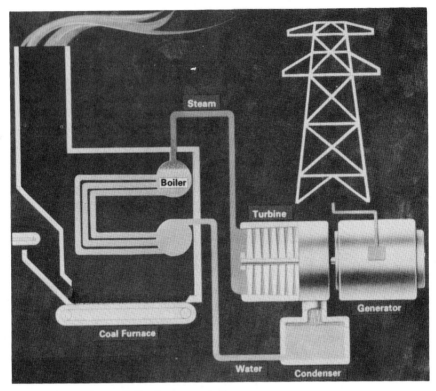

Figure 8.1　The coal-fired power station

The intensity of the fission process sustained by the chain reaction, and hence the amount of heat produced in the core and boiler, is adjusted by neutron absorber control rods which are raised and lowered automatically to keep the power steady at the required level. An adequate supply of additional absorber rods is also held in readiness should an emergency shut-down become necessary.

Virtually all nuclear power stations throughout the world use reactors which are fuelled with either natural or slightly enriched uranium and which require the presence of a moderator to slow down the neutrons produced in the core. Such reactor types are known as 'thermal reactors' because the function of the moderator is to slow down the neutrons until their speeds are comparable with those which occur at normal temperatures. Such neutrons are said to have been 'thermalized' and to have 'thermal' energies.

A very small number of reactors in various parts of the world – used

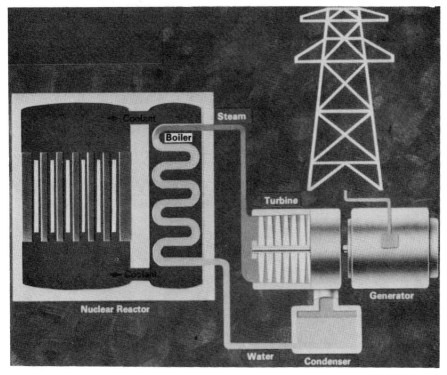

Figure 8.2   The nuclear power station

mostly for research – employ plutonium as fuel and in consequence do not require the presence of a moderator. Such reactor types are known as 'fast reactors' since they are able to sustain a chain reaction with the fast, that is un-moderated, neutrons produced in the core. To avoid unnecessary repetition in this book all references to nuclear reactors should be understood as being to thermal reactors unless the word 'fast' is specified in the description.

GAS-COOLED REACTORS

Magnox reactors

The first generation of nuclear power stations in Britain was based upon what are known as Magnox reactors, so called because the fuel elements they use are clad in a special type of magnesium–aluminium alloy which was given the proprietary name Magnox. A total of eleven Magnox stations were built in Britain between the years 1956 and 1971 and all are still operating. Two are owned and operated by BNFL and nine by the generating boards. Between them they have a combined effective generating capacity of about 4130 MWe (megawatts electrical). British-designed Magnox power stations were also constructed by British companies in Japan, in 1963, and in Italy in 1964; both are still operating. Gas-cooled reactors are also currently used in France and the USSR, although these countries have now standardized on the water-cooled types.

The fuel elements used in the British Magnox reactors are similar in appearance to that shown in Figure 6.2. Each element comprises a solid rod of natural uranium metal which has a heat energy equivalent of about 150 tonnes of coal. Many thousands of such elements are used in a typical reactor. The elements are cooled by pumping pressurized carbon dioxide gas (CO_2) through them and the neutrons which they generate are moderated by many tonnes of graphite in the form of large blocks containing holes to accommodate the fuel elements.

The core of the reactor comprising the fuel elements, the coolant gas, the moderator blocks and the control rod assembly are contained in a gas-tight pressure vessel surrounded by a thick concrete container which provides radiation shielding for the protection of the operating staff. In the earlier types of Magnox reactors the pressure vessel was made of thick steel plate

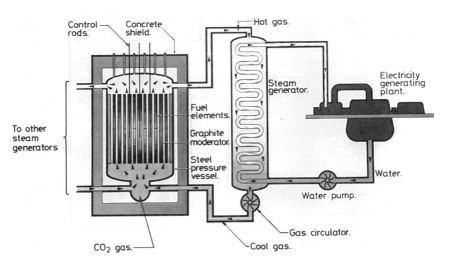

Figure 8.3   Operating principles of the early Magnox reactors

and the steam generators – typically four per reactor – were located outside the concrete shield. A simplified diagram illustrating the operating principles of this type of reactor is shown in Figure 8.3. In later designs, however, due to increases in coolant pressure, operating temperature and physical size, the functions of pressure vessel and radiation shield were combined in the form of a pre-stressed concrete pressure vessel.

Wylfa

The Wylfa station in Anglesey (Figure 8.4) is typical of power stations using this type of reactor. Completed in 1971 this is the largest of the Magnox stations and the last of the type to be built.

The power station comprises two independently-operated nuclear reactors housed in a building which is 167 m (550 feet) long, 70 m (230 feet) wide and 55 m (180 feet) high, an inside view of which is shown in Figure 8.5. The steam generated by the two reactors is fed to four turbo-generator sets which, after allowing for the electricity used on-site, actually supply 840 MW of electricity to the National Grid.

The concrete pressure vessel of each reactor is, in places, up to 3.6 m (12 feet) thick and weighs 80 000 tonnes. It contains a graphite core weighing 3800 tonnes, throughout which are dispersed more than 49 000 Magnox fuel elements contained in 6156 channels, with eight elements per channel. Access holes to the channels from the reactor top for loading and unloading purposes are clearly visible in Figure 8.5. The total weight of the natural uranium fuel in each reactor is nearly 600 tonnes. The graphite core also contains nearly 200 channels which accommodate boron–steel neutron-absorbing control rods, each of which is 7.9 m (26 feet) long and 76 mm (3 inches) diameter.

Each pressure vessel also contains four gas circulator pumps for the CO_2 coolant and four steam generators which convert ordinary water into steam at a temperature of 335°C and at a pressure of 555 psi (about 38 atmospheres). The maximum temperature of the coolant gas is 360°C and is maintained at a pressure of 400 psi (about 27 atmospheres).

Having passed through the turbines and done its job, the steam, now at very low pressure, is re-converted into water and returned to the steam generator in the reactor core for re-use. The re-conversion to water takes place in a condenser – one for each of the four turbine sets – which is supplied with cooling water from the Irish Sea at a rate of 53 million gallons per hour. Having passed through the condensers the warmed cooling water is returned to the sea where it is dispersed in deep water. The operation of the steam cycle section of the generating system is shown, much simplified, in Figure 8.6.

Figure 8.4   The Magnox nuclear power station at Wylfa, Anglesey

Figure 8.5   Inside the reactor building at the Wylfa nuclear power station
In the foreground is shown the top of one of the two reactors and some of the 6156 access holes to the fuel channels. In the background is the fuel loading and unloading machine which serves both reactors whilst they are operating at full load.

125

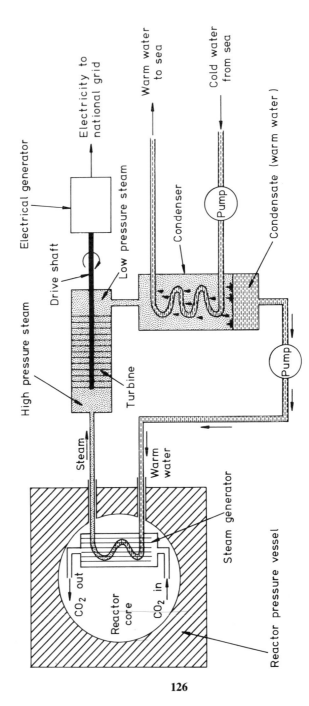

Figure 8.6 **Steam cycle section of the generating system**

Thermal efficiency   In any steam-producing power station, whether the source of heat be from coal, oil, gas or nuclear fission, only a small fraction of the total heat produced is actually converted into electrical power; the rest is wasted in the form of low pressure steam from the turbines and is eventually dissipated in the condenser cooling water.

The *thermal efficiency*, sometimes called the *steam cycle efficiency*, of any steam-producing power station is defined by the ratio:

$$\text{net efficiency } (\%) = \frac{\text{electricity available for sale}}{\text{total heat produced}} \times 100$$

or

$$\text{gross efficiency } (\%) = \frac{\text{total electricity generated}}{\text{total heat produced}} \times 100$$

The gross efficiency figure takes into account the electricity consumed by the generating station itself.

The Wylfa nuclear power station, for example, generates 840 MW of electricity for the National Grid and, in so doing, produces 3080 MW of thermal power in the form of heat. The net thermal efficiency of the station is therefore:

$$\frac{840 \text{ MW}}{3080 \text{ MW}} \times 100\% = 27.3\%$$

This figure compares favourably with many of the older coal-fired power stations but unfavourably with the efficiencies achieved by the modern coal and nuclear stations which have figures in excess of 40 per cent.

Advanced gas-cooled reactors

The advanced gas-cooled reactor, or AGR as it is more usually known, was developed in the 1960s to improve the relatively poor thermal efficiency of the Magnox type of reactor. The AGR operates at a much higher coolant temperature than the Magnox – typically more than 600°C – and in consequence achieves thermal efficiencies which match those of the most modern coal-fired power stations.

The essential differences between the two types of reactors are to be found in the chemical and physical composition of the uranium fuel and the material used for its cladding. Graphite is still used for neutron moder-

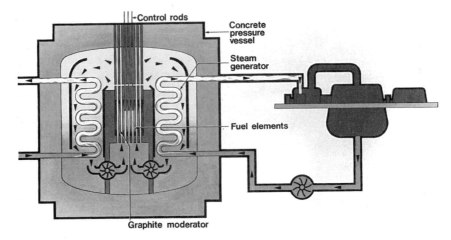

Figure 8.7   Advanced gas-cooled reactor (AGR)

ation, and CO_2 gas – albeit at higher temperatures and pressures – is still used as the coolant material. The concrete pressure vessel is also similar to that used in the later types of Magnox stations, although much thicker, and contains the reactor core, the steam generators and the gas circulator pumps. The operating principles of the AGR are summarized in the simplified Figure 8.7.

The high density and good thermal conductivity of natural uranium metal makes it eminently suitable for use in low-temperature reactors of the Magnox type. At temperatures above about 600°C, however, many unpleasant things begin to happen to uranium metal, partly because of irreversible changes which take place in its crystal structure when it is subjected to temperature cycling, and partly because of the physical damage, for example swelling and creep, caused by high energy neutrons and fission products. These limitations, coupled with the low melting point of Magnox cladding (about 600°C), made existing Magnox fuel totally unsuitable for use in the AGR and a completely new type of fuel assembly was therefore necessary.

The fuel assemblies currently used in commercial AGRs were developed after exhaustive tests in the Windscale AGR (WAGR), a small prototype reactor (Figure 8.8) which operated at the plant now known as Sellafield from 1962 to 1981, during which time it regularly supplied 33 MW of electricity to the National Grid. This was the first AGR in the world.

It was explained on page 99 and illustrated in Figures 6.4–6.6 that fuel

Figure 8.8   The Windscale AGR: the World's first

129

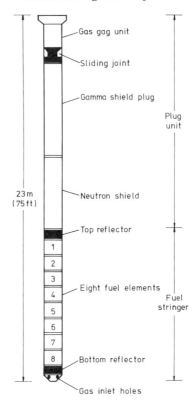

Figure 8.9   AGR fuel assembly arrangement (not to scale)

for the AGR takes the form of small ceramic pellets of enriched uranium dioxide (UO_2) which are packed into thin stainless steel tubes called 'pins'. Thirty-six pins are then clamped together and fixed inside a graphite sleeve to form a fuel element. The final form of a complete fuel assembly for an AGR is shown, much simplified, in Figure 8.9. It comprises seven or eight fuel elements clamped end-to-end and a number of other components, all held firmly together by a central tie bar. A commercial AGR would contain more than 300 such assemblies.

The U-235 content of AGR fuel has to be enriched to a level between 1.4 per cent and 2.6 per cent, depending upon its position in the core and the time of loading, to compensate for the greater number of neutrons which are lost in the stainless steel cladding because of its higher neutron capture cross section than Magnox metal.

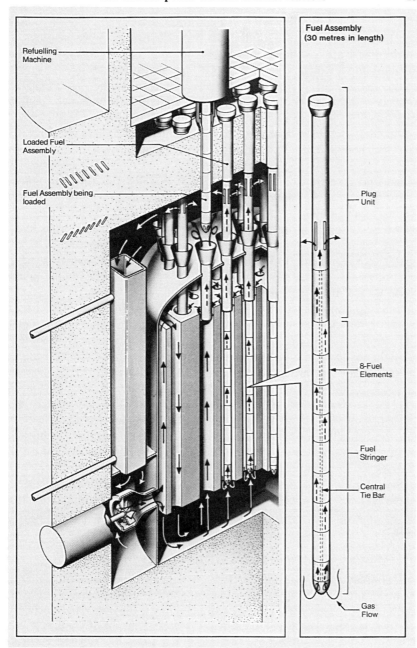

Refuelling Machine

Loaded Fuel Assembly

Fuel Assembly being loaded

Fuel Assembly
(30 metres in length)

Plug Unit

8-Fuel Elements

Fuel Stringer

Central Tie Bar

Gas Flow

Figure 8.10   Cross section through the AGR reactor

The fuel assembly illustrated in Figure 8.9 is the type currently used in the Hinkley Point-B and Hunterston-B AGR nuclear power stations. Its overall size is approximately 23 m (75 feet) length and 240 mm (9½ inches) diameter and it weighs about 2.5 tonnes. The section of the assembly which resides deep in the core of the reactor is known as the *fuel stringer*. It comprises eight (sometimes seven) fuel elements and two neutron reflectors. The reflectors – one above and one below the string of elements – reflect back into the core some of those neutrons which might otherwise have escaped and, in so doing, improve the neutron efficiency of the reactor.

The upper section of the fuel assembly is known as the *plug unit*. It comprises two main parts, each of which functions as a radiation shield. The part immediately adjacent to the fuel stringer is made from a neutron absorbing material, for example cadmium, which prevents neutrons reaching the reactor top where they could harm the operating staff. The part closest to the reactor top consists of a concrete-filled steel tube which functions as a gamma radiation shield. This part of the assembly resides inside the lid of the concrete pressure vessel.

A sliding joint immediately above the gamma shield enables the overall length of the assembly to be adjusted to accommodate either seven or eight fuel elements. The *gas gag unit* at the top of the assembly provides a gas-tight seal with the pressure vessel and enables the amount of cooling gas (CO_2) flowing through the assembly to be adjusted.

Figure 8.10 illustrates how the fuel assemblies are arranged in the graphite moderator blocks and how the CO_2 cooling gas is circulated through them and the heat exchanger boiler. The diagram also illustrates how the refuelling machine changes one of the fuel assemblies whilst the reactor is still operating at full power. This technique is also used on the Magnox reactors and is known as *on-line refuelling*.

The refuelling machine for the AGR is itself an enormous structure as can be seen in Figure 8.11. The machine weighs nearly 600 tonnes and runs on two side rails between two reactor tops; it is more than 29 m (90 feet) high.

The Hinkley Point power station

The Hinkley Point nuclear power station near Bridgwater in Somerset, shown in Figure 8.12, comprises two Magnox reactors (Station-A) with a combined net electrical output of 450 MW and, a short distance away, two AGRs (Station-B) with a combined net electrical output of 1250 MW. The A-station commenced operation in 1965, the B-station in 1976; both are

Figure 8.11 Refuelling machine for the Hinkley Point AGR shown in position above the reactor top plate (courtesy of CEGB)

owned and operated by the CEGB. The following brief description summarizes the salient features of the B-station.

The two AGRs and the turbo-generating plant are housed under the
same roof with a central block for fuel handling, instrumentation and
control. Both reactors are served by a single refuelling machine like that
shown in Figure 8.11. The core of each reactor takes the form of a 16-sided
stack of graphite (the moderator) containing 308 fuel channels arranged on
a square lattice. Immediately outside the core are four boiler units (the
heat exchangers) each of which is supplied with hot coolant gas from two
gas circulator pumps.

Each boiler functions as a fully independent unit complete with its own
water feed regulating equipment, integral pipework and valves, and may be
taken out of service whilst the reactor is still operating at full load.
Defective tubes can be blanked-off from outside the pressure vessel whilst
the reactor and other boilers remain on full load. Similarly, each of the
eight gas circulators functions as a fully independent unit with its own
electric drive motor and control gear, and is housed inside its own 'tunnel'
in the pressure vessel wall (see Figure 8.10). The entire unit can be
removed and replaced whilst the reactor remains at full load.

The prestressed concrete pressure vessel of each reactor is cylindrical in
shape with walls 5 m (16.5 feet) thick. Its top (the so-called *pile cap*) is 5.5 m
(18 feet) thick and its base is 8.5 m (27.75 feet) thick. Its internal dimensions
are 19.4 m (63 feet 8 inches) height and 18.9 m (62 feet) diameter, and it
contains the graphite core weighing 1800 tonnes – complete with control
rods and 122 tonnes of UO_2 fuel – plus four boiler units and eight gas
circulator pumps.

The operating temperature of the CO_2 coolant passing through the
boilers is 645°C and its pressure is maintained at 576 psi (about 40
atmospheres). The steam produced by the boilers for the turbines is at a
temperature of 541°C and at a pressure of 2419 psi (about 167 atmospheres). The steam produced by the two reactors is fed to two 660 MW
turbo-generator sets which generate electricity at 22 000 volts. This voltage
is then stepped up to 400 000 V and coupled into the National Grid. The
net station electrical output is 1250 MW and the total thermal output is
3000 MW; the net thermal efficiency of the station is therefore equal to:

$$\frac{1250}{3000} \times 100\% = 41.7$$

Cooling water for the turbine condensers is taken from the Bristol Channel
at a point about 500 m (0.3 mile) off shore and at a rate of about 109 million

Figure 8.12 The CEGB's Hinkley Point nuclear power station complex (courtesy of CEGB)
The two Magnox reactors which form Station-A are in the background of the picture. The two AGRs which form Station-B are in the foreground and are housed under the same roof

litres (24 million gallons) per hour. The water is returned to the sea at a point about 800 m (0.5 mile) east of the intake with a temperature rise of about 9°C. Some of this warm water is used by a nearby fish farm where fish are found to thrive on it.

HEAVY-WATER COOLED REACTORS: CANDU

The Canadian CANDU reactor is a thermal reactor whose operating principles are the same as any other type of thermal reactor, that is its core contains a critical mass of fissile material; it employs a moderator for slowing down the neutrons produced during fission; it employs neutron absorbers to control its power output; and the heat generated in its core is removed by a circulating coolant. Where the CANDU differs from the British Magnox and AGR reactors and from the American BWRs and PWRs described below, is that it uses heavy water (D_2O) as both a coolant and a moderator.

Heavy water is the best moderating material available (see Chapter 15), but is also the most expensive, costing many hundreds of pounds per gallon (£700 per gallon in 1983). It is not surprising, therefore, that Canada, with its large heavy water manufacturing facilities, is the only country in the world to have developed commercial-size nuclear power stations using heavy water reactors. At one time Britain thought seriously about adopting heavy water reactors for its second-generation reactor programme and went so far as to construct the prototype Steam-Generating Heavy-Water Reactor (SGHWR) at the UKAEA research establishment at Winfrith, in Dorset, where it has been operating successfully since 1968 putting 100 MW of electricity into the National Grid. However, the complexity of the SGHWR, coupled with its large D_2O inventory, caused it to be dropped in favour of the AGR.

Heavy water is an attractive moderating material to the reactor designer because its very low capture cross section to thermal neutrons brings about good neutron economy (little wastage) which, in turn, means that the reactor like the Magnox, can be fuelled with natural uranium instead of the more expensive enriched uranium. This also means that more of the U-235 content can be 'burnt up' before the fuel is removed from the reactor. Spent fuel from a typical CANDU reactor contains about 0.2 per cent U-235, compared with the 0.7 per cent U-235 content of fresh fuel, plus about 0.3 per cent Pu-239. In fact, the CANDU reactor, with its D_2O moderator, extracts almost twice as much energy from each gram of U-235 consumed than does a light-water moderated reactor; mainly because of the greater amount of Pu-239 created from the U-238 content of the fuel (about 2.5 g Pu-239 per kg U-238), much of which is incinerated in the core to add to the overall power produced (see Table 9.5) for relative Pu-production rates). This is another way of saying that the conversion ratio of the CANDU reactor, that is *the ratio of fissile material of all types produced to that destroyed* is greater than that of the light-water reactor. A typical value for conversion ratio for the CANDU reactor is 0.79 whereas for a light-water reactor the value is nearer 0.6.

Another attractive and very important characteristic of heavy water is its *neutron lifetime* value, that is the time spent by a neutron in the moderator before it escapes or is captured by an atom of fuel, by the moderator itself, or by structural materials or control absorbers. This is often as much as 50 ms, a period which is some 500 times longer than of a light-water reactor. This means that the response of the heavy-water reactor to a rapid transient is relatively sluggish, making it more easy and less critical to control; a very important safety feature.

Pressure tubes

Apart from the differences in types of moderator and coolant materials used in the heavy-water and light-water types of reactors, there exist other major differences regarding the way in which heat is extracted from the core and in which steam is generated for the turbo-generators.

In light-water reactors the fuel elements are grouped close together in a single large-diameter pressure vessel and completely immersed in a large pool of light water which functions as both moderator and coolant. The water is pressurized to between 1030 and 2280 psi (70 to 155 atmospheres) to stop it boiling so that it can be circulated through the fuel elements as a liquid coolant at temperatures in excess of 300°C. Steam is produced either directly from the coolant and fed straight to the turbo-generators or indirectly by way of a small number (typically four) of heat exchangers. This will be explained in much greater detail when the BWR and PWR reactors are described.

In the CANDU type of heavy-water reactor many hundreds of individual channels containing bundles of fuel elements are grouped closely together to form the reactor core and arranged to pass through a large tank filled with heavy water. The tank is known as a *calandria* and the heavy water it contains functions as the moderator. The arrangement is illustrated in Figure 8.13.

The heavy water moderator is operated at near-atmospheric pressure and is maintained at a temperature within the range 43°C to 71°C by a circulating pump and a small heat exchanger with a maximum thermal

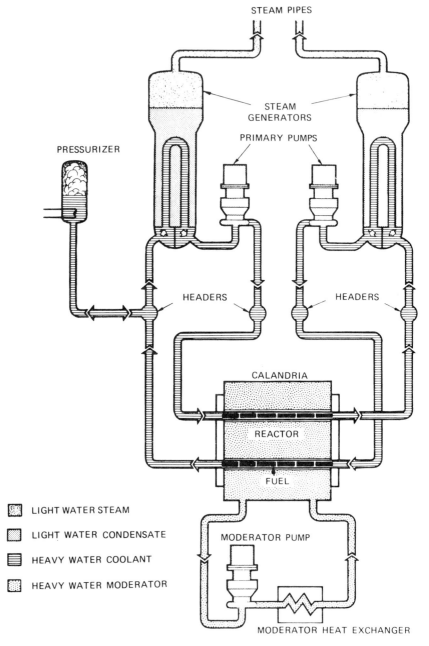

Figure 8.13　CANDU reactor: simplified flow diagram

capacity of nearly 150 MW (about 6 per cent of the reactor's thermal output). The total amount of heavy water in a typical calandria is in the region of 300 tonnes (about 67 000 gallons).

In an emergency the control rods and standby neutron absorbers are dropped under the influence of gravity into the reactor core in a number of channels passing vertically through the calandria. Should insufficient absorbers drop to shut down the reactor then the moderator itself can be forced out of the calandria – using Helium gas under pressure – into a standby tank or, in more recent designs, it can be temporarily 'poisoned' by the rapid injection of a liquid neutron absorber such as gadolinium nitrate. The advantage of retaining a poisoned moderator in the calandria is that much of the reactor's decay heat can be removed by the moderator heat exchanger should all of the main heat exchanger circuits fail; a most unlikely event!

The calandria is cylindrical in shape and fabricated from 317 mm (1.25 inch) stainless steel. Its overall size varies with design but is typically 8.5 m (28 feet) in diameter and 6 m (19.7 feet) in length. The 480 or so fuel channels passing through the calandria are made from Zircaloy-2, a zirconium-based alloy possessing a very low neutron absorption cross section. The channels have a diameter of about 130 mm (5.1 inch) and a wall thickness of 1.37 mm (0.054 inch). The total quantity of fuel contained by a typical calandria is in the region of 108 tonnes of natural uranium in the form of UO_2 pellets.

Fresh fuel is loaded and unloaded at both ends of the channels and in order to provide adequate protection for the operating staff the end faces of the calandria are each fitted with radiation shields weighing nearly 230 tonnes and measuring 1.06 m (3 ft 6 in) in length. These form an integral part of the calandria shell and provide support for the fuel channels.

Core heat removal

Heat is removed from the core of the CANDU reactor by circulating pressurized heavy water through the individual fuel channels and through a number of steam generating heat exchangers where heat from the D_2O is transferred to a light water circuit which is allowed to boil. The steam so produced is fed to the turbo-generators and, after it has done its work and been cooled by the condensers, is returned to the steam generators as a warm-water condensate for re-boiling.

Heavy water rather than light water is used as a coolant since it effectively forms part of the D_2O moderator during the time when it passes through the calandria. The coolant D_2O is pressurized to stop it boiling and to allow it to operate as a liquid at temperatures which are typically as high as 300°C. The pressure is maintained by an electrically-heated pressurizer at typically 1322 psi (about 90 atmospheres), a figure which lies midway between the coolant pressures of the BWR and PWR reactors. For this reason the CANDU reactor is often described as being a pressurized heavy water reactor (PHWR). A typical CANDU coolant circuit would comprise four circulating pumps feeding eight steam generators and producing a coolant flow rate (per channel) of 89 tonnes per hour, or about 334 gallons per second.

The total quantity of heavy water coolant in a typical CANDU reactor is approximately the same as the heavy water moderator contained in the calandria, that is 300 tonnes (67 000 gallons). This means that the total heavy water inventory of such a reactor is about 600 tonnes (134 000 gallons). Such a large amount represents a great deal of money – in fact, about 15 per cent of the capital cost of the reactor.

CANDU fuel

Fuel for the CANDU reactor consists of natural uranium dioxide (UO_2) in the form of small ceramic pellets packed into thin-wall Zircaloy tubes. The tubes are about 500 mm (19.7 inches) in length and 15 mm (0.6 inches) diameter and are clamped together to form what is known as a *fuel bundle*, like that shown in Figure 8.14. Design varies from reactor to reactor but a typical fuel bundle would consist of 37 tubes and have an overall diameter of about 102 mm (4 inches). The total UO_2 content of such a bundle would be about 21 kg and a 800 MWe reactor would contain 6240 bundles with 13 of them placed end to end in each of 480 pressurized fuel channels. The total weight of fuel in such a reactor would be about 108 tonnes of natural uranium. Typical burn-up rating of the fuel would be in the region of 8125 MW-days per tonne of uranium and up to 15 bundles per day would be changed during the on-line re-fuelling operations.

Pros and cons

The main advantage of the CANDU reactor, from a safety point of view, is that there is no single large pressure vessel to fail, as there is with the BWR and PWR types of reactors, and hence no possibility of a massive loss-of-coolant-accident (LOCA). Any leakage which *does* occur is likely to be confined to a few series-connected fuel channels which form part of a separate coolant circuit and which can quickly be isolated from the undamaged channels and, if necessary, flooded with gravity-fed light water

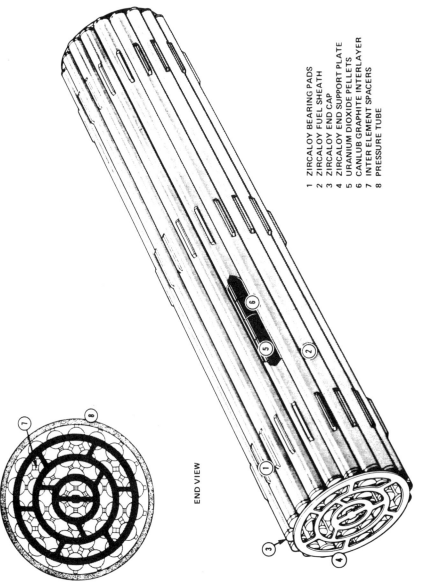

END VIEW

1 ZIRCALOY BEARING PADS
2 ZIRCALOY FUEL SHEATH
3 ZIRCALOY END CAP
4 ZIRCALOY END SUPPORT PLATE
5 URANIUM DIOXIDE PELLETS
6 CANLUB GRAPHITE INTERLAYER
7 INTER ELEMENT SPACERS
8 PRESSURE TUBE

Figure 8.14   Typical fuel bundle for the CANDU reactor

141

from a so-called 'dousing tank' built into the roof of the reactor containment building. As this tank is emptied, water collecting in the sump of the containment building is picked up by a pump and returned to the dousing tank via a heat exchanger.

Any significant LOCA would, of course, automatically initiate a reactor trip so that after a very short space of time only decay heat would have to be removed from the core. This would be removed mainly by the undamaged heat exchangers of the intact D_2O coolant circuits, partly by the moderator heat exchangers, and partly by the emergency core cooling water from the dousing tank.

Another, less obvious, advantage of the CANDU reactor is that scaling-up the power rating in a new design need not involve any new technology but simply more of the same, that is more channels of the same well-proven type.

Another advantage of the CANDU reactor, which has an important bearing on its running costs, is the fact that there is easy access to both ends of the individual fuel channels, making it relatively easy to re-fuel whilst the reactor is operating at full load; this, as will be discussed below, cannot be done on the BWR and PWR reactors, both of which must be shut down during the refuelling process. You will recall that both the Magnox and AGR reactors can be re-fuelled whilst 'on-load'.

On the debit side the most significant disadvantages of the CANDU reactor are the vulnerability to failure of the many hundreds of pressure tubes (and even more tube welds), the construction costs of such a complex system and the very large cost of the heavy water inventory. Nevertheless, the CANDU reactor is acknowledged throughout the world as being one of the most reliable of all types with operating load factors often in excess of 90 per cent. In 1980 CANDU reactors occupied the first four positions in world performance rankings and had eight listed in the first 24; a remarkable and enviable achievement.

One other disadvantage of the CANDU reactor, and one which applies to all water-cooled reactors, is that its thermodynamic efficiency is relatively poor – typically less than 31 per cent; the British AGR and PFR reactors have efficiencies in excess of 42 per cent.

The Bruce CANDU

A good example of a modern CANDU reactor installation is the nuclear power station located at Bruce Township, Ontario. Owned and operated by Ontario Hydro and designed jointly by Ontario Hydro and Atomic Energy of Canada Limited, the station comprises four completely separate

CANDU reactors, each of which generates up to 791 MW of electrical power. The first reactor (Unit 1) began operation in 1977; the last (Unit 4) in 1979.

The reactor containment buildings are rectangular in shape and constructed from reinforced concrete 1.83 m (6 feet) thick. Each is 31.7 m (104 feet) long, 28.04 m (92 feet) wide and 14.18 m (46.5 feet) high, and is maintained with a small negative pressure so as to prevent the release of radioactive gases to the atmosphere in the event of an accident.

A large cylindrically-shaped building, separate from but close to the reactors, is known as the *vacuum building*. It is made from reinforced concrete with a wall thickness of 1.14 m (3 ft 9 in). Its internal height is 45.4 m (149 feet) and its internal diameter is 49 m (160 feet). Its internal volume is 62 297 m^3 (2.2 million cubic feet) and, in its roof space, is a water storage capacity of 10 000 m^3 (2.2 million gallons). The vacuum building serves all four reactors and functions as an emergency storage capacity in the event of a large build-up of pressure within any of the reactor containment buildings due, for example, to a major breach of the primary coolant circuit. In such an event, pressure relief valves would automatically open and the steam–air mixture would be routed to the vacuum building where it would be safely contained. *The vacuum building is capable of accommodating all the steam which would be generated from boiling of the entire heavy water coolant inventory*, and is assisted in doing so by a cold water spray from the roof tank which is actuated automatically at a preset pressure build-up. This cools the air within the building and condenses the steam.

The combination of the reactor containment building and the vacuum building is designed so as safely to contain the total energy which would be released in the worst conceivable rupture of the primary coolant circuit.

LIGHT-WATER COOLED REACTORS

About 70 per cent of the world's nuclear reactors are of the light-water type (LWR) in which ordinary water is used as both moderator and coolant. There are essentially two main types of LWR, one (the most predominant) is the pressurized water reactor (PWR), the other is the boiling water reactor (BWR). Both types originated in the USA but are now manufactured in other parts of the world, principally in France, Germany and the USSR.

Development of the LWR began in the USA in the late 1940s when that country embarked upon a plan to construct a fleet of nuclear submarines. The decision to use light water instead of heavy water or graphite, for example, was because its good neutron moderating properties coupled with its high density and excellent heat transfer ability meant that it could readily fulfil the dual role of moderator and coolant and that, unlike heavy water, it was cheap and plentiful. Its big drawback was its relatively high neutron capture cross-section (see Chapter 15) which meant that any reactor using light water as a moderator would have to use for its fuel uranium which had had its U-235 content artificially enriched to a level about four times above its natural abundance of 0.7 per cent. *It is impossible to sustain a neutron chain reaction using natural uranium in a light-water moderator, no matter how much uranium is used.* Fortunately for the USA, this requirement presented no problems because of that country's plentiful supply of enriched uranium from its diffusion enrichment plants constructed during the war for military purposes. To Canada, Britain and the countries of Western Europe, however, who were embarking on a programme of building nuclear power stations, the requirement meant that LWR reactors were out of the question because there were no enrichment plants in operation outside the USA and the technology for building them was denied them by the US McMahon Act of 1946. *It is for this reason that Canada, Britain and France opted for natural uranium burning reactors for their first-generation nuclear power programmes and why the CANDU and Magnox reactors came into being.*

Work on the design of a nuclear-powered submarine began in the USA in 1948 and a land-based prototype known as STR-1 (Submarine Thermal Reactor Mark-1) began operating in Idaho in 1953. It used a PWR-type of reactor and generated electricity using a light-water heat exchanger and steam-driven turbine. STR-1 was quickly followed by the launching in 1954 of USS Nautilus, the world's first nuclear powered submarine, which also used a PWR type of reactor.

The pressurized water reactor (PWR)

Having successfully demonstrated the ability of the PWR in its fleet of nuclear submarines, it seemed natural for the USA to look favourably upon this type of reactor when it began designing nuclear power stations. And so it was that the first prototype nuclear power station to be built in the USA made use of a Westinghouse-designed PWR reactor which began operating on 2 December 1957 at Shippingport, Pennsylvania. Its full power output was initially 60 MW electrical (230 MW thermal) but after some re-design of its core this was raised to 150 MW electrical (505 MW thermal). This reactor was eventually shut down for ever in 1982. Modern PWRs operate at power levels in the region of 1000 MW to 1300 MW

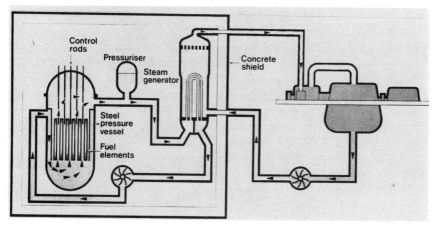

Figure 8.15 Simplified operating diagram of the pressurized water reactor (PWR)

electrical (3000 MW to 3425 MW thermal) and there are more than 100 of them operating throughout the world with many more under construction.

The operating principles of a modern PWR reactor are illustrated in the simplified diagram shown in Figure 8.15. A cylindrically-shaped pressure vessel containing the fuel elements and control rods is completely filled with ordinary (light) water which functions as both moderator and primary coolant. This water is circulated in a closed circuit through the fuel, where it becomes very hot, and also through a heat exchanger in an external steam generator where it gives up much of its heat to a completely separate light water circuit: the secondary coolant.

The primary coolant water which circulates through the fuel is prevented fom boiling by an external *pressurizer* and is allowed to reach temperatures typically as high as 325°C. The secondary coolant water in the steam generator, however, is not artificially pressurized and is allowed to boil and produce steam for driving the turbo-generator set. Prevention of the primary coolant from boiling is an essential requirement of the PWR. If this water is allowed to boil then a two-phase (steam and water) mixture will be created and its efficiency as a coolant will be severely reduced, leading rapidly to overheating of the reactor core. On the other hand, it is advantageous to keep the outlet temperature of the primary coolant as high as possible because this, in turn, produces high-temperature steam in the steam generator and yields a good thermodynamic efficiency.

The pressurizer forms part of the primary coolant circuit and is partially filled with primary coolant water. Its operation is very similar to the master brake cylinder in a motor car except that pressure is created by locally-generated steam above its water level rather than by a mechanically-operated piston. The steam is generated in the pressurizer by an electrical immersion heater and its pressure carefully controlled so as to maintain a constant pressure in the pressure vessel. Part of the control process involves a high-pressure injection nozzle located immediately above the surface of the boiling water. This is used to spray cold water into the pressurizer whenever it is necessary to reduce pressure quickly. The pressure created in this way is typically 2200 psi (about 150 atmospheres), a figure which ensures that the primary coolant remains in the liquid phase when operated at temperatures in the region of 325°C.

The pressurizer is itself quite large: typically 16 m (52 feet) in height and 2.2 m (7.2 feet) outside diameter, containing over 30 m³ (6600 gallons) of water. Its total capacity is more than 50 m³ (about 11 000 gallons). The pressurizer is fitted with a *pressure-operated relief valve* (PORV) which is designed to blow off excess pressure once it exceeds about 2500 psi (170 atmospheres); domestic pressure cookers are fitted with a similar safety valve to prevent them exploding.

PWR layout

Most PWRs in the USA have been designed by the Westinghouse Electric Corporation of America and the remainder by Construction Engineering or Babcock & Wilcox, both American companies. Westinghouse has also granted manufacturing licences to organizations outside the USA and which include Kraftwerkunion of Germany, Framatome of France, and Mitsubishi Heavy Industries of Japan.

Although designs have changed over the years a great degree of standard-ization has been agreed upon by various design authorities and has resulted in the formation of what has become known as the SNUPPS project (Standardized Nuclear Power Plant Systems). This was formed by a group of five US organizations so as to produce a standard design which would reduce the amount of time required in licensing applications and in design and construction.

Westinghouse designs are based on a modular concept in which the Nuclear Steam Supply System (NSSS) comprising the coolant circulating pumps, the steam generators, the pressurizer, the reactor pressure vessel and its internals, the fuel and control rod assemblies, and the coolant loop layouts are the same for all plants of the same size. Westinghouse produces standardized designs for two-, three- and four-loop layouts, where each loop represents one steam generator and its associated circulating pump. The simplified diagram shown in Figure 8.16 illustrates the physical layout of a modern four-loop PWR NSSS designed by the Westinghouse corporation. Such a system would produce 3425 MW of heat, from which would be generated about 1200 MW of electricity. A single pressurizer serves all four steam generators, each of which has its own coolant circulating pump.

Reactor pressure vessel

The reactor pressure vessel of the NSSS system illustrated in Figures 8.16 and 8.17 is constructed from carbon steel with an internal cladding of stainless steel to reduce corrosion (ordinary water at 325°C is highly corrosive). It is about 13 m (43 feet) high, 5 m (16 feet) outside diameter and with a wall thickness of up to 215 mm (8.6 inches) in the region surrounding the core. Its overall weight when empty of fuel and water is nearly 400 tonnes.

Under normal operating conditions the vessel contains about 100 tonnes of UO_2 fuel, in the form of 193 fuel assemblies, each containing 264 fuel rods and 25 neutron absorber control rods arranged in a 17×17 matrix, like that shown in Figure 6.10. The vessel also contains 334 m^3 (almost 74 000 gallons) of light water pressurized to 2200 psi (150 atmospheres)

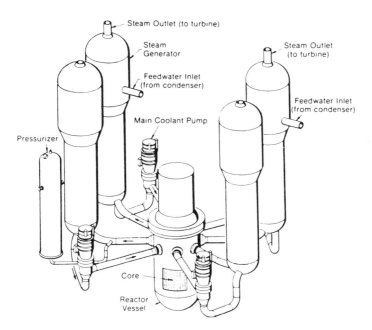

Figure 8.16   Physical layout of a four-loop Westinghouse PWR NSSS (courtesy of Westinghouse Electric Corporation)

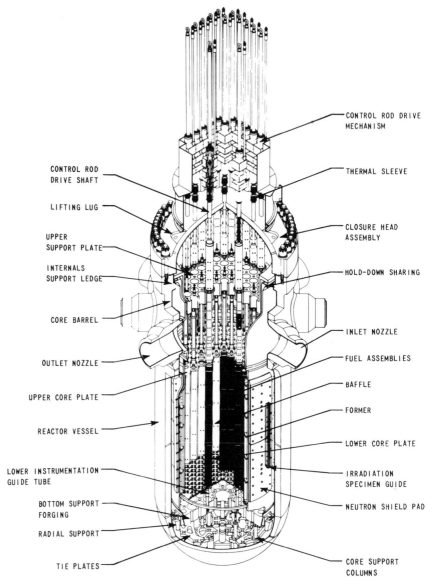

CONTROL ROD DRIVE MECHANISM

THERMAL SLEEVE

CONTROL ROD DRIVE SHAFT

LIFTING LUG

UPPER SUPPORT PLATE

INTERNALS SUPPORT LEDGE

CORE BARREL

OUTLET NOZZLE

UPPER CORE PLATE

REACTOR VESSEL

LOWER INSTRUMENTATION GUIDE TUBE

BOTTOM SUPPORT FORGING

RADIAL SUPPORT

TIE PLATES

CLOSURE HEAD ASSEMBLY

HOLD-DOWN SHARING

INLET NOZZLE

FUEL ASSEMBLIES

BAFFLE

FORMER

LOWER CORE PLATE

IRRADIATION SPECIMEN GUIDE

NEUTRON SHIELD PAD

CORE SUPPORT COLUMNS

Figure 8.17   The reactor pressure vessel of a Westinghouse PWR (courtesy of Westinghouse Electric Corporation)

and circulated by four pumps at a combined rate of nearly 19 tonnes per second (about 4200 gallons per second). Each circulating pump is rated at 6 MW (8000 hp) and weighs more than 94 tonnes.

The upper head of the pressure vessel is removable for maintenance and refuelling and is retained by a bolted flange and sealed with metal 'O-rings'. Refuelling takes place approximately once per year and takes between five and six weeks, during which time about one-third of the fuel assemblies are replaced and the remainder rearranged in position so as to occupy a more favourable position in the core. The residence time for each fuel assembly in the core is therefore about three years and burn-up ratings of about 36 000 MWd/t (equivalent to nearly 4 per cent heavy atoms) are achieved. All structures within the pressure vessel are removable once the upper head has been removed.

The volumetric ratio of water-to-UO_2 fuel in the PWR is almost 2:1, that is the light water content of the reactor pressure vessel occupies about twice as much space as does the UO_2 in the core.

The specific thermal power rating of the fuel in a modern PWR is typically 33 kW/kg of uranium, a factor which is about twice that of the AGR. However, because of the much higher specific heat of the light water coolant used in the PWR, compared with the much less efficient CO_2 gas used in the AGR, less coolant is required in the PWR for a given power rating and its fuel assemblies can therefore be placed much closer together giving a small, compact core with a very high volumetric thermal power density, typically as high as 102 MW/m^3; this is about thirty-six times that of an AGR of comparable power output with its much larger core size.

A high volumetric thermal power density is most desirable from an economic point of view because of the relatively small physical size of the very expensive pressure vessel (costing tens of millions of pounds), and the lower fabrication and construction costs of the associated containment building and pipework, etc. From a safety point of view, however, a low volumetric power density means a large core with a large thermal capacity, lower coolant pressures, and therefore much less likelihood of abnormally high temperatures being generated during an accident involving loss of coolant.

For comparison, an existing PWR core producing 3800 MW of heat requires a core volume of approximately 37 m^3; equivalent in size to a room measuring 3.9 m (12.8 feet) × 3.9 m (12.8 feet) × 2.44 m (8 feet). An AGR core producing the same power would require an impractical core volume of 1367 m^3. In fact, each of the two AGR reactors at Hinkley Point produces 1500 MW of heat and each has a core volume of 540 m^3.

The PWR steam generator

Each of the steam generator units shown in Figure 8.16 comprises a tightly packed bundle of more than 5600 inverted U-shaped tubes through which the primary coolant from the reactor pressure vessel is circulated. These are surrounded by a second light-water circuit which absorbs heat from the U-tubes and is allowed to boil. The resulting steam produced in this way is used to drive the turbo-generator set, after which it is condensed and returned to the steam generator as feedwater for re-boiling. A cutaway diagram of a modern Westinghouse steam generator is illustrated in Figure 8.18.

The steam generator is a very large component which is nearly 21 m (69 feet) in height and with a maximum diameter of about 4.5 m (14.75 feet). Its weight when empty is 325 tonnes and it is fabricated from carbon steel lined with stainless steel.

The primary coolant enters the U-tube bundle at a temperature of 325°C and leaves at a temperature of 293°C. On the secondary side of the generator steam is produced at a temperature of 285°C, at a pressure of about 1000 psi (68 atmospheres), and the temperature of the feedwater condensate is 227°C. The mass flow rate of the steam fed to the turbines is nearly 480 kg/s. Steam dryers and centrifugal moisture separators located near the top of the steam generator ensure that the steam leaving the vessel contains no more than 0.25 per cent wetness.

The containment building

The whole of the NSSS of a PWR, and some of its emergency core cooling (ECC) system, are housed inside a structure known as the *containment building*. Its purpose is to contain the whole of the primary coolant in the event of a LOCA, and withstand the expected overpressures, and to prevent unacceptable releases of radioactivity to the environment. A cross-sectional diagram of a Westinghouse PWR containment building is shown in Figure 8.19.

The building is fabricated from steel-lined reinforced concrete with walls up to 1.3 m (4.3 feet) thick and with an overall height of about 53 m (175 feet); its outside diameter is approximately 38 m (125 feet).

Not shown in Figure 8.19 is a stainless steel storage vessel known as the accumulator. This is half-filled with borated water (water dosed with boron – a very good neutron absorber) which is held under pressure by nitrogen gas occupying the other half of the vessel. The pressure on the borated water is approximately 650 psi (44 atmospheres). A modern four-loop PWR has four accumulators in the containment building – one for each loop – each containing about 40 m³ (8800 gallons) of borated water. The

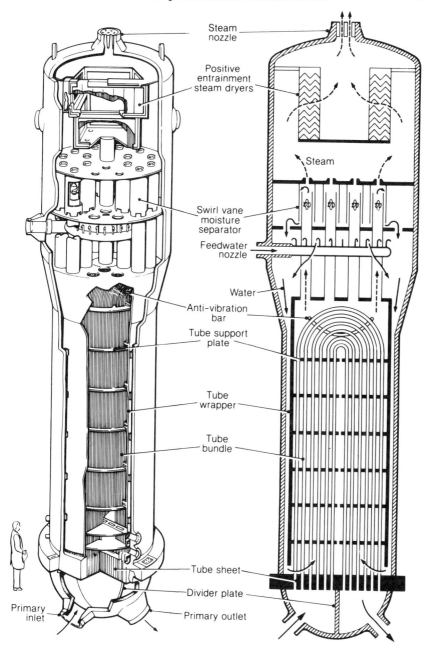

Steam nozzle

Positive entrainment steam dryers

Steam

Swirl vane moisture separator

Feedwater nozzle

Water

Anti-vibration bar

Tube support plate

Tube wrapper

Tube bundle

Tube sheet

Divider plate

Primary inlet

Primary outlet

Figure 8.18   Cutaway view of a Westinghouse PWR steam generator (courtesy of Westinghouse Electric Corporation)

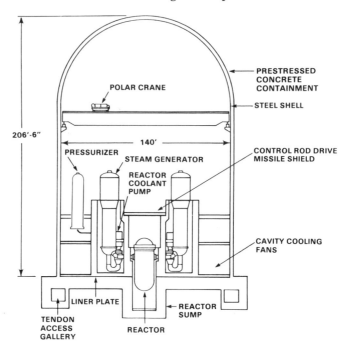

Figure 8.19   Cross-section of a Westinghouse PWR containment building (courtesy of Westinghouse Electric Corporation)

accumulators form part of what is known as the passive emergency core cooling system and are triggered into action as soon as the primary coolant pressure drops below about 1000 psi (69 atmospheres) due, for example, to a fracture in one of the primary coolant loop pipes. When this happens the borated water is injected at high pressure into the reactor pressure vessel, either directly or through the coolant loop pipes, thereby providing without delay the essential decay-heat cooling required in the early stages after shut-down and, at the same time, providing back-up neutron absorption for the shut-down rods which should also have been triggered into operation by the reduction in primary coolant pressure.

The action of the accumulators is supplemented by a containment spray system located in the roof area of the primary containment shell. The system is supplied from cold water header tanks and is triggered into operation by the onset of a LOCA. Its function is to condense steam and prevent the build-up of excessive pressures.

Two pumps located in the sump of the containment shell are used to return water to the spray header tanks by way of a heat exchanger. Four heat exchangers are also employed to remove heat from the containment building under both normal and fault conditions.

The roof sprays ensure that a serious energy release from the NSSS can safely be accommodated indefinitely without exceeding the design working pressure of the containment building.

Additional safety features

The four accumulators in the containment building provide immediate first-aid for any major LOCA in the PWR NSSS but their limited and non-replenishable combined capacity of borated water – about 160 m^3 (35 200 gallons) – coupled with the very high injection flow rate, means that once operated they are soon exhausted. The accumulators are therefore backed up by a number of *active* systems which supply a continuous (pumped) flow of emergency cooling water which is designed to provide long-term compensation for water being lost from the pressure vessel.

Designs vary but a typical active ECC system would comprise four high-head safety pressure injection (HHSI) pumps and two low-head safety injection (LHSI) pumps which also feed water to two heat-removing heat exchangers for the ECC water. These pumps, operating together, are able to provide more than enough cooling water for the NSSS circuit to ensure indefinite long-term cooling of the core in the most severe LOCA.

Water for the HHSI and LHSI pumps is initially supplied from a large high-head storage tank containing 1740 m^3 (almost 400 000 gallons) of borated water which, under normal operating conditions, is used to flood the reactor pressure vessel when its top is removed during the refuelling operation. When used for ECC purposes, however, it is emptied by the HHSI and LHSI pumps and finally ends up in the sump of the containment building. Here it is picked up again by the same pumps and recirculated indefinitely past heat exchangers and through the NSSS circuit.

In addition to some of the safety systems so far described for the primary side of the reactor coolant circuit there are a number of standby pumps and feedwater storage tanks for the secondary side of the steam generators, and also means for venting surplus steam in the event of a defective turbine.

Sizewell B

On 11 January 1983 a Public Inquiry began to consider the pros and cons of a planning application submitted by the CEGB for the construction of a nuclear power station based upon a pressurized water reactor (PWR) at Sizewell in Suffolk. The proposal is for the new station, to be known as

Sizewell B, to be built alongside the CEGB's existing Magnox station (Sizewell A) which began operating there in 1966. The inquiry was led by the Government-appointed Inspector, Sir Frank Layfield QC, and took place in the town of Leiston, Suffolk, about 4 miles from Sizewell.

The Inquiry, the longest ever in Britain, eventually ended on 7 March 1985 after sitting for 340 working days spread over two years and hearing evidence and opinions from 200 witnesses. The eight-volume report of the Inquiry, comprising some 3000 pages, was published on 28 January 1987 and was debated in the House of Commons on 23 February 1987.

On 12 March 1987 the Secretary of State for Energy (Mr Peter Walker) informed members of the House that after careful consideration of the Layfield Report, and of the points raised in the subsequent debate – in particular those relating to the Chernobyl accident which occurred after the Inquiry had ended – he had decided to grant permission for the CEGB to construct the Sizewell B power station, subject to the CEGB incorporating all the recommendations contained in the Report and, most importantly, the acquisition of an operating licence from the Nuclear Installations Inspectorate (NII).

Among the many recommendations contained in the Report are:

- Loading of fuel into the new reactor shall not start until at least one year after a simulator for Sizewell B has been installed and is ready for use.
- At least two members of the Radioactive Waste Management Advisory Committee (RWMAC) should be appointed specifically to provide an independent expert environmental contribution to the handling of radioactive waste management.
- In making proposals for future power stations the CEGB should provide a probabilistic analysis of the costs of generation from new capacity (the net effective cost), together with a clear statement of the probabilities it has assumed.
- In view of the CEGB's proposals for the Sizewell site the Suffolk County Council, in consultation with the CEGB, should undertake a study of alternative routes for heavy goods vehicles traffic from the A12 road to the Sizewell site to identify the best route, its cost, effects and practicality.
- Any reprocessing in the UK of spent fuel from Sizewell B or later PWRs should be carried out in buildings separate from those used for military reprocessing, and the separated plutonium shall be stored apart from plutonium from military reactors.
- The revised pre-construction safety report and the final safety report

should be published by the CEGB as soon as possible after they have been submitted to the Nuclear Installations Inspectorate.

- The Health & Safety Executive (HSE) should publish consultation and policy documents on subjects that concern nuclear safety policies. Those papers should be designed to be comprehensible to the interested public and should, so far as appropriate and practicable, be prepared in consultation with other regulatory or advisory organizations.

The main points contained in the CEGB's case for wishing to build a PWR power station at Sizewell are as follows:

- The Board possesses a good, safe and economic design for the PWR which has been adapted to meet British conditions and which is suitable for replication.
- The PWR would help to keep the cost of electricity as low as possible. It would have lower total costs than a new coal, oil or AGR type of power station and would more than pay for itself over its lifetime from fuel saving alone.
- The PWR would help to improve the security of fuel supplies by further diversifying the types and sources of fuel used for electricity generation at a time when the price of oil requires its minimum use and the Board is over-reliant on coal.
- It makes good sense to build the PWR now to reap the earliest benefits from displacing older, less economic plant.
- Since the CEGB's step-by-step approach to future development seems likely eventually to lead to the ordering of further nuclear capacity, the Board must ensure that it has available the best technology and it is therefore necessary to establish the PWR as an alternative in the UK to the AGR.
- The Sizewell site is the right place for helping to balance electricity supply and demand in south-east England; no new transmission lines would be needed and the site is already owned by the Board. The use of seawater for cooling also means that no new cooling towers would be required.
- The PWR type of power station is preferred to that based on the AGR because:
  - it costs 17 per cent less per kilowatt to build;
  - it will have a ten-year longer life;
  - it will be marginally cheaper to run;
  - it will have only slightly less availability;

- it will have a lifetime advantage over the AGR by saving electricity consumers more than £600 million;
- it will boost export opportunities;
- 90 per cent of the total construction cost (£1600 million at 1986 prices) will be spent in the UK;
- it will create 10 000 jobs over seven years, 300 of which will be on the construction site.

The Sizewell B reactor The following is a summary of some of the technical and physical features of the PWR reactor to be built at Sizewell:

Electrical power output	1155 MW (nett)
Thermal power output	3425 MW
Total fuel inventory	101 tonnes UO_2
Number of fuel rods	51 000
Number of fuel assemblies	193
Pressure vessel	13.55 m high; 4.39 m diameter; 385 tonnes weight
Steam generators	four; each 20.6 m high and weighing 325 tonnes
Coolant	light water pressurized to 158 atmospheres (2250 pounds per square inch)
Containment building	cylindrical, steel-lined pre-stressed concrete envelope 1.3 m thick, 45.7 m internal height.

The boiling water reactor

The boiling water reactor (BWR) is a development of the pressurized water reactor (PWR) and is in direct commercial competition with it. About 36 per cent (1985 figures) of all light-water reactors (LWRs) throughout the world are of the BWR type and most of them have been designed and constructed by the US General Electric company. The total installed capacity for BWRs throughout the world (1985 figures) is about 37 GWe, a figure which is a little below half that of the PWR (about 81 GWe).

The first BWR to be constructed was a test reactor facility known as BORAX-1, built in Idaho (USA) in 1953. This was followed by the power-producing Experimental Boiling Water Reactor (EBWR) which was built at the Argonne National Laboratory (ANL) and began operating on 1 December 1956. It began generating electricity on 23 December 1956 and achieved its full design power output of 5 MWe six days later. The first commercial BWR to be built was the 210 MWe Dresden-1 reactor at Morris, Illinois. This was constructed by the General Electric company and went critical in October 1959; full commercial operation began in August 1960.

The BWR is essentially very similar to the PWR. It uses light water for both moderator and coolant and it uses enriched uranium fuel in the form of small ceramic pellets of UO_2 contained in Zircaloy tubes (fuel pins). The main difference between the two reactor types is that in the PWR the light water coolant is artificially pressurized to stop it boiling whereas in the BWR it is not. In consequence the coolant is allowed to boil as it passes through the core and steam is produced above the water level in the steel pressure vessel which also contains the core. In the PWR the pressure vessel is completely filled with water; in the BWR it is only partially filled. There is, however, always enough water in the pressure vessel to ensure that the core is fully immersed at all times. The basic operating principles of the BWR are summarized in the simplified diagram shown in Figure 8.20.

Heat produced by the core causes the coolant to boil and steam collects in the upper region of the pressure vessel above the water level. The steam is conveyed to the turbo-generator electricity generating plant, after which it is condensed and returned to the pressure vessel as warm water condensate. About 14 per cent (by weight) of the water contained in the pressure vessel is converted to steam, the remainder is re-circulated through the

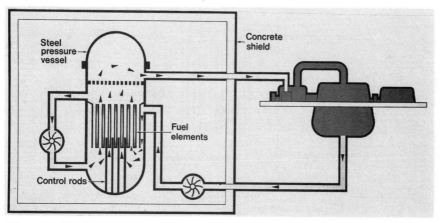

Figure 8.20   Simplified operating diagram of the boiling water reactor (BWR)

core by a number of so-called 'jet pumps'. These force the water in a downward direction through the annular space formed between the wall of the pressure vessel and the core shroud, and then up through the fuel elements which make up the core where some of it is converted to steam. This action is more clearly illustrated in Figure 8.21. Typical flow rate for the BWR coolant water is about 14 tonnes per second (about 3140 gallons per second).

Operation of the BWR is very similar to that of a domestic pressure cooker where the boiling point of the water content is raised above the normal temperature of 100°C by preventing the steam escaping and allowing the pressure to build up. This is precisely what happens in the

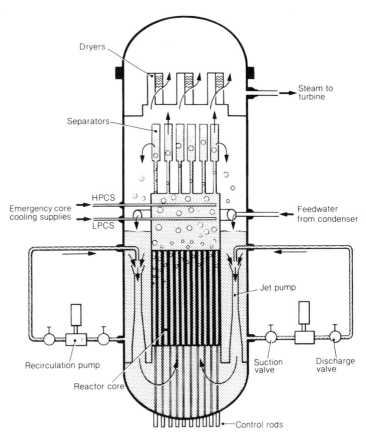

Figure 8.21 Coolant flow and steam generation in the BWR (courtesy of General Electric)

BWR; the steam created in the top of the pressure vessel is kept in a closed circuit formed by the vessel itself and the turbo-generator set and the pressure builds up to a level of about 1040 psi (about 71 atmospheres). This self-generated pressure is sufficient to raise the boiling point of the coolant, and the temperature of the resulting steam, to about 288°C, a figure which is about the same as the temperature of the steam produced in the PWR steam generator (see page 150). It is because of this similarity in steam temperatures that the thermodynamic efficiencies of the PWR and BWR are about the same and close to 33 per cent.

The pressure vessel

Since the coolant in the BWR is operated at a self-generated pressure of about 1040 psi (71 atmospheres) the coolant must be contained in a thick-walled pressure vessel similar to that used for the PWR. However, since the coolant pressure of the BWR is a little less than half that of the PWR, which is about 2200 psi, its pressure vessel thickness is correspondingly less. Figure 8.22 illustrates a cutaway view of a modern BWR pressure vessel for a core producing 3800 MW of thermal power. It is made from low alloy steel and has a diameter of about 6.5 m (21.3 feet); its height is approximately 22 m (72 feet) and it has a wall thickness of 152 mm (6 inches).

The steam separator assembly shown in Figure 8.22 separates the steam from the steam–water mixture produced as the coolant passes through the core and directs it into the steam dryer assembly located immediately above. The separation is achieved by swirling the mixture in long vertical tubes so that the water forms a central vortex (like a whirlpool) through which the steam passes to the dryers. The separated water flows downward through the annular space between the core shroud and the pressure vessel wall where it joins the water being circulated through the core. Any moisture remaining in the separated steam is removed by the steam dryer assembly, after which the dried steam is conveyed to the turbines. This is a vital operation since wet steam can cause rapid errosion of the turbine blades due to the formation of high velocity water droplets which literally 'wear away' the blades.

It is at this stage that another major difference is observed between the PWR and BWR reactors. In the PWR the steam produced in the steam generator is derived from a secondary water loop which is allowed to boil. The heat which causes the boiling is derived from a completely separate primary water loop which passes through the reactor core and which is *not* allowed to boil. The steam which passes through the turbines is thus free from radioactivity since it is derived from water which has not been

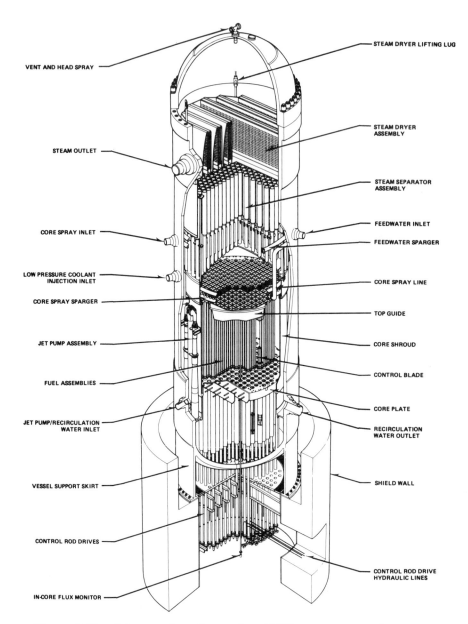

VENT AND HEAD SPRAY

STEAM DRYER LIFTING LUG

STEAM OUTLET

STEAM DRYER ASSEMBLY

STEAM SEPARATOR ASSEMBLY

CORE SPRAY INLET

FEEDWATER INLET

FEEDWATER SPARGER

LOW PRESSURE COOLANT INJECTION INLET

CORE SPRAY LINE

CORE SPRAY SPARGER

TOP GUIDE

JET PUMP ASSEMBLY

CORE SHROUD

FUEL ASSEMBLIES

CONTROL BLADE

JET PUMP/RECIRCULATION WATER INLET

CORE PLATE

RECIRCULATION WATER OUTLET

VESSEL SUPPORT SKIRT

SHIELD WALL

CONTROL ROD DRIVES

IN-CORE FLUX MONITOR

CONTROL ROD DRIVE HYDRAULIC LINES

Figure 8.22   Cutaway view of a modern BWR pressure vessel (courtesy of General Electric)

circulated through the core. This method of steam generation is known as the *indirect cycle* system and was illustrated in Figure 8.15.

As explained above, the steam fed to the turbines in the BWR reactor is generated directly from water which has passed through the core and is therefore radioactive due to neutron activation of its constituent parts. It also carries with it gaseous and volatile fission products which included the radioisotopes of iodine, krypton, xenon and tritium; these are always present in the coolant to some extent – albeit at very low levels – due to the inevitable pin-hole imperfections which occur from time to time in all types of fuel cladding. Also carried along with the steam are tiny fragments which have become detached from the surfaces of internal pipework and structural materials due to corrosion. These fragments, in circulating through the core, will also be made radioactive due to the process of neutron activation.

The coolant activation products are mostly the nitrogen isotopes N-16 and N-17 produced by n,p reactions on the oxygen content of the water:

$$^{16}_{8}O \ (n,p) \ ^{16}_{7}N \ (7.1 \ seconds)$$

$$^{17}_{8}O \ (n,p) \ ^{17}_{7}N \ (4.1 \ seconds)$$

plus the oxygen-19 isotope produced by the n,γ reaction:

$$^{18}_{8}O \ (n,\gamma) \ ^{19}_{8}O \ (29 \ seconds).$$

Because of their very short half-lives, the main activation products present no special hazards to operating personnel once the reactor is shut-down. On the other hand they give rise to very high levels of radioactivity in the vicinity of the turbines whilst the reactor is operating, necessitating massive shielding (concrete) and making routine on-load maintenance more difficult. This is the main disadvantage of the BWR's *direct cycle* of steam generation. The main advantage is an economic one because of the absence of expensive steam generators and their associated pipework.

Power density and burn-up

For a given power rating the physical size of the core in a BWR reactor is almost twice that of a PWR and its total fuel inventory is about one-and-a half times as much: typically 170 tonnes for 1300 MWe rating. The individual fuel rods (or pins) are also larger in diameter than those of the PWR and the fuel assemblies are spaced much further apart. All this amounts to a much larger core volume and a correspondingly lower volumetric power density (see page 149) when compared with the PWR and which, in turn, requires a much larger pressure vessel.

The volume of a BWR pressure vessel is approximately three times that of a similarly-rated PWR and the volumetric thermal power density of its core is a little more than half: typically 56 MW/m^3 for the BWR compared with 102 MW/m^3 for the PWR.

The specific thermal power rating (see page 149) for BWR fuel is about 20 per cent less than that for the PWR and is typically 26 kW/kg of uranium; this compares with 33 kW/kgU for the PWR. The U-235 enrichment of fresh fuel for the BWR is typically 2.6 per cent, which is somewhat lower than the 3.3 per cent fresh-fuel enrichment for the PWR. The burn-up rating of BWR fuel is also less than that of the PWR and is typically 27 500 MWd/t; this compares with 36 000 MWd/t for the PWR. However, because of the larger fuel inventory of the BWR its in-core residence time is greater than that of the PWR (for a given burn-up) and is, in some designs, as long as $4\frac{1}{2}$ years. Refuelling procedures vary considerably with design but typically one-third of the fuel is removed and replaced every 18 months, the whole procedure taking about 8 days.

The refuelling procedure involves removal of the top section of the pressure vessel (see Figure 8.22) and also of the steam dryer and steam separator assemblies. This operation allows access to the individual fuel channels by the refuelling machine which can then remove the spent fuel assemblies and replace them with fresh ones, the whole operation being conducted through the open top of the pressure vessel. Before this operation is started, however, the pressure vessel is completely filled with water which is allowed to overflow into a circular-shaped surrounding well to a depth of a few metres. Spent fuel is then taken out of the reactor and transported through water-filled passage-ways to a spent fuel storage pond where it remains under water until ready for reprocessing. Spent fuel is similarly loaded under water.

When the refuelling operation is completed, the top section of the pressure vessel and the surrounding well is drained of water and the components which were removed from the vessel interior are replaced. The top cap of the pressure vessel is then bolted back into place and the reactor is ready for 'start-up'.

Spent fuel assemblies are removed from the central regions of the core and replaced by partially-spent fuel assemblies from the outer regions; fresh fuel assemblies are loaded into the outermost regions of the core. The whole purpose of this so-called 'fuel shuffling' is to maintain a relatively flat power distribution across the core.

The actuating mechanisms for the control rod assemblies used in the

BWR core penetrate the pressure vessel from the bottom and are manipulated hydraulically. This method of control is made necessary because the top section of the vessel is occupied by the steam dryer and steam separator assemblies.

Fuel assemblies

A fuel rod assembly for a modern BWR takes the form of a square-section cluster of fuel rods (or pins) arranged in an 8 × 8 matrix like that shown in Figure 8.23. Unlike the PWR the BWR fuel clusters are enveloped by a

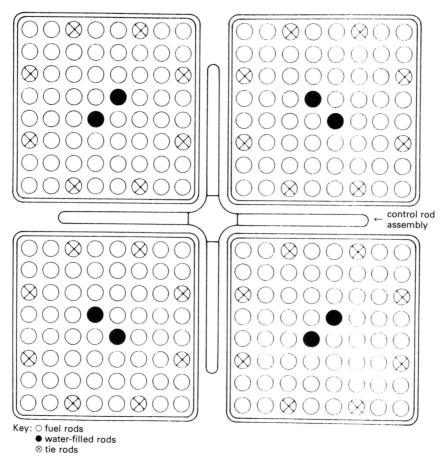

Key: ○ fuel rods
    ● water-filled rods
    ⊗ tie rods

Figure 8.23   Sectional diagram showing the physical arrangement of four fuel-rod and one control-rod assemblies in the core of a modern BWR (courtesy of General Electric)

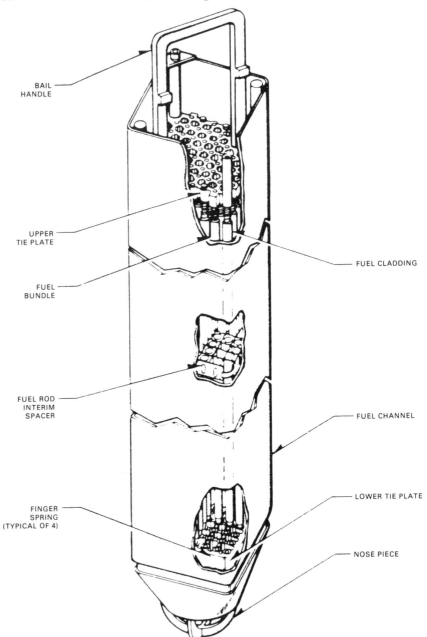

BAIL
HANDLE

UPPER
TIE PLATE

FUEL CLADDING

FUEL
BUNDLE

FUEL ROD
INTERIM
SPACER

FUEL CHANNEL

LOWER TIE PLATE

FINGER
SPRING
(TYPICAL OF 4)

NOSE PIECE

**Figure 8.24   Typical fuel-rod assembly used in a BWR
(courtesy of General Electric)**

Zircaloy metal wrapper (or channel) which prevents lateral flow of the cooling water and ensures that it flows in an upward direction *only* between the fuel rods. The bottom end of the wrapper is tapered to form a nose connector which locates with a supporting structure and enables the flow of coolant through the assembly to be regulated. A complete fuel assembly for a BWR is illustrated in Figure 8.24.

The fuel rods are about 4 m (13 feet) in length and are made from thin-walled Zircaloy tubing. Each is filled with UO_2 pellets of cylindrical shape 18 mm (0.71 inches) high and 10.6 mm (0.42 inches) in diameter. Not all of the 64 spaces in the assembly are occupied with fuel rods (see Figure 8.23); some are occupied by tie rods which hold the assembly together and some by water-filled rods which help to level out the power distribution in the assembly by providing additional moderation in the centre. A modern BWR reactor would contain nearly 800 fuel assemblies and more than 48 000 fuel rods containing a total of 170 tonnes of UO_2.

Control rod assemblies

The neutron-absorbing control-rod assemblies used in the BWR take the form of a cruciform (cross-shaped) container filled with stainless steel tubes packed with boron carbide powder. Each assembly contains about 72 tubes and there are nearly 200 assemblies in a modern BWR. Figure 8.25 illustrates a typical BWR control-rod assembly; its overall length is about 4.4 m (14.4 feet).

The assemblies occupy positions in the core between groups of four fuel-rod assemblies (see Figure 8.23) and are driven hydraulically into the core from below. The advantage of this type of control mechanism is that the assemblies remain operational even when the top of the pressure vessel is removed during the refuelling procedure.

The bottom of the assembly, and its outer surfaces, are fitted with rollers which act as spacers and also allow rapid up-and-down movements of the assemblies within the core for control purposes. The 'velocity limiter' at the bottom of the assembly shown in Figure 8.25 limits the rate of descent of the assembly, should the hydraulic support mechanism fail, and thereby allows full control to be re-established by the remaining assemblies. This arrangement prevents sudden power transients in the reactor core.

Another advantage resulting from manipulation of the control-rod assemblies from below the core is that they may be withdrawn during operation by an amount which compensates for the lack of moderation in the upper regions of the core brought about by the formation of steam pockets (voids) as the water boils. Steam pockets reduce the density of the water and hence its moderating ability. This facility assists in maintaining a uniform heat generation in the vertical (axial) direction of the core.

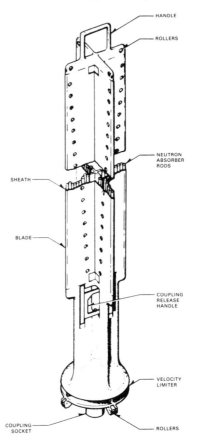

Figure 8.25   One of the many control-rod assemblies used in a BWR (courtesy of General Electric)

Containment building

Like the PWR, the BWR and its associated equipment are housed in a containment building equipped with safety features to combat the effects of a LOCA. The layout of a typical BWR containment building is illustrated in Figure 8.26.

The pressure vessel and its coolant circulation pumps, etc., are completely surrounded by a concrete structure known as a *dry well*, a lid in the roof of which can be opened for refuelling purposes. Surrounding this is a large steel structure which functions as the primary containment. This is designed to withstand the excess temperatures and pressures which could

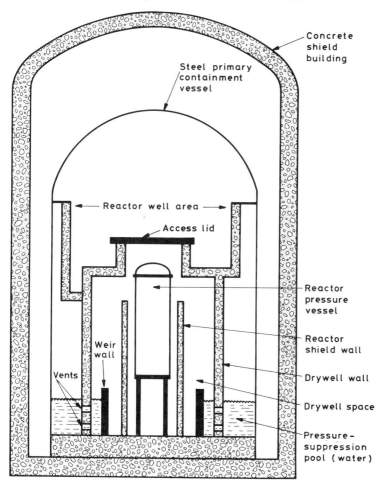

Figure 8.26   Sectional diagram of a BWR containment building (courtesy of General Electric)

arise during a LOCA and provides a sealed barrier against the release of radioactive materials to the environment.

The annular space formed between the base of the primary containment and the dry well is partially filled with water. This water is allowed to enter the bottom of the dry well through a number of vents around its circumference but is prevented from reaching the pressure vessel by a circular wall which functions as a weir. This volume of water is known as the *suppression pool* and it surrounds the pressure vessel in the same way as a water-filled moat once surrounded a medieval castle.

In the event of a LOCA, water and steam would escape from the pressure vessel and pressure would quickly build up in the dry well. This would eventually force its way down through the water in the bottom of the dry well and thence through the vent holes into the suppression pool. In bubbling-up through the suppression pool it would then be trapped in the confines of the primary containment. However, not much steam would find its way here since most of it would have been condensed and its pressure reduced to zero as it passed through the suppression pool.

The *reactor well* is a concrete structure surrounding the top cap of the pressure vessel and which is flooded during the refuelling operation when the cap is removed. The secondary containment is a structure known as the *shield building*. This is fabricated from reinforced concrete and serves as a second-line defence against the release of radioactive materials to the environment and also affords some protection against missiles such as falling aircraft, ice blocks, etc. and extremes of weather.

A number of emergency core cooling systems are also installed inside the pressure vessel to combat the effects of a LOCA and take the form of high- and low-pressure spray systems and coolant injection facilities similar to those used on the PWR.

SUMMARY OF REACTOR TYPES

Table 8.1 summarizes the main features of the commercial reactors which have been described in this chapter. However, because of differences between reactors of the same generic type the table should not be looked upon as a definitive summary but more as a guide which highlights the essential differences between the different types. Nevertheless, the details presented are representative of actual nuclear power stations and have been derived from information published by the station operators.

The Magnox reactor details given in the table are those for the Wylfa nuclear power station in Anglesey, North Wales, which is the largest of the British Magnox power stations and the last of its type to be built (see page 123 for detailed description of Wylfa). The AGR details are those for the Hinkley Point-B power station in Somerset which was described in detail on page 132. In each case the details refer to *one* reactor unit only.

Details for the Canadian CANDU reactor are those relating to the Bruce nuclear power station in Ontario, Canada, which was described in detail on page 142.

The two power stations used to represent the PWR and BWR types of

Table 8.1
Operating characteristics of some commercial nuclear power stations

Reactor Type	Thermal power generated (MW)	Net electrical power generated (MW)	Net thermal efficiency (%)	Type of fuel	Fuel inventory (tonnes)	Average fuel enrichment (% U-235)	Type of moderator	Type of coolant	Coolant pressure (atmos) (psi)	Coolant outlet temp. (°C)	Core volume (m³)	Volumetric thermal power density (MW/m³)	Specific thermal power density (kW/kg)	Discharge burn-up (MWd/t)	Refuelling sequence
Magnox (Wylfa)	1540	420	27	U-metal rods	600	Natural	Graphite	Carbon dioxide	27 (400)	360	2565	0.60	2.6	5500	On-load
AGR (Hinkley Point)	1500	615	42	UO_2 pellets	112	1.57	Graphite	Carbon dioxide	40 (576)	630	540	2.8	13.4	24 000	On-load
CANDU (Bruce)	2855	740	31	UO_2 pellets	117	Natural	Heavy water	Heavy water	90 (1322)	300	340	8.4	24.4	7160	On-load
PWR (Trojan)	3411	1095	32	UO_2 pellets	98	2.6	Light water	Light water	153 (2247)	325	33	103.5	38.5	30 000	33.3% per year
BWR (Browns Ferry)	3293	1067	32	UO_2 pellets	142	2.2	Light water	Light water	70 (1024)	294	65	56	23	19 760	26% per year

reactors were deliberately chosen because of their similarity in electrical and thermal power outputs, which made comparisons between them much easier. The PWR station chosen was the Trojan station in Prescott, Oregon, USA; this achieved full commercial operation in May 1976. The BWR station chosen was the Browns Ferry station at Decatur, Alabama, USA; this first achieved commercial operation in August 1974.

REFERENCES

Layfield, F., Sir, 'Sizewell B Public Inquiry: Report on Application by the Central Electricity Generating Board for Consent for the Construction of a Pressurized Water Reactor'. 8 volumes. Department of Energy, London. HMSO, 1987. £30. ISBN 0114-1157-53.

Layfield, F., Sir, 'Sizewell B Public Inquiry: Summary of Conclusions and Recommendations from the Inspector's Report on the Central Electricity Generating Board's Application for Consent for the Construction of a Pressurized Water Reactor'. Department of Energy, London. HMSO, 1987. 20 pp. £4.95. ISBN 0114-1157-61.

9 The nuclear power station: fast breeder reactors

The thermal reactors which have so far been described are fuelled with either natural uranium containing 0.7 per cent U-235 or enriched uranium containing up to about 3 per cent U-235, and they require a moderator to slow down the neutrons produced during fission. Since such reactors use mainly the U-235 content of the fuel then most of the uranium, in the form of U-238, is actually wasted. It was the need to reduce the magnitude of this waste which gave rise to the so-called *fast reactor*.

FAST BREEDER REACTOR THEORY

A fast reactor is one which uses no moderator to slow down the neutrons produced in its core and the fuel it uses is fissioned directly by the fast neutrons emitted during the fissioning process; hence the name.

The fast reactor concept is not new; in fact the first reactor *of any type* in the world to be used for generating electricity was a fast reactor and it began operating in the USA in 1951; it was known as EBR1 (Experimental Breeder Reactor 1) and it used pure U-235 for its fuel. Its electrical power output was a modest 200 kW.

Fast reactor research and construction has also been undertaken for many years in Britain, France and the USSR. The first fast reactor in Britain to be used for the generation of electricity was constructed at the UKAEA's fast reactor research establishment at Dounreay, Scotland, where it began operating in November 1959. It was known as DFR

Figure 9.1   View of the UKAEA's Dounreay research establishment show-ing the 250 MWe PFR (foreground) and the 14 MWe DFR (spherical con-tainment) which operated for almost 18 years

(Dounreay Fast Reactor) and was initially loaded with fuel composed of highly enriched uranium containing 75 per cent U-235; it was later fuelled with a mixture of U-238 and Pu-239. DFR regularly supplied the National Grid with 14 MW of electrical power for almost 18 years and yielded valuable information which was later used when the much larger Prototype Fast Reactor (PFR) was designed and later built alongside DFR (Figure 9.1). DFR was eventually closed down as a working reactor in March 1977.

Nuclear cross-section

The usually non-fissile U-238 isotope which makes up most of naturally-occurring uranium *can* be made to fission, although not very readily, with highly energetic fast neutrons whose energies are in excess of about 1.1 MeV – the so-called *fission threshold* of U-238. Although many neutrons emitted during the fast fissioning of U-238 have energies well in excess of this figure (the average fission neutron energy is about 2 MeV), most of them are involved in multiple non-fission-producing collisions with other U-238 atoms and, in consequence, lose much of their initial energy. The result of this is that even if an external source of fast neutrons were used to cause fissioning of a few atoms of U-238, the average energy of the resulting fission neutrons would be below the critical 1.1 MeV threshold required to initiate and sustain a chain reaction. *This is why it is impossible to maintain a fast neutron chain reaction in a mass of pure natural uranium, no matter how much is used.*

The reason why so few neutron collisions with atoms of U-238 result in fission is because of the very small *fission cross-section* of U-238. (The term 'cross-section' as applied to neutron capture was explained on p. 44.)

Another nuclear event which plays an important role in the study of reactor physics is that known as *neutron scattering*. This takes into account those neutron–nuclei collisions which result in neither capture nor fission and it is this which brings about the progressive loss of kinetic energy of the fast fission neutrons produced in a mass of natural uranium. The greater the likelihood of neutron scattering taking place in a particular material, the greater is the *scattering cross-section* of that material.

The concept of nuclear cross-section is a theoretical one in which the willingness of a target nucleus to undergo a particular nuclear event when struck by a neutron is expressed in terms of the size of the apparent area it presents to the neutron. The greater the likelihood of the event occurring, the greater is the apparent area presented by the target nucleus and the greater is its associated cross-section.

Nuclear cross-section is represented by the greek letter sigma (σ), with subscripts to indicate which event is being represented. Fission cross-section is represented by σ_f, capture cross-section by σ_c and scattering cross-section by σ_s. The total *absorption cross-section* of an individual nucleus is known as the *microscopic cross-section* and is represented by the symbol σ_a; its value is equal to the sum of the capture and fission cross-sections:

$$\sigma_a = \sigma_c + \sigma_f$$

So far only the cross-sections of individual nuclei have been considered. If a beam of neutrons is arranged to irradiate a volume of material equal to 1 cm³, then the total effective area of the target presented to the incident neutrons is equal to $N\sigma_a$, where N is the total number of nuclei (or atoms) per cubic centimetre. This product is known as the *macroscopic cross-section* of that particular material and is represented by the greek capital letter sigma (Σ), so $\Sigma = N\sigma_a$.

The unit of measurement of nuclear cross-section is the *barn* (abbreviated to the letter b), where 1 barn is equivalent to a target area of 10^{-24} cm^2. Thus, a nuclear cross-section of, say, 3.2×10^{-26} cm^2 would be written as 0.032 b, or as 32 mb (32 millibarns).

All nuclear cross-sections vary widely in value between different elements and are strongly dependent on the energy of the incident neutron. The fission cross-section of U-235 for thermal neutrons, for example, is quite large (about 580 b) whereas for U-238 it is zero. Similarly, the capture cross-section of Pu-239 for thermal neutrons is more than two and a half times that for U-235.

Table 9.1 lists the capture and fission cross-sections of the nuclides U-235, U-238 and Pu-239 for thermal and fast neutrons, and also the average number (ν) of neutrons produced from each fission event. Also shown is the ratio (α) of capture-to-fission cross-sections. Thermal neutrons, in this instance, are assumed to possess a mean kinetic energy of 0.025 eV and travel at about 2200 ms^{-1}. Corresponding figures for fast neutrons are about 2 MeV and 2×10^4 kms^{-1}.

The values indicated in the table are approximate and intended for comparison purposes only. The important points to observe from the table are:

1   As far as thermal neutrons are concerned, U-238's only value is its ability to produce the more useful fissile isotope Pu-239 by neutron capture (see page 70).

2   The α-value of U-235 is less than half that of Pu-239 for thermal

Table 9.1

Thermal- and fast-neutron cross-sections and fission neutron yields for natural uranium isotopes and Pu-239

Nuclide	Thermal neutrons				Fast neutrons			
	σ_c	σ_f	α	ν	σ_c	σ_f	α	ν
U-235	100	580	0.17	2.4	0.08	1.2	0.06	2.7
U-238	2.7	0	∞	0	0.09	0.3	0.03	2.8
Pu-239	270	745	0.36	2.9	0.05	1.8	0.03	3.2

neutrons and U-235 would therefore appear, at first sight, to be a better fissile material. On the other hand, Pu-239 has a higher fission neutron yield than U-235 which, as you will shortly see, makes it the better of the two.

3　The fission neutron yield (ν) for all of the nuclides listed is higher for fast neutrons.

4　The fission and capture cross-sections for both U-235 and Pu-239 are very much less for fast neutrons.

5　The α-value for Pu-239 is less than that of U-235 for fast neutrons. This fact, coupled with its higher neutron yield, makes Pu-239 superior to U-235 as a fissile material in a fast neutron flux.

6　Significant fissioning of U-238 occurs in a fast neutron flux and its α-value is comparable with that of Pu-239.

7　The fast fission neutron yield of U-238 is less than that of Pu-239 but higher than that of U-235.

Fast fission

When a fissile nucleus absorbs an incident neutron it either fissions and emits neutrons, or it is transformed into a higher-order isotope by the process of neutron capture. From the point of view of the nuclear reactor it is the fission process which is the more important of the two and it is the likelihood of this happening, coupled with the magnitude of the fission neutron yield, which determines the usefulness of the material as a source of reactor fuel.

A parameter which indicates the usefulness of a fissile material is represented by the Greek letter eta (η) and expresses the average number of fission neutrons produced per neutron absorbed by fission and capture processes in the material. Eta is related to the fission (σ_f) and capture (σ_c) cross-sections and the fission neutral yield (ν) by the expression:

$$\eta = \frac{\nu.\sigma_f}{\sigma_f + \sigma_c}$$

$$= \frac{\nu}{1 + \sigma_c/\sigma_f}$$

$$= \frac{\nu}{1 + \alpha}$$

The expression shows that the magnitude of η increases both as v becomes larger and as α becomes smaller.

Table 9.2
Values of η for the fissile materials U-235
and Pu-239

Fissile material	η	
	Thermal neutrons	Fast neutrons
U-235	2.07	2.50
Pu-239	2.15	3.09

Table 9.2 lists the values of η for the fissile materials U-235 and Pu-239 for both fast and thermal neutrons and shows that the greatest values of η are obtained, for both materials, for fast neutrons. The table also shows that Pu-239 is the superior of the two materials for both thermal and fast neutrons, especially so for fast neutrons where η is 24 per cent larger than the corresponding figure for U-235.

- *In a fast reactor more of the available neutrons result in fission of the nuclear fuel than in a thermal reactor. This results in good neutron economy.*
- *Pu-239 is a better nuclear fuel than U-235, for both thermal and fast reactors, because its capture-to-fission cross-section ratio is much lower and its neutron yield much higher.*
- *The superiority of Pu-239 over U-235 is most evident when used as fuel in a fast reactor.*

Fertile material

It has been demonstrated that although U-238 is virtually non-fissile in a thermal reactor it *can* be made to fission in a fast reactor, although to an extent which is far below that of U-235 and Pu-239. On the other hand the relatively high capture cross-section of U-238, especially for thermal neutrons, makes it an ideal material from which Pu-239 can be produced for use in the fast reactor. *In fact, without U-238 there would be no Pu-239 since this is the only material from which it can be made.*

A non-fissile material like U-238, which can be transmuted into a fissile material such as Pu-239, is said to be *fertile*; U-238 is therefore a *fertile material*. Another fertile material is naturally-occurring thorium-232 (Th-232) which, on capturing a neutron, is transmuted into the fissile material U-233 via Th-233 and Pa-233. Much research has been and continues to be done on the use of Th-232/U-233 as a source of nuclear fuel but as yet there have been no large-scale commercial developments. It is of interest to note that U-233 is significantly superior to both U-235 and Pu-239 as a source of fuel in a thermal reactor (η-value = 2.29), whereas in a fast reactor it is inferior to Pu-239 but better than U-235 (η-value = 2.61).

Fertile U-238 is used to produce Pu-239 in a fast reactor by packing depleted uranium (see page 81) around the core in such a way that the core is completely enveloped by it. In this way, stray neutrons which escape from the core are captured by the surrounding material and used to transmute U-238 into Pu-239. The process is known as *breeding* and the U-238 deployed in this way is known as the *breeding blanket*. That which appears above and below the core is incorporated in the individual fuel sub-assemblies and is known as the *axial breeder*. That which is packed around the sides of the core is known as the *radial breeder*.

Figure 9.2. illustrates the core and blanket arrangement of a fast reactor.

The fact that a fast reactor uses unmoderated (fast) neutrons and is able to breed fissile Pu-239 from virtually non-fissile U-238 gives rise to the

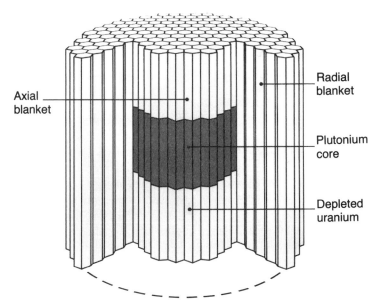

Axial blanket

Radial blanket

Plutonium core

Depleted uranium

Figure 9.2   The core and breeding blanket arrangement in a fast reactor

name *fast breeder reactor* (FBR). It should be remembered, however, that *a fast reactor does not 'operate faster' (a meaningless phrase) than any other type of reactor, nor does it breed fast, and it only breeds when required to do so.* In fact, most thermal reactors breed Pu-239 considerably faster than does a fast reactor for a given operating power. In the fast reactor, however, the breeding takes place *outside* the core and makes use of surplus neutrons which would otherwise have escaped and been wasted. Breeding in a fast reactor is therefore a bonus to what already is a more efficient way of using nuclear fuel.

Fissile concentration

The fast reactor has been shown to be more efficient than the thermal reactor in using nuclear fuel, in spite of the much smaller fission cross-section of its fuel for fast neutrons. Unfortunately, this low fission cross-section also means that many of the neutrons produced during fission are likely to escape from the core without causing further fission, making it difficult to create and sustain a chain reaction. This is why a greater concentration of fissile material is required in the fast reactor than in the thermal reactor whose fuel has a much higher fission cross-section because it is irradiated by slow neutrons.

In a virtually loss-free core (from which no neutrons escape) of infinite size it is possible to sustain a chain reaction in a fast reactor with fuel containing about 7 per cent U-235 or 5 per cent Pu-239. In a practical arrangement, however, the core of a fast reactor does not behave as though it were of infinite size and many fission neutrons do escape. This in itself is not so important because most of these will be captured by the fertile blanket and put to good use in creating Pu-239. What *is* important is that enough neutrons must be created and retained in the core to sustain a chain reaction. This can be achieved with large quantities of fuel with relatively low fissile concentration, or with smaller quantities of fuel with a high fissile concentration. The choice is a complicated one dictated partly by economics – large quantities of fissile material are expensive – and partly by operating conditions and the problems associated with a high power density and rapid heat transfer.

Representative of fissile concentrations in fast reactor fuel is that adopted for the 250 MWe British PFR. The fuel used in this reactor was described on page 102 and was said to comprise a homogeneous mixture of PuO_2 and UO_2. The fissile concentration varies from about 21 to 27 per cent, depending upon the position occupied by the fuel in the core, and averages about 25 per cent. The actual volume occupied by the core region of the fuel sub-assemblies (see Figure 9.2) is quite small – less than 2 m^3 – and contains about 1.1 tonnes of PuO_2 and 3 tonnes of UO_2. The power density of the core is approximately 142 kW (heat) per kg of PuO_2/UO_2. Bearing in mind that 1 kg of PuO_2/UO_2 is similar in size to a tennis ball (about 56 mm diameter), and that 142 kW of heat is the same as that produced by 142 one-bar electric fires, the power density of the PFR is seen to be very high indeed. For comparison, the power densities of the Wylfa Magnox and Hinkley Point AGR reactors are approximately 3 kW/kg of uranium and 13 kW/kg of UO_2, respectively.

It is the very high power density of the PFR which presents problems of heat transfer and the need for a special type of coolant, as you will shortly see.

Conversion ratio

It has been shown that the average number of neutrons produced for each neutron absorbed by a fissile nucleus is represented by eta (η) and that for Pu-239 its value is 3.09. Of the 3.09 neutrons produced in this way, one is required to initiate fission in another Pu-239 nucleus and so sustain the chain reaction, leaving 2.09 neutrons with the potential ability of producing 2.09 atoms of Pu-239 in the breeder blanket, or in the UO_2 content of the fuel. In practice, of course, there are the inevitable losses in a reactor and some of the surplus neutrons are lost by total escape or by capture in the coolant and structural materials; this latter phenomenon is known as *parasitic absorption*. Some of the surplus neutrons, especially those of high energy, may cause fissioning of U-238. This is a most desirable event because the energy it releases contributes to the overall power output produced by the reactor and makes the best possible use of the available fuel. In fact, as much as 20 per cent of a fast reactor's power is typically due to fast-neutron fission of U-238.

Taking into account the various losses and other events which reduce the number of surplus fission neutrons available for breeding purposes, figures of between 1.2 and 1.4 are generally assumed as being typical. This means that for every 100 nuclei of Pu-239 fuel destroyed by fission in the core, between 120 and 140 nuclei of fresh Pu-239 will be created in the core and breeding blanket. The ratio expressing the rate at which new fissile material is being created to that at which the fissile content of the fuel is being consumed is known as the *conversion ratio* (CR):

$$CR = \frac{\text{rate of production of new fissile material}}{\text{rate of consumption of fissile material in fuel}}$$

In a thermal reactor, fissile Pu-239 is created from fertile U-238 at a rate which is always less than that at which the U-235 content of the fuel is being consumed; the CR value for such a reactor is therefore always less than unity (typically 0.76). In a fast breeder reactor, on the other hand, the rate at which Pu-239 is being created in the breeding blanket always exceeds that at which the Pu-239 content of the fuel is being consumed and the CR value is always greater than unity (typically 1.2). It is usual when referring to fast breeder reactors to use the term *breeding ratio* instead of conversion ratio.

Plutonium isotopes

Table 9.1 showed that the fast neutron capture cross-section (σ_c) for Pu-239 is about one-thirtieth of its corresponding fission cross-section (σ_f). This means that about one in 30 of all Pu-239 nuclei which manage to capture a fast neutron will be transformed by the process of parasitic capture into Pu-240. This particular isotope has a half-life of 6540 years and a fission threshold of 0.6 MeV (about half that of U-238); *in a thermal reactor, therefore, Pu-240 is virtually non-fissile*. In a fast neutron flux, however, Pu-240 fissions better than does U-235 and U-238 and its η-value of 3.1 is the same as that of Pu-239 (see Table 9.3). Thus, although some Pu-239 is lost because of its transformation to Pu-240 by parasitic capture, this loss is compensated to a large extent by the fast fissioning of the Pu-240 created. Further compensation occurs when some of the Pu-240 formed from the Pu-239 is itself transformed by parasitic capture into Pu-241, some of which is then readily fissioned. Pu-240 thus behaves as a fertile material.

Table 9.3
Thermal- and fast-neutron cross-sections and fission neutron yields for some plutonium isotopes

Nuclide	Thermal neutrons					Fast neutrons				
	σ_c	σ_f	α	ν	η	σ_c	σ_f	α	ν	η
Pu-241	370	1000	0.36	2.9	2.1	0.040	1.66	0.024	3.2	3.1
Pu-240	256	0	∞	0	0	0.09	1.24	0.07	3.3	3.1
Pu-239	270	745	0.36	2.9	2.1	0.05	1.8	0.03	3.2	3.1
U-235	100	580	0.17	2.4	2.0	0.08	1.2	0.06	2.7	2.5

Pu-241 is a highly fissile nuclide with a half-life of 14 years. Its thermal fission cross-section is in excess of 1000 barns (almost twice that of U-235), making it eminently suitable for use in thermal reactors, and its η-values for both thermal and fast neutrons are superior to U-235, Pu-239 and Pu-240.

Table 9.3 summarizes the nuclear properties of Pu-240 and Pu-241 and, for comparison purposes, includes data presented earlier for U-235 and Pu-239. The values listed are approximate and intended for comparison purposes only.

Resonance capture

In a working reactor, processes are never as clearly defined as explained above; the neutrons in the core are neither *all fast* nor *all thermal* but are more likely to exist over the whole energy spectrum.

Neutrons emitted during fission have energies which range typically from 1 MeV to 20 MeV with an average energy of around 2 MeV. In a thermal reactor these neutrons will have to travel enormous distances – in atomic terms – before getting out of the fuel itself and entering the moderator where thermalization can take place. In traversing the fuel many fast neutrons will cause fast fissioning of U-238, some will be captured by U-238 and transformed into Pu-239, some will then fission the Pu-239 created by other neutrons, some will escape from the core altogether or be captured by the structural materials, and some will experience multiple collisions with other nuclei and have their energies reduced to intermediate levels lying between what has been defined as 'fast' and 'thermal'. It is the effect which these intermediate-level neutrons have on the nuclear cross-sections of various materials which must be considered when studying the behaviour of a working reactor.

To say that U-235 has a very large fission cross-section to thermal neutrons and a very small fission cross-section to fast neutrons is not to imply that this cross-section to intermediate neutron energies is non-existent, nor that it decreases smoothly with increasing neutron energy. In fact, a graphical plot of fission cross-section versus neutron energy is far from smooth for all the fissile and fertile nuclides so far considered and contains many abrupt discontinuities, as illustrated for U-235 by Figure 9.3. The abrupt transitions in the curve are caused by the preference of U-235 nuclei to be fissioned by neutrons of certain energies in the same way that a radio-frequency filter has a preference for certain *resonant* frequencies. For this reason, the individual peaks in the curve are known as *resonant captures*. Note that these occur at neutron energies in the inter-

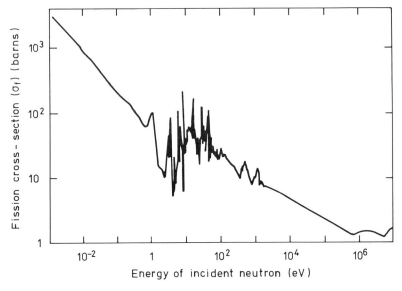

Figure 9.3   Fission cross-section for U-235 over a wide range of incident neutron energies

mediate range and, depending upon the nuclide being considered, may therefore have a significant effect on the operation of the reactor.

Equally important when studying fission cross-sections is the fact that the fission neutron yield (ν) of a nuclide, and hence the magnitude of the all-important eta-factor (η), is also strongly dependent on neutron energy and therefore varies in a manner very similar to that of the cross-section. This is vividly illustrated in Figure 9.4 by the graphical plot of η versus neutron energy for the nuclides U-233, U-235 and Pu-239, and demonstrates why the values given in Tables 9.1–9.3 are intended for guidance only.

Several important observations can be made from Figure 9.4:

1   At neutron energies below about 0.01 eV the η-values for all three nuclides are fairly steady and that U-233 has the highest value; U-235 has the lowest value.

2   Although some fluctuations occur between 1 eV and 200 eV, the η-value for U-233 is fairly steady at about 2.25 over a wide range of neutron energies extending from about 0.001 eV to 100 keV.

3   The η-values for all three nuclides are less than 2.5 over the entire thermal and intermediate energy regions.

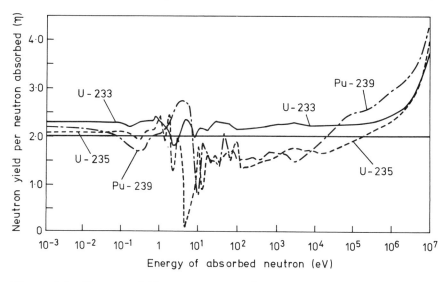

Figure 9.4   Neutron yield versus energy of absorbed neutron for the nuclides U-233, U-235, Pu-239

4   The η-values begin to increase steeply with increasing neutron energy above about 100 keV, starting with Pu-239.

5   Both U-235 and Pu-239 experience considerable resonant fluctuations over the intermediate energy regions of the curves and that for much of this range their η-values are substantially less than 2.

Doubling time

The total fissile fuel inventory of a single fast reactor is made up of two distinct parts. The first part is that fuel actually contained in the core of the reactor itself; the second part is that fuel which is outside the reactor and at various stages in the nuclear fuel cycle. This out-of-core material comprises spent fuel in the 'cooling pond', spent and recovered fuel held and being used at the reprocessing plant, fuel held and being used at the fuel manufacturing plant, and fresh fuel in the form of sub-assemblies en route to, and held in store at, the reactor site. The total quantity of such material is typically 80 per cent of that present in the core.

The time taken for a fast breeder reactor to produce enough plutonium to provide a complete fuel inventory for an identical reactor is known as the *doubling time* and is typically between 12 and 25 years, depending upon many factors such as operating load factor, burn-up, breeding gain, etc.

Burn-up

The residence time of the fuel in a nuclear reactor before it has to be removed for reprocessing is determined by what is known as *burn-up*. Burn-up is a term used to describe the amount of thermal energy extracted from a specified quantity of fuel and is usually expressed as so many *thermal megawatt-days per tonne of fuel* (abbreviated MWd/t). For example, a specified burn-up of 10 000 MWd/t could mean that the reactor, in consuming one tonne of fuel, has operated at a thermal power rating of 1000 MW for 10 days, or at 100 MW for 100 days, or at 25 MW for 400 days.

A more recent, but as yet little used method of expressing burn-up is to specify the thermal energy (in joules) produced during the residence time of the fuel in the reactor for each kilogram of fuel consumed. The abbreviations used are joules per kilogram (j/kg), megajoules (10^6 joules) per kilogram (MJ/kg), or terajoules (10^9 joules) per kilogram (TJ/kg). The relationship between the two methods is given by:

$$1 \text{ MWd/t} = 86.4 \text{ MJ/kg}$$

When the fuel used in a reactor is in pure metallic form, for example the uranium metal rods used in the Magnox reactors, the burn-up refers to the quantity of uranium *metal* consumed. In oxide-fuelled reactors, however, such as the AGR and PFR, the oxygen content of the fuel plays no part in the fissioning process and so the burn-up rating is then intended to refer to the plutonium or uranium content of the oxide. This is usually referred to as the *heavy atom* content (abbreviated ha) so that it covers both single oxide (UO_2) and mixed oxide (PuO_2/UO_2) fuels. The burn-up would then be specified as so many megawatt-days per tonne of heavy atoms and abbreviated MWd/tha.

Yet another method of expressing burn-up is to quote the percentage of heavy atoms fissioned during the in-core residence time of the fuel. This method is usually used when talking about fast reactors which contain mixtures of both fissile and fertile fuel within each fuel pellet. The heavy atoms referred to include the fissioning of U-238 directly or using Pu-239. 1 per cent heavy atom burn-up is approximately equal to 10 000 MWd/t, or to about 0.87 Tj/kg.

The greater the burn-up rating of a fuel element the longer can it stay in

Table 9.4
Burn-up ratings

Reactor type	Approximate burn-up	
	MWd/t	% heavy atoms
Magnox	5500	0.55
AGR	18 000	1.8
	(24 000)	(2.4)
PFR	70 000	7.0
	(150 000)	(15)
PWR	33 000	3.3

the reactor and the less often must it be removed for reprocessing and re-fabrication. These are expensive operations in the fuel cycle and it is the aim of every reactor designer to ensure that the fuel can remain in the reactor for long periods of time and give up as much as possible of its available energy.

At the time of writing, approximate burn-up values for some of the various reactor types used throughout the world are presented in Table 9.4; figures shown in parentheses are target figures which the designers hope to attain.

THE PROTOTYPE FAST REACTOR

Without any doubt the most outstanding feature of the fast breeder reactor is that it allows up to about 60 per cent of natural uranium to be burnt up in its core compared with less than 1 per cent of that allowed by thermal reactors. This is because thermal reactors operate almost exclusively on the tiny 0.7 per cent U-235 content of natural uranium whereas fast breeder reactors are able to convert much of the remaining – and otherwise useless – 99.3 per cent U-238 into fissile Pu-239, which can then be used as fuel. The importance of this may be appreciated by considering the fact that Britain has more than 20 000 tonnes of depleted uranium stockpiled as residue from its fuel reprocessing and uranium enrichment processes. If this were to be used as feedstock for the breeder blanket in fast reactors it would, on eventual conversion to Pu-239, yield a heat energy equivalence of more than 40 000 million tonnes of coal; an amount greater than

Britain's existing coal reserves, currently estimated at 300 years supply, or more than five times its North Sea oil reserves.

Put another way, if Britain were to initiate a programme of building fast reactors to complement its existing thermal reactors, the importation of less than half a million tonnes of uranium Yellow Cake would provide enough energy to supply all of Britain's electrical requirements for hundreds of years to come. Without the contribution from fast reactors, however, the same quantity of Yellow Cake would have to be imported every 20 years or so and would have to be acquired in competition with other countries fighting for a share of the world's dwindling supplies of uranium.

An adequate installation of fast reactors would therefore substantially reduce Britain's dependence on uranium imports and hence reduce future costs of electricity generation.

Plutonium production

Although the fast reactor is fuelled by plutonium, and is able to breed additional plutonium in its breeding blanket, it is in fact much less efficient as a net producer of plutonium than most thermal reactors. This is because much of the substantial amount of plutonium created in the fuel and blanket of the fast reactor is incinerated almost as quickly as it is formed – the fast reactor has been specially designed to burn plutonium.

Table 9.5
Net plutonium production for various
thermal reactors operating for
1 GWe(net)-year

Reactor type	Net Pu production (kg)
Magnox	617
CANDU	493
PWR	270
AGR	173
PFR	190

Table 9.6
Plutonium content of a fast breeder reactor operating for
1 GWe(net)-year

Pu content of core (kg)	Pu created in core (kg)	Pu inciner- ated in core (kg)	Pu content of blanket (kg)	Pu created in blanket (kg)	Pu inciner- ated in blanket (kg)	Net Pu produc- tion (kg)
1936	558	789	0	455	34	190

Tables 9.5 and 9.6 summarize the differences between the plutonium-producing capabilities of the various reactor systems. Table 9.5 illustrates the relative differences between the thermal reactor systems and lists the net quantity of plutonium created in the fuel after one year's operation of generating 1 GW (net) of electricity, that is 1 GWe(net)-year. The Magnox reactor is seen to be a prolific producer of plutonium when compared with the PWR and AGR. The reasons for this are complex but are due in part to the fact that the Magnox reactor, in being fuelled with natural uranium, contains a larger quantity of fertile U-238 atoms than the PWRs and AGRs which are fuelled with enriched uranium and that it has a larger moderator-to-fuel mass ratio.

The in-situ incineration of plutonium in a thermal reactor is highly desirable because it makes better use of available fuel and compensates for the progressive consumption of the U-235 content; in short it extends the burn-up rating of the fuel. For example, well over half of the plutonium created in the fuel of a modern PWR is incinerated *in situ* and is responsible for about one-third of the total heat generated in the core. Of the plutonium remaining, about one-sixth is lost through neutron capture into the non-fissile Pu-240 isotope and the rest appears in the fuel when it is removed from the reactor; this is the net quantity listed in Table 9.5 and is recoverable at the reprocessing plant.

The figures listed in Table 9.6 give a comprehensive breakdown of what happens to the plutonium created and incinerated in the core and breeding blanket of a fast reactor after operating for 1 GWe(net)-year. The core is assumed to be that of the PFR containing mixed oxide fuel composed of 25 per cent PuO_2 + 75 per cent UO_2.

The initial fuel loading of 1936 kg of plutonium is supplemented during the operating period of the reactor by the creation of 558 kg of plutonium through neutron capture in the UO_2 content of the fuel, making a total plutonium inventory of 2494 kg. Of this amount, 789 kg are incinerated during the operating period of the reactor leaving an in-core residue of 1705 kg.

The initial plutonium content of the breeder blanket is assumed to be zero and that 455 kg are created during the operating period of the reactor. Of this, 34 kg are incinerated in the blanket itself, leaving an in-blanket residue of 421 kg. The total amount of plutonium remaining in the reactor at the end of its working period is thus 1705 kg + 412 kg = 2126 kg; an effective increase of 190 kg on the initial amount supplied to the core. This gain is substantially less than that produced by either the PWR, CANDU and Magnox reactors operating under the same conditions, which demonstrates that the fast reactor is a good incinerator of plutonium but a relatively poor net producer.

Construction of the Prototype Fast Reactor (PFR) began in 1967 and was completed in 1974. It reached full thermal power of 600 MW in 1977, since when it has been putting 250 MW of electrical power into the National Grid. The PFR is described as being a 'pool type' of reactor because its fuel elements are immersed in a pool of liquid sodium which functions as the coolant; a simplified operating diagram of the PFR is shown in Figure 9.5.

The sodium coolant is pumped through the fuel elements from bottom to top and also through the intermediate heat exchanger where it gives up the heat it has extracted from the fuel. The heat transferred in this way is taken

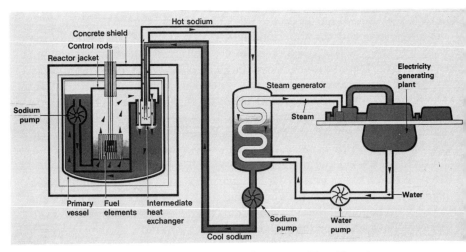

Figure 9.5   Operating diagram of the Prototype Fast Reactor

away by another – completely independent – sodium loop which forms part of the steam generator located outside the core. The reason for having two heat exchange steps is to ensure that the highly radioactive sodium flowing through the fuel is confined to the core and isolated from the external water loop and steam which flows through the turbines. The sodium which flows through the fuel is made radioactive by neutron capture and transformed into the sodium-24 (Na-24) isotope. This is a very powerful gamma emitter but with a short half-life of only 15 hours. The big advantage of the short half-life is that should the reactor need to be shutdown then its primary coolant would lose most of its induced radioactivity within a week or two.

The secondary sodium loop, in being remote from the fuel and shielded from the intense neutron flux which surrounds it, is not subjected to very much neutron radiation and is therefore not made significantly radioactive. The total inventory of sodium coolant in the fast reactor is approximately 1000 tonnes.

Normal control of the PFR is achieved by five boron carbide absorber rods which are raised and lowered within the core so as to maintain a steady neutron flux. Six sub-assemblies containing clusters of boron carbide absorber rods are also used for emergency shut-down purposes.

The fuel inventory of the PFR comprises 78 sub-assemblies which are arranged in the core in the manner shown in Figure 9.2. The diameter of the area occupied by the sub-assemblies is 1524 mm (about 5.5 feet). Each sub-assembly comprises three main sections as illustrated in Figures 9.6 and 9.7. The upper section contains a cluster of what are called mixer-breeder pins. There are 19 of these and they are filled with either natural or depleted uranium which is employed to breed plutonium by the process of neutron capture. The outer surfaces of the pins are fluted in a spiral formation so that the liquid coolant is thoroughly mixed as it passes out through the top of the sub-assembly. The centre section of the sub-assembly is made up of a cluster of 325 fuel pins shown being assembled in Figure 6.8. Each of these contains fuel and breeder pellets arranged in the manner illustrated in Figure 9.7. The central region of the pin, which represents the central core of the reactor, is packed with solid and sometimes hollow PuO_2/UO_2 pellets extending over a length of 914 mm (about 3 feet). The upper region of the pin is filled with solid pellets made from depleted uranium and forms part of what is called the *upper axial breeder* of the reactor core. The region of the pin immediately below the fuel pellets is packed with pellets made from depleted uranium and forms part of the so-called *lower axial breeder*. The pellets are made with a central hole so that fission product gases such as krypton and iodine can flow easily into the hollow region immediately below the lower axial

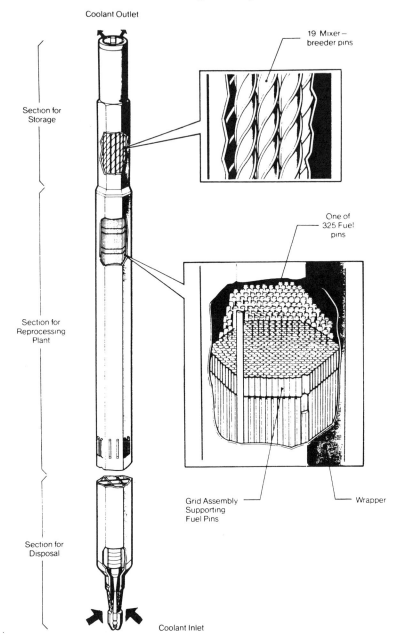

Figure 9.6   PFR fuel sub-assembly

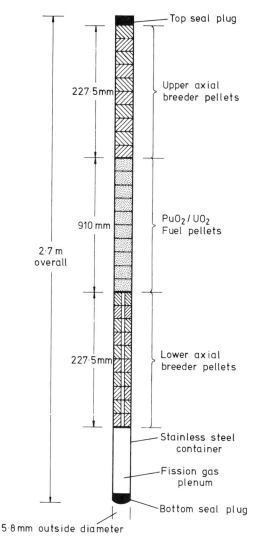

Figure 9.7   Arrangement of fuel and breeder pellets inside a PFR fuel pin

breeder where room is made available for their expansion and build-up. This region is known as the fission gas plenum, without which the pressure build-up from the gases would cause swelling and buckling of the fuel pin.

The total combined weight of fuel and breeding material in the sub-assemblies is approximately 4 tonnes, of which about 1 tonne is in the form of PuO_2. The maximum temperature achieved at the surface of any fuel pin is approximately 700°C; a temperature corresponding to red heat.

Although the illustration of Figure 9.5 shows only one intermediate heat exchanger and only one steam generator, there are actually six intermediate heat exchangers connected to three sodium pumps and to three steam generators; one for each pair of intermediate heat exchangers. The steam generators supply steam at a temperature of approximately 540°C and at a pressure of 2352 psi (about 160 atmospheres) to a single turbo-generator set which supplies 250 MW of electricity to the National Grid.

The hottest temperature reached by the primary sodium loop is approximately 560°C.

Radial breeder

The radial breeder which surrounds the vertical sides of the fuel sub-assemblies, as illustrated in Figure 9.2, consists of sub-assemblies containing long stainless steel tubes filled with depleted uranium. This is illustrated in Figure 9.8 which also shows two workers carrying out a fuel loading rehearsal. Both men are standing inside the stainless steel containment tank which houses the liquid sodium coolant, the reactor core sub-assemblies, the radial breeder sub-assemblies, the radial shield rods, the intermediate heat exchangers and the sodium circulating pumps. The dimensions of the tank are 12.1 m (39.7 feet) diameter and 15.2 m (49.8 feet) in depth. One man is shown standing on top of the core and breeder sub-assemblies and looking up at his colleague who is guiding down a sub-assembly which is supported by a large pulley. The vertical tubes which surround the workers are those which form the radial shield.

The radial breeder serves three very important purposes. Firstly, it functions as a first-stage neutron shield, thereby reducing the amount of concrete required outside the containment tank. Secondly, it functions as a neutron reflector and ensures that some of those neutrons which have escaped from the core are reflected back into the core where it is hoped they will cause additional fissioning of the plutonium fuel; this serves to improve the overall neutron economy of the core. Thirdly, it functions as a fertile blanket for the production of Pu-239 from neutrons which have escaped from the core.

The reason for locating the radial breeder within the containment tank – and immersed in the sodium coolant – is that many of its U-238 atoms are directly fissioned by the neutrons escaping from the core; this causes considerable heat to be produced in the blanket. The amount of heat

Figure 9.8   Fuel element loading rehearsal in the containment tank of the Dounreay Prototype Fast Reactor

produced in this way represents about 10 per cent of that produced by the core material and makes a useful contribution to the total heat supplied to the heat exchangers.

Sodium coolant

Because no neutron moderating material is required in a fast reactor its fuel elements can be placed in close proximity to one another, thereby leading to a small, densely-packed core. Although this results in the advantages of a physically small reactor for a given power output, and a short doubling time, it also results in a very high power density for the core. This was explained on page 179 where it was shown that the power density of the PFR core is about 142 kW of heat per kg of PuO_2/UO_2 fuel; a figure which is ten times greater than that of the Hinkley Point AGR and *forty-seven times that of the Wylfa Magnox reactors*! It is the rapid removal of this heat which presents one of the biggest problems in fast reactor design, although the power density could quite easily be reduced.

For various very important reasons, which will be explained in Chapter 16, the type of coolant which has been adopted for fast reactors worldwide is liquid sodium (Na). Sodium is a soft, silver-white metal which has a very low neutron capture cross-section, a density about the same as that of water, a high thermal conductivity and specific heat, a melting point which is a little below the temperature of boiling water and a boiling point which is much higher than the normal operating temperature of the reactor core, which means that pressurization to suppress boiling of the coolant is unnecessary. As explained below, these are nearly ideal characteristics for a fast reactor coolant. It is not without its disadvantages, however, the most important of which is that it reacts violently with water, virtually exploding when it comes in contact with it at high speed. Naturally this presents considerable problems in the pipework outside the reactor core and in the heat exchanger where the external surfaces of the pipes carrying hot sodium are in contact with water. This is why two completely separate heat exchange steps are used in the PFR; one (sodium to sodium) located within the primary containment tank, the other (sodium to water) outside the tank (see Figure 9.5).

Safety features

In addition to the double heat exchanger principle adopted for the PFR there is also an outer tank which surrounds the primary vessel containing the liquid sodium. The presence of this tank ensures that in the unlikely event of a leakage in the primary tank the sodium would be retained and the fuel prevented from becoming uncovered.

The large volume of sodium coolant (about 1000 tonnes), coupled with its relatively high heat capacity (specific heat), ensures a uniform heat distribution throughout the reactor core and thermal stresses in the primary tank are therefore avoided. Another important advantage resulting from the large volume of sodium and its high specific heat, is its ability to absorb the heat from the core should there be a complete failure of the sodium pumping system including, most improbably, failure of all auxiliary motors. That this is so has been demonstrated on the PFR by deliberately stopping the sodium pumps whilst the reactor was operating at full power so that forced circulation of the sodium ceased completely. Under normal circumstances this situation – or even part of it – would result in a reactor 'trip' (automatic shut-down of the reactor by rapid insertion of the control rods), and the only heat to be generated in the core would be the so-called 'decay heat' from decay of the short-lived fission products in the fuel; in practice a few tens of megawatts instead of the normal 600 MW produced when the reactor is working at full power. It has been demonstrated on the PFR that this decay heat is removed by natural convection currents within the static sodium and that no danger whatsoever is brought about by complete failure of the sodium circulating pumps. This fail-safe feature is unique to the pool-type of sodium-cooled fast reactor. All other types of reactor, for example Magnox, AGR, BWR, PWR, CANDU, depend absolutely on the efficient forced removal of their decay heat, as demonstrated so vividly to the world when the PWR at Three Mile Island 'tripped' in March 1979.

To think the unthinkable, what would happen if the reactor failed to 'trip' when all sodium circulating pumps failed (a situation which could only arise if *five completely separate fail-safe systems were to fail simultaneously*)? The answer is that the reactor would continue to operate and its core would rapidly start to overheat. However, calculations based upon earlier experiments with decay heat removal by non-circulating sodium have shown that natural convection within the sodium would remove a considerable proportion of the heat and that, under the most favourable conditions, the rise in temperature and corresponding increase in power would not result in the development of a dangerous runaway situation. On the other hand, if less favourable fault conditions are assumed then some localized damage to the core is expected to occur – due to overheating – and would result in the need to remove some fuel sub-assemblies before their allotted residence time. Naturally, any experiment likely to result in damage to the core would never be carried out in practice but would instead be assessed theoretically using experimental data acquired from less risky experiments.

Fire

The risk of a sodium/water fire or explosion in the PFR core is reduced to a negligible level by the use of two completely separate heat exchangers. It is further reduced by the technique of double containment of vulnerable pipes and by a 'blanket' of the inert gas argon above the surface of the sodium pool. The argon is maintained at a slight pressure so that atmospheric air is prevented from entering the sodium containment tank and the coolant circuit.

Sodium burns very slowly in air, with a low flame temperature, and considerable experience in dealing with sodium fires has been acquired over many years from its use in the chemical industry.

Although sodium is not pleasant to work with it has not presented any serious problems to fast reactor design, as has been demonstrated by successful operation of the DFR for 18 years and, since 1977, of the PFR.

Doppler effect

An extremely important phenomenon which contributes to the inherent safety of the fast reactor is that known as the *Doppler effect*. A full description of this effect is beyond the scope of this book but, in essence, it describes the way in which the neutron multiplication in the core of a nuclear reactor varies with operating temperature. The magnitude of the effect depends on the temperature differences between the fuel, the coolant and the moderator (if present) and on the composition of the fuel, that is whether metal, oxide or mixed oxide, all of which affect the number of useful neutrons in the core which are lost through neutron capture and neutron scattering. Most nuclear reactors of all types are designed in such a way that the Doppler effect gives rise to what is known as a negative temperature coefficient for the rate at which neutrons are created. This ensures that in the event of the core overheating for any reason then the resultant reduction in the production rate of core neutrons would act in such a way as to oppose (but not completely compensate for) the original temperature excursion. To this extent, a reactor (such as the PFR) with a negative temperature coefficient behaves in a self-stabilizing manner; this contributes to its overall stability and its ability to operate with loss of coolant flow.

THE FAST REACTOR FUEL CYCLE

The fuel cycle described for the British Magnox and AGR reactors is essentially the same for the PFR, with the difference that the fuel is fabricated at Sellafield instead of at Springfields and that it is reprocessed on-site at Dounreay instead of at Sellafield. Figure 9.9 illustrates the operation of the PFR fuel cycle.

Mixed oxide fuel comprising 25 per cent PuO_2 plus 75 per cent depleted UO_2 is fabricated at Sellafield from material supplied from the Capenhurst uranium enrichment plant and the Dounreay PFR reprocessing plant. The material from Capenhurst comprises depleted uranium taken from its 20 000 tonne stockpile; that from Dounreay comprises both plutonium and depleted uranium extracted from the reprocessing of spent PFR fuel and some of the uranium breeder blanket. The plutonium is in the form of plutonium nitrate liquor ($PuNO_3$), a solution consisting of plutonium dissolved in nitric acid.

On completing its scheduled residence time in the PFR the spent fuel is removed from the core and stored in a cooling pond containing liquid sodium. When most of the short-lived fission products have decayed the now much cooler fuel is removed from the pond and transferred to the PFR reprocessing plant, which is also located on the Dounreay site. Here the fuel is dissolved in nitric acid and the plutonium separated from the uranium and from the fission product waste. The recovered plutonium, in the form of a plutonium nitrate solution, is then transported to Sellafield where it is used in the fabrication of more PFR fuel elements.

A more detailed description of fuel reprocessing is given in Chapter 11.

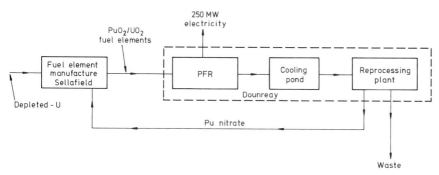

Figure 9.9   The PFR fuel cycle

THE COMMERCIAL FAST REACTOR

Although the PFR continues to operate satisfactorily and to yield performance figures even better than expected, it is, nevertheless, no more than a
prototype whose main purpose has always been to yield experimental data
and operating experience which would assist in the design and construction
of a much larger commercial-size demonstration fast reactor; the so-called
CDFR.

At the time of writing no firm date has been proposed for starting the
construction of the CDFR but preliminary plans already exist for its
essential components. Present thinking envisages an electrical power output of about 1320 MW for the CDFR, a scaling-up factor of 5.3 when
compared with the PFR. The number of fuel sub-assemblies would
increase from 78 to 349 there will be four instead of three primary sodium
circulating pumps and eight sodium/water heat exchangers in four secondary circuits instead of the nine heat exchangers in three circuits used in the
PFR. The importance of this conception is that although the power output
of the CDFR is more than five times that of the PFR, the physical size of its
essential components require to be increased very little since the additional
work load has been achieved by using 'more of the same' of tried and tested
components.

Tables 9.7 and 9.8 summarize the essential differences between the PFR

Table 9.7
Differences between the PFR and proposed CDFR

Main parameters	PFR	CDFR
Electrical power output	250 MW	1320 MW
Gross thermal power	600 MW	3300 MW
Number of reactors	1	1
Number of fuel sub-assemblies	78	349
Number of primary sodium pumps	3	4
Number of secondary heat exchanger vessels	9	8
Number of secondary heat exchanger circuits	3	4
Number of turbo-generators	1	2
Fuel cladding temperature	700°C	670°C
Core sodium outlet temperature	560°C	540°C

Table 9.8
Scaling-up factors for the CDFR

Electrical power output	5.3
Core containment tank diameter	1.6
Fuel sub-assembly length	1.1
Fuel sub-assembly power	1.14
Primary sodium pump flow rate	2.5
Primary heat exchanger power	4.0

and the proposed CDFR. Note that operating temperatures for the CDFR have been reduced slightly below those of the PFR. This has been done to ease thermal stress problems and to allow the use of ferritic steel in certain areas instead of stainless steel used previously, thereby resulting in greater reliability for a small sacrifice in thermodynamic efficiency.

THE FAST REACTOR WORLDWIDE

There are at present nine countries throughout the world actively engaged in development of the fast reactor and total R&D expenditure outside the communist bloc is currently running at about £1000 million per year. Table 9.9 summarizes the location and number of fast reactors – either operating or under construction – having a thermal power output in excess of 200 MW. All are fuelled with mixed PuO_2/UO_2 and all use liquid sodium as a coolant.

There are also a number of low-power fast reactors operating for test purposes in the USA, France, West Germany, Japan and the USSR, and two are under construction in Italy and India.

Construction and funding of the Super Phenix Fast Reactor at Crays-Malville, near Lyon, was a joint undertaking by France and Italy and a consortium formed by West Germany, Belgium and the Netherlands. The consortium, known as DEBENE (an acronym derived from the names of the member countries), is at present also building the SNR-300 Fast Reactor in West Germany.

After many years of 'going it alone' in fast-reactor research, Britain eventually entered into partnership with its European neighbours when, on 10 January 1984, it signed a joint research and development agreement with France, West Germany, Italy and Belgium. The aims of the agreement are to pool the vast wealth of knowledge possessed by these countries in

Table 9.9
Fast reactors in other countries

Country	Name of reactor	Thermal power (MW)	Electrical power (MW)	Year of start-up
France	Super Phenix	3000	1240	1984
France	Phenix	560	250	1974
USSR	BN 600	1470	600	1980
USSR	BN 350*	1000	135	1973
UK	PFR	600	250	1976
USA	FFTF	400	0	1979
USA	CRBR	975	350	1990
W. Germany	SNR 300	762	300	1986
Japan	MONJU	714	250	1988

*Thermal power currently limited to 700 MW. The reactor is also used to produce desalinated water.

the field of fast reactors and to join together in the work needed to develop the fast reactor into a proven commercial power system.

THE FUTURE

The unexpected world recession has led to a slowdown in the growth of electricity demand and has had a similar effect on the expansion of electricity generating plant; including, of course, nuclear. Predictions made during the 1960s of a worldwide shortage of uranium in the 1980s have therefore failed to materialize and the world's known reserves of economically-recoverable uranium are now expected to last well into the next century. All this, of course, has had a profound effect on the likely future of the fast reactor since its ability to burn uranium much more efficiently than any of the thermal types of reactors is now less important than it was 20 years ago. The situation is not helped by the fact that the fast reactor is more expensive to build than any of the many well-proven thermal types, mainly because of the sodium containment vessel and associated pipework and pumps; plus the double-circuit (primary and secondary) heat exchanger requirements. Nevertheless, the outstanding advantages of the fast reactor in uranium utilization (sixty-times better than any thermal reactor) have not been overlooked; it is only that the pace of fast reactor development has been slowed down to keep pace with the anticipated needs of the future.

10 Spent nuclear fuel: temporary storage and transportation

THE NEED FOR REMOVAL OF SPENT FUEL

It would be of great economic value if the fuel of a nuclear reactor could remain in the core until all of its fissile atoms had been fissioned (split in two). Unfortunately, there are numerous reasons why this is not possible although considerable research is being devoted worldwide into finding ways in which the in-core residence time of nuclear fuel can be extended.

Reactor poisons

The impurities which are picked up by human blood as it is circulated through the body are removed by the kidneys, or by a kidney machine. If this is not done the body becomes progressively more lethargic until it eventually stops working altogether and dies. So it is with a nuclear reactor. Each time an atom of uranium or plutonium is fissioned it breaks up into two completely different elements – the so-called fission products (see page 63). Many of these elements – xenon for example – have a very high neutron capture cross-section and if present in sufficient quantities will reduce the reactivity of the reactor by capturing neutrons which would otherwise be maintaining the chain reaction in the core. If they are allowed to build-up to too large an extent they will eventually stop the reactor working altogether; they are said to have *poisoned* the reactor. This is the main reason why it is necessary to remove the fuel from the reactor periodically and cleanse it of its fission product impurities at the reprocessing plant.

Fuel swelling

The density of uranium is approximately 19 g/cm^3, so one gram of uranium occupies a volume of about one-nineteenth of a cubic centimetre (0.053 cm^3). However, when an atom of uranium fuel is fissioned in the core of a nuclear reactor it breaks into two completely different elements, each of which has a mass approximately equal to half that of the uranium atom from which it was formed and each has a density which is very much lower than that of uranium, especially the gaseous fission products such as krypton and xenon. This means that the total volume occupied by the fission products as they build up in the nuclear fuel is considerably greater than that of the uranium (or plutonium) atoms from which they were formed, thereby causing the fuel to increase in size. This increase may take the form of elongation, localized swelling, or a combination of both, and if allowed to continue may eventually cause rupture of the fuel cladding; a most undesirable event.

Rupture of fuel cladding allows gaseous and volatile fission products to escape into the coolant and find their way into areas outside the reactor core where they give rise to abnormally high levels of nuclear radiation. Cladding rupture also makes handling of the fuel difficult when it is removed from the reactor for intermediate storage and transportation since it causes radioactive contamination of everything it comes in contact with. The phenomenon of fuel swelling is, of course, well understood by manufacturers of nuclear fuel and allowances are made in the design of the fuel element to accommodate the expected increase in volume. Nevertheless, there may be a limit to the swelling which can be tolerated and it is one of the many features which govern the maximum residence time of the fuel in the core.

To illustrate the effects of fuel swelling, consider what happens when one atom of uranium splits into the two fission products caesium-137 ($^{137}_{55}$Cs) and rubidium-97 ($^{97}_{37}$Rb). The density of caesium is 1.9 g/cm^3 and that of rubidium is 1.5 g/cm^3. One gram of caesium atoms would therefore occupy a volume of $\frac{1}{1.5}$cm^3 = 0.53 cm^3, and one gram of rubidium would occupy a volume of $\frac{1}{1.9}$cm^3 = 0.67 cm^3. If, for the purpose of this example, it is assumed that one gram of uranium, occupying a volume of 0.053 cm^3, splits evenly into half a gram of caesium and half a gram of rubidium, then the volume occupied by the caesium would be 0.265 cm^3 and that occupied by the rubidium would be 0.335 cm^3. The total volume occupied by the one gram of combined fission products would therefore be 0.6 cm^3, a figure which is more than *ten times* that of the one gram of uranium from which they formed. Although this example is perfectly valid, things are not quite as bad as they might at first sight appear, since fissile uranium, that is U-235, comprises only a small percentage of the total mass of fuel and fission product swelling only affects that part.

Radiation damage

Yet another factor which may limit the residence time of fuel in a nuclear reactor – certainly in a fast reactor – is the effect of radiation damage to the fuel cladding. Prolonged bombardment by fast neutrons causes many of the atoms which make up the cladding to be knocked out of their normal

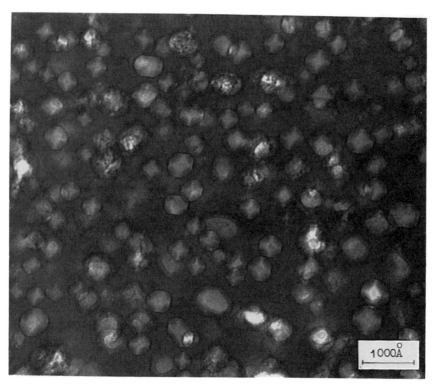

Figure 10.1   Voids produced in stainless steel by irradiation in the Harwell Variable Energy Cyclotron

positions in the atomic structure of the material and to leave behind an increasing number of atomic vacancies, some of which may combine with other vacancies to form large voids. These have the effect of reducing the density of the cladding (just like air bubbles do in a bar of chocolate) which, in turn, causes it to swell.

The effect of radiation-induced swelling is a very complex one and depends on many factors. It is, for example, very severe in stainless steel at temperatures between 350°C to 700°C but less severe at temperatures outside this range. It is also dependent upon the amount of mechanical stress to which the material is subjected and to the number of gaseous atoms present in the material. Gaseous atoms exist in the material as original impurities (hydrogen, for example), or as fission products (such as krypton and xenon) or are formed from neutron capture (for example helium). Certain types of steel contain substantial amounts of nickel in the form of the stable isotope Ni-58. On capturing a neutron this is transformed into the radioactive isotope Ni-59 which has a 75 000 year half-life and a relatively high neutron capture cross-section of 13 barns. On capturing a neutron Ni-59 undergoes an n,α reaction and is transmuted into the stable iron isotope Fe-56, the reaction being accompanied by the emission of an alpha particle. The alpha particle, being the nucleus of a helium atom, quickly captures the necessary number of orbital electrons and becomes an atom of the gas helium.

Figure 10.1 demonstrates the appearance of radiation-induced voids in stainless steel. The voids were created by irradiating the steel with charged particles derived from the Harwell Variable Energy Cyclotron (VEC) so as to simulate the effects of prolonged neutron irradiation.

Irradiation creep

Prolonged irradiation of nuclear cladding can also bring about the effect known as *irradiation creep*. This term describes the phenomenon whereby hitherto stable material undergoes plastic deformation under the influence of relatively low mechanical stresses and temperatures after it has been subjected to prolonged irradiation. Irradiation creep is therefore yet another of the many factors which could limit the residence time of nuclear fuel in a reactor core.

INTERIM STORAGE

Spent fuel which has recently been removed from an operating reactor is intensely radioactive, mainly because of the radiation emitted by those

Table 10.1
Fission products commonly found in spent fuel

Fission product	Approx. $\frac{1}{2}$-life	Radiation emitted
Iodine-135	6.7 h	$\beta\gamma$
Xenon-135	9.2 h	β,γ
Xenon-133	5 d	β,γ
Iodine-131	8 d	β,γ
Cerium-141	32.5 d	β,γ
Niobium-95	35 d	β,γ
Strontium-89	51 d	β,γ
Zirconium-95	65 d	β,γ
Sulphur-35	88 d	β
Cerium-144	284 d	β,γ
Ruthenium-106	1 y	β
Promethium-147	2.65 y	β
Krypton-85	10.6 y	β,γ
Strontium-90	28 y	β
Caesium-137	30 y	β,γ

fission products which have a relatively short half life; typical of these are given in Table 10.1, along with an indication of the type of radiation which they emit. All of them are seen to emit beta (β) radiation and some of them to emit gamma (γ) radiation as well. It is the intensity of this radiation which prevents the immediate transportation of the spent fuel to the reprocessing plant and which necessitates a period of interim storage on the site of the reactor installation. The actual period of interim storage varies considerably with different types of fuel and its burn-up but is seldom less than 90 days and is typically 150 days in many parts of the world.

Because of the exponential way in which a radioactive nuclide decays (see page 57), about 97 per cent of its initial radioactivity will have decayed after a period of five half-lives and 99.9 per cent of it after ten half-lives. If a period of one year is assumed for interim storage of spent fuel then after this period of time the Xe-133, I-131, Ce-141 and Nb-95 nuclides listed in Table 10.1 will have virtually disappeared from the fuel and also much of the Zr-95 and S-35. Ce-144 will have decayed to a little less than half and Ru-106 to exactly half. Pm-147 will have decayed to about 23 per cent of its initial activity whereas Kr-85, with its 10.6-year half-life, will still retain about 94 per cent of its initial activity. Sr-90 and Cs-137 will each have decayed by little more than 2 per cent during the one-year storage period because of their relatively long half-lives. The residual activity of a nuclide can be calculated:

$$\text{Residual activity } (\%) = 100e^{\frac{-0.693t}{t_{\frac{1}{2}}}}$$

where $t_{\frac{1}{2}}$ is the half-life of the nuclide being considered and t is the storage period expressed in the same units as $t_{\frac{1}{2}}$.

Heavy elements

Although the vast majority of the radioactivity associated with freshly-removed fuel is attributable to the short-lived fission products, some small amount is due to radiation emitted by those nuclides which have been created in the reactor due to neutron capture by the original uranium, and by successive neutron captures of the nuclides created in this way. Typical of these are the plutonium isotopes 238 to 242, the americium isotopes 241 and 243 and the curium isotopes 242 and 244. Most of these have long half-lives ranging from 14 y to 387 000 y and they are all alpha emitters with the exception of Pu-241 which emits beta radiation only. Both of the curium isotopes are also spontaneously fissile, that is they fission without any external stimulus and in fissioning release neutrons and create fission products of their own. Curium-242 is of special interest since its half-life is unusually short (164 d) and its initial activity is comparable with that of some of the normally-produced fission products in the fuel. However, after one year's interim storage its activity will have fallen to 21 per cent of its initial value and its presence is therefore less significant.

A somewhat unusual source of radioactivity from spent fuel is that associated with the uranium isotope U-237. This artificially-made nuclide is simultaneously formed in two completely different ways; firstly from the U-235 content of the fuel by two successive neutron captures:

$$\text{U-235} \xrightarrow{\text{n}} \text{U-236} \xrightarrow{\text{n}} \text{U-237}$$

and secondly from the U-238 content of the fuel via a single n,2n reaction:

$$\text{U-238} \xrightarrow{\text{n}} \text{U-239} \longrightarrow \text{U-237} + 2\text{n}$$

U-237 is a gamma emitter with a very short half-life of 6.75 days and its presence is therefore of some significance in the early stages of spent fuel removal.

Methods of storage

There are two methods in use throughout the world for the interim storage of spent nuclear fuel: wet storage and dry storage.

Wet storage

In the 'wet' storage method the fuel is stored in large deep ponds filled with demineralized light water (ordinary water which has had its natural salts removed). The water functions as an inexpensive radiation shield and also as a very efficient coolant which may be circulated through a heat exchanger to remove the residual heat generated by the fuel as its radioactivity decays. In addition to these advantages the transparency of the water allows the fuel elements to be observed, their radiation to be monitored, and their position in the pond to be re-arranged by the use of remote manipulator tools. Virtually all methods used throughout the world for the interim storage of spent nuclear fuel are based upon the wet method. It is safe, reliable and has been proven over many years.

A typical CEGB spent fuel storage pond at a Magnox power station would measure about 65 m long, 12 m wide and 7.5 m high. It would contain about 3500 m³ (772 415 gallons) of demineralized water to a depth of about 6 m. The water would have its upper temperature limited to about 15°C – using a chilling plant if necessary – and it would be dosed with sodium hydroxide (caustic soda; NaOH) to yield a concentration of 200 mg NaOH per litre of water. The presence of the sodium hydroxide reduces the acidity of the water and raises its pH value to about 11.6. This, coupled with the relatively low maximum operating temperature of the water, greatly reduces the corrosion rate of the Magnox cladding during its period of storage in the pond. The depth of water is about twice that considered desirable to ensure a safe radiation working environment for the operating staff.

The storage time of the spent fuel in the power station pond before it is transported to the reprocessing plant depends upon many factors but is never less than 100 days and is often as long as 150 days.

Storage ponds for AGR spent fuel are similar to those designed for Magnox fuel but are somewhat smaller and have a greater depth of water because of the higher radiation levels emitted by the enriched fuel which will have been subjected to a very high burn-up. A typical AGR pond would be 25 m long, 9 m wide and 9 m high and it would contain 900 m^3 (nearly 200 000 gallons) of demineralized water to a depth of 7 m.

The maximum operating temperature and chemical composition of the water used for AGR fuel storage is different from that used for Magnox fuel because of the much higher corrosion resistance of the stainless steel cladding and the greater radioactivity of the spent fuel. Maximum operating temperature is typically 25°C and its pH value is raised to about 7 by the addition of sodium hydroxide. Also present in the water, at a concentration of about 0.7 per cent, is boric acid (H_3BO_3). This is added to give the water a high neutron absorption (provided by the boron) so that there is no possibility of a chain reaction being initiated by the delayed neutrons emitted by the highly-irradiated spent fuel.

In both Magnox and AGR storage ponds the irradiated fuel is stored in open-topped skips on the pond floor. In a typical Magnox pond each skip contains up to 200 fuel elements loosely stored and positioned horizontally one on top of the other. In AGR ponds each skip contains 20 fuel elements positioned vertically, each comprising a cluster of 36 fuel pins contained in a graphite sleeve (see Figures 6.5 and 6.6). The elements are separated from one another by neutron-absorbing boron steel metal inserts, rather like the separators in a milk crate, so that there is no possibliity of a chain reaction beginning. The unintentional initiation of a chain reaction is known as a criticality event.

A question frequently asked following a description of wet storage is 'what happens to the water; does it become radioactive?'. The simple answer is 'no, it does not, provided the fuel cladding remains intact'. The reason for this being so is due to the fact that most of the radiation emitted by the fuel is in the form of gamma rays and that it is not possible to make water radioactive by irradiating it with gamma rays of the energies emitted by spent fuel. On the other hand, should a hairline crack or pinhole develop in the fuel cladding then radioactive gases and fission products will be released into the surrounding water. Although such occurrences are rare, they nevertheless happen from time to time and were anticipated when the pond was designed. This is why the water in the pond is continually circulated through filters on its way to the heat removal plant. The filters vary with the type of fuel being stored. For Magnox storage they take the form of deep sand beds through which the water is passed under pressure. For AGR storage the sand filters are preceded by filters formed from a diatomaceous-earth material (a siliceous deposit which occurs as a whitish powder; it is also widely used in fireproof cements and furnace walls, etc.).

Figure 10.2 shows the spent fuel storage pond at Harwell.

Figure 10.2   The spent fuel storage pond at Harwell

Decay heat   The radioactivity of a spent fuel element on being lowered into the storage pond depends upon many factors and it is therefore difficult to be precise. However, it can be said that this radioactivity causes the element to generate considerable heat due to the radioactive decay processes of the many fission products created in the fuel when it was being irradiated in the reactor core and that it therefore acquires a relatively high surface temperature. This form of heat is known as *decay heat* and is typically 200 W for each Magnox fuel element at the time of entering the pond; for a 36-pin AGR fuel element the figure can be as high as 5 kW, assuming there has been no pre-cooling.

Again, depending on the length of storage time and other important factors, it is equally difficult to be precise about the residual decay heat of a particular fuel element at the time when it is removed from the pond to be transported to the Sellafield reprocessing plant. However, typical figures for decay heat following pond storage would be 25 W for a Magnox fuel element and 1 kW for a 36-pin AGR fuel element.

Dry storage

Although water-filled ponds could be designed to meet the required safety standards of certain types of fuel (but not the Magnox clad type), a more attractive method for long-term storage, from both the safety and economic points of view, is that known as *dry vault storage*. There is nothing particularly new about this method; in fact it has been used continuously for interim storage of spent fuel at the largest of the Magnox nuclear power stations at Wylfa, in Anglesey, since it began operating in 1971 (see Figure 8.4) and it could be used equally satisfactorily for other types of reactor fuel.

The Wylfa station was originally equipped with three dry storage cells but the construction of two much larger ones was completed in 1981. The original cells each had a storage capacity of 83 tonnes of spent fuel comprising 6840 fuel element assemblies stacked one above the other in a matrix formation of vertical storage tubes. The physical arrangement of one of the cells is shown in Figure 10.3. Each storage tube is closed at its bottom end and fuel is lowered into it from the top. A single canopy covers the tops of all the tubes and is filled with carbon dioxide (CO_2) gas at a pressure of about 3 pounds per square inch. In this way each of the tubes is filled with pressurized CO_2. A removable plug at the top of the canopy, and an internal storage shute, allows fuel elements to be loaded into storage tubes.

The tubes and the canopy of each cell are surrounded by air at normal atmospheric pressure and are located inside a cylindrically-shaped concrete containment vessel which functions as a radiation shield; cooling takes

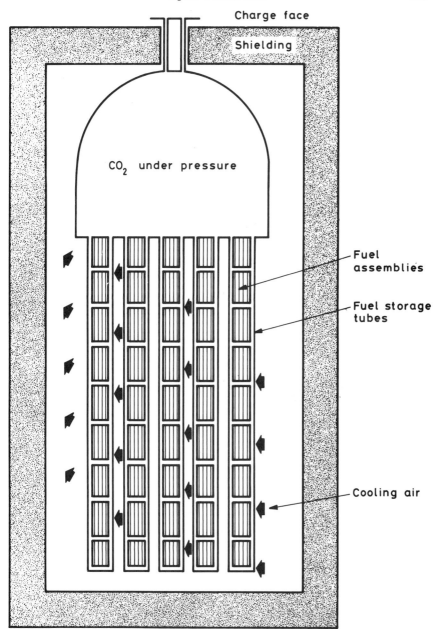

Figure 10.3   Physical arrangement of one of the three original dry-store cooling cells at the Wylfa nuclear power station (courtesy GEC Energy Systems)

place by natural convection. Decay heat from the fuel is conducted away by the surrounding CO_2 gas to the walls of the storage tubes and thence via the air to the atmosphere. Cool air enters the cell via ducting at the bottom and leaves at the top after being heated where it is conveyed by more ducting to a tall stack which serves all three cells. The hotter the fuel the greater is the draught up the stack and the greater is the quantity of cool air drawn in at the bottom of the cell.

The effectiveness of the original dry storage cells at Wylfa was clearly demonstrated when spent fuel which had been stored in them for four years was removed and found to be in perfect condition. In fact, permission was granted by the Licensing Authority for it to be re-loaded into the reactor for further irradiation, should this be considered desirable. Naturally, it is most unlikely that the Wylfa station operators would ever want to do such a thing but the fact that they were given permission to do so reflects the very high level of confidence placed in this method of storage.

Additional capacity   The only disadvantage of the original dry storage cells at Wylfa was their relatively low capacity when compared with other Magnox stations which used cooling ponds for spent fuel storage. To give an example, the average storage capacity of Magnox cooling ponds is about 1 tonne of spent fuel per MWe of station output whereas the corresponding figure for the Wylfa dry store is only 0.3 tonnes.

Bearing in mind that Wylfa is the largest of the nine Magnox power stations and that it is responsible for about 25 per cent of all the spent fuel they produce, the storage capacity of the original dry cells was thought to be inadequate should a hold-up occur at the reprocessing plant which would prevent the transportation of spent fuel. It was for this reason that additional on-site storage capacity was proposed in 1976 for the Wylfa station.

Two new storage cells for the Wylfa station were designed and constructed by the British company GEC Energy Systems Limited and were completed in 1981. The new cells, like the original ones (which were also built by GEC Energy Systems), are of the dry vault type but differ from them in relying entirely upon air for cooling instead of using CO_2 as well. *These cells represent the first in the world to make use entirely of air only for spent fuel storage.*

The new cells are designed to complement the original cells which are still used for the first-stage storage of freshly-removed spent fuel for at least 150 days. It is only after the decay heat has fallen to less than 60 W per element that the fuel is transferred to the new cells and elaborate monitoring systems ensure that it is mechanically impossible to transfer spent fuel to the new cells whose power rating is in excess of 60 W per element. Other precautions ensure that the moisture content of the naturally-convected air is maintained at a low level so that in the unlikely event of the uranium core becoming exposed through defective cladding it would not lead to the formation of uranium hydride. This is very important because under certain fault conditions, which might be brought about by physical damage and high temperature, the presence of uranium hydride could lead to spontaneous ignition of the highly reactive fuel.

An internal view of one of the new dry storage cells is shown in Figure 10.4. Each cell has a total storage capacity of 350 tonnes of uranium representing nearly 29 000 Magnox fuel elements. The elements are stored individually in vertical tubes which are closed at the bottom and open at the top. The storage tubes are linked together physically in groups of 192 in what is known as a 'skip'. The tubes in each skip are arranged in a 16×12 formation and are mounted as a separate unit on a wheel-driven conveyor system which runs the full length of the cell. The total capacity of the cell is 151 skips (28 992 fuel storage tubes) arranged in six rows of 25 skips, plus one spare.

The cell is essentially a concrete box structure 60 m long, 11 m wide and 4.5 m high with walls 2 m thick. Its interior is maintained at slightly below atmospheric pressure so that in the unlikely event of a breach in the cell sealing system, coinciding with a release of radioactivity, the leakage would be inwards instead of outwards, thereby preventing the release of radio-active material to the external atmosphere.

Whilst designing the new air-cooled cells for Wylfa, GEC Energy Systems realized that similar types of cells would be suitable for the long- and intermediate-term storage of AGR- and LWR-types of oxide fuel, and also of vitrified high-level radioactive waste (see Chapter 12), and have therefore devoted considerable research into perfecting cells suitable for these purposes.

TRANSPORTATION

After the spent fuel has been stored for the requisite period of time at the nuclear power station it is transported by road, rail or sea to the reprocess-ing plant in massive shielding flasks usually filled with water and weighing between 45 and 100 tonnes. The flasks vary in design according to the type of fuel they are designed to carry. The type used for transporting Magnox spent fuel is illustrated in Figures 10.5 and 10.6. It is cuboid in shape and, when empty, weighs 48 tonnes. It can carry up to 200 fuel elements representing about 2 tonnes of spent fuel. Its dimensions are approximately

Figure 10.4   Inside one of the new dry storage cells for spent fuel at the Wylfa nuclear power station
(courtesy CEGB)

Figure 10.5   Spent fuel from a Magnox power station cooling pond begins its journey to the Sellafield reprocessing plant (courtesy CEGB)

217

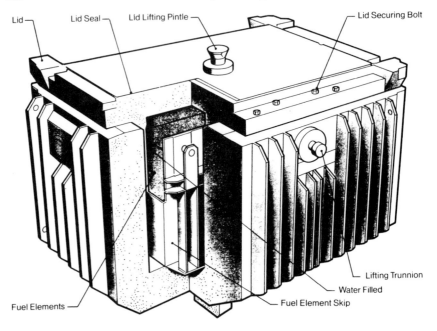

Lid — Lid Seal — Lid Lifting Pintle — Lid Securing Bolt

Lifting Trunnion
Water Filled
Fuel Elements — Fuel Element Skip

Figure 10.6   Cutaway diagram of a Magnox spent fuel transport flask

2.4 m (8 feet) cube and its walls are made from steel 368 mm (14.5 inches) thick. The lid, which weighs 9 tonnes, forms a water tight seal with the body of the flask and is held in position by 16 high-tensile steel bolts. The great thickness of the flask walls ensures that the maximum radiation level measured at a point 1 m from any surface is never more than 10 μSv per hour, a figure which is one-tenth of the legally permitted level.

The Magnox elements within the flask are stored in a single skip and, in the case of pond storage, are loaded into the flask under water in the pond. When its sealing lid has been secured, the fully laden flask is removed from the pond and washed down to ensure that it has not picked up any surface contamination. After further testing, inspection and radiological monitoring by health physics staff the flask is loaded onto the transporter ready to begin its journey.

The water within the flask conducts the heat away from the spent fuel elements and conveys it to the thick walls of the flask where it is radiated to the atmosphere by the large surface area presented by the cooling fins.

The reprocessing plant also receives spent fuel from British AGR nuclear power stations and from light water reactors. This latter type of

Figure 10.7   Typical flask of the type used to transport spent oxide-type nuclear fuel from LWRs

fuel is transported in cylindrically-shaped water-filled flasks like the one shown in Figure 10.7. British flasks of this type are fabricated from 90 mm (3.5 inch) thick steel and the walls are lined with 190 mm (7.5 inch) of lead. The unladen weight of such a flask is typically 70 tonnes and it can carry up to 3 tonnes of spent oxide-type fuel. The fuel is loaded into the flask through a removable end cap which is then bolted into position to form a water tight seal.

Safety

Since reprocessing began at Sellafield more than 30 years ago, over 25 000 tonnes of spent fuel have been reprocessed. This has involved the transport of spent fuel over many millions of route miles of road and rail track in Britain and although a small number of minor accidents have occurred during that time involving the flasks, *not one has resulted in a breach of the flask's integrity and none has resulted in the release of any radioactive material*.

The flasks are designed to withstand a free fall of 9 m onto an unyielding target of steel plate on massive foundations when oriented in the worst possible impact attitude, followed (or preceded) by a free fall of 1 m onto a steel punch of 150 mm diameter mounted on the target, followed by an all-round fire producing temperatures up to 800°C for not less than 30 minutes. The regulations governing the design of flasks used for the transportation of spent fuel have been prepared by international specialists under the direction of the International Atomic Energy Agency and are intended to ensure that even the worst possible credible accident will not result in unacceptable damage to the flask.

Proof of the design was demonstrated in public in 1984 when the CEGB dropped one of its 48-tonne flasks from a height of 9 m onto an unyielding metal plate anchored into solid rock in Cheddar, in Somerset. The flask reached a speed of 30 mph during its 1½-second fall but afterwards was found to be still fully sealed. The purpose of the test was part of a demonstration to the public that flasks meeting regulations used for transporting spent nuclear fuel would be able to survive a high-speed train crash without releasing unacceptable levels of radioactive material.

An even more drastic test was staged by the CEGB in July 1984 on a British Rail test track near Melton Mowbray, Leicestershire. A 48-tonne Magnox transportation flask of the type used to take spent nuclear fuel from a cooling store to the reprocessing plant at Sellafield was filled to the normal level with water and the remaining gas volume was pressurized. It was then strapped to its rail vehicle and set on its side across the line at an

Figure 10.8　The rail crash staged by the CEGB at Melton Mowbray to demonstrate the safety of its nuclear fuel transport flasks (courtesy CEGB)

221

angle designed to ensure the most damaging impact. Eight miles away an elderly (and unmanned) diesel locomotive weighing 140 tonnes was coupled to three 35-tonne coaches and launched along the track towards the flask. Before an audience of 1500 people the train hit the flask at an estimated speed of 100 mph (Figure 10.8). Fragments from the locomotive were flung several hundred metres through the air and the flask itself was pushed 100 metres down the track. Half an hour later, it was examined by engineers and found to be effectively intact with the pressure inside it fully maintained. The worst apparent damage was a 230 mm (9-inch) gouge in its 370 mm (14.5-inch) thick steel surface.

In the extremely unlikely event of an accident involving a flask which resulted in the release of radioactive material, well-rehearsed emergency procedures would at once be initiated and would involve, among others, the regional generating board (CEGB/SSEB), British Rail and its Railway Police, the Civil Police, the Fire Brigade, the Ambulance Service and the local authority responsible for the area in which the accident occurred.

11 The reprocessing plant

Britain's main nuclear fuel reprocessing plant is located at Sellafield in West Cumbria. The plant functions as a purification process in which valuable unburnt fissile uranium (as much as 50 per cent of the original for enriched fuels) is separated from the unwanted fission products which constitute the impurities in the spent fuel. The process also segregates small quantities of valuable plutonium (mostly Pu-239) and other, but unwanted, transuranic elements such as neptunium, americium and curium. Plutonium is an extremely valuable source of fuel for fast reactor systems and is currently used to supply Britain's PFR at Dounreay in the north of Scotland.

There is nothing unique about the reprocessing of spent nuclear fuel; the USA, Britain and France have been doing it for more than 30 years but only Britain and France have plants of sufficient reprocessing capacity to support their own commercial nuclear power programmes and, at the same time, offer a reprocessing service to operators of other reactor systems throughout the world.

Reprocessing of spent nuclear fuel first started at Sellafield (then known as Windscale) in 1952. Since then more than 25 000 tonnes of Magnox fuel and about 100 tonnes of oxide fuel from British AGRs and LWRs from other countries have been processed and it has been estimated that by the turn of the century a further 20 000 tonnes of Magnox fuel and 6000 tonnes of oxide fuel will have been processed.

Since 1971 the Sellafield plant has been owned and operated by British

Figure 11.1　An aerial view of BNFL's reprocessing plant at Sellafield, Cumbria (courtesy BNFL)

Nuclear Fuels plc (BNFL), a nationalized organization operating as an autonomous commercial company. Prior to 1971 it was part of the United Kingdom Atomic Energy Authority (UKAEA). An aerial view of the Sellafield plant is shown in Figure 11.1. To the right of the picture can be seen the 200 MWe Calder Hall nuclear power station with its four cooling towers. First commissioned in 1956 this station is still supplying electrical power 30 years later to the National Grid as well as electricity and process heat to the Sellafield plant.

The first of the Sellafield reprocessing plants began operating in 1952 and was designed specifically to reprocess spent fuel from the early Windscale 'piles' to derive the plutonium required to support Britain's nuclear weapons programme. The Windscale piles were actually two air-cooled, graphite-moderated reactors fuelled with natural uranium. The first began operating in 1951, the second eight months later. The sole purpose of the piles was to produce plutonium. The cooling chimneys of these piles can be seen towards the centre rear of the photograph shown in Figure 11.1. Following a very serious fire in one of the piles in 1957 both were later de-fuelled and sealed up and have never operated since. This particular incident is covered in more detail in Chapter 19.

The plant later began reprocessing spent fuel from the eight Calder Hall and Chapelcross reactors, all of which were designed for the dual purpose of plutonium production for nuclear weapons and as prototypes for the commercial Magnox nuclear power stations. In 1964 a much larger repro-cessing plant began operating at Sellafield and the smaller plant was shut down. The newer plant was designed to handle the large quantities of spent fuel which would eventually arise from Britain's own Magnox nuclear power stations and from similar types of stations operating overseas; this plant is still operating.

Modifications to the first reprocessing plant in 1969 enabled it to pre-treat spent oxide fuel arising from the experimental Windscale AGR (WAGR) and from overseas LWRs, and to feed the partially-separated fuel to the newer plant. This combination of the two plants worked well until 1973 when after an accidental release of radioactivity into a working area it was decided to discontinue using the older plant in this way.

The existing Magnox reprocessing facilities at Sellafield are at present undergoing considerable improvement with replacement in some plant areas in order to cope with a continuing demand for capacity. Construction of a completely new plant for reprocessing oxide fuel started in 1985. To be known as THORP (*Th*ermal *O*xide *R*eprocessing *P*lant) it will have a design capacity of 1200 tonnes per year of spent AGR and LWR fuel and a guaranteed capacity for contractual purposes of at least 600 tonnes per year. Large-scale test rigs have been constructed at Sellafield to examine various operating features of the plant and it is hoped to have the plant fully operational by the early 1990s.

DOUNREAY PLANTS

The reprocessing facilities at Sellafield are used solely for the treatment of spent fuel arising from British and overseas thermal reactors in which the uranium enrichment seldom exceeds about 3 per cent. In 1958, however, a small plant was constructed at Dounreay in Scotland for the reprocessing of highly enriched spent fuel arising from the Dounreay Materials Testing Reactor (DMTR) and from the DIDO and PLUTO reactors at Harwell which were nearly identical to the DMTR. Although the DMTR, having outlived its usefulness, was shut down in 1969 the plant is still used to reprocess spent fuel from the Harwell reactors.

A second reprocessing plant was built at Dounreay for the treatment of spent fuel from the experimental Dounreay Fast Reactor (DFR). This small plant began full operation in 1962 and continued operating until the DFR was closed down in 1975. It was then decontaminated and completely rebuilt so that it could reprocess plutonium-based spent fuel and depleted uranium arising from the much larger Prototype Fast Reactor (PFR) at Dounreay. This rebuilt plant commenced the reprocessing of PFR fuel in 1980 and has an approximate throughput capacity of 6 tonnes per year. The Dounreay reprocessing plants, like the site itself, are owned and operated by the UKAEA.

THE REPROCESSING PROCEDURE

A general summary of the function of a spent nuclear fuel reprocessing plant is illustrated in Figure 11.2. For Magnox fuel, after the transport flask carrying the spent fuel arrives at Sellafield, it is unloaded from the transporter and transferred to a shielded core in the Fuel Handling Plant where its lid is removed (Figure 11.3). The spent fuel is then transferred to a special container which is placed under water in fixed geometric positions until required for reprocessing; during this additional waiting time further radioactive decay takes place.

The next stage in the operation is removal of the fuel cladding. This is carried out in shielded caves in air with water sprayed over the fuel to prevent self ignition of either the uranium or the swarf. Automatic stripping machines are used as illustrated in Figure 11.4. The operation results in the accumulation of solid debris like that shown in Figure 11.5 and is

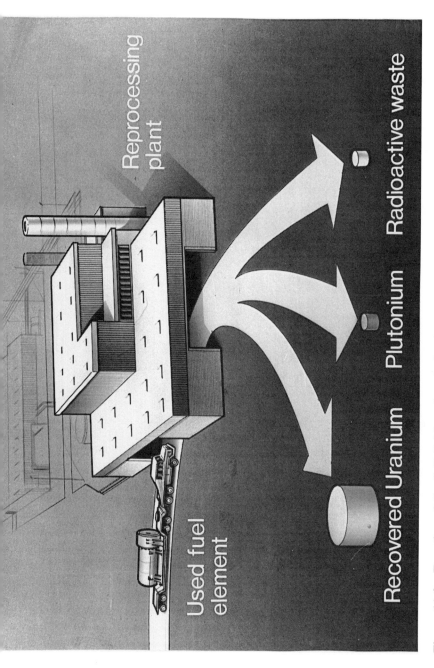

Figure 11.2   Function of the Sellafield nuclear fuel reprocessing plant

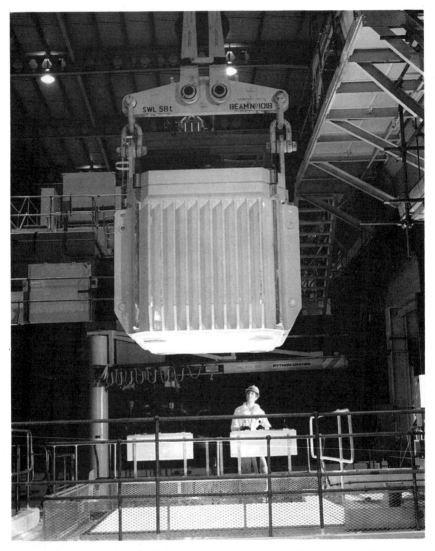

Figure 11.3 A transit flask being lowered through the ante-chamber roof hatch in a sub-pond at BNFL's Sellafield site
(courtesy BNFL)

Figure 11.4 Decanning spent Magnox fuel in the fuel handling plant at BNFL's Sellafield site (courtesy BNFL)

categorized as 'intermediate-level' nuclear waste; this and other forms of nuclear waste will be discussed in more detail in Chapter 12.

After the cladding has been removed from the Magnox fuel elements, the solid uranium rods are transferred in shielded containers to the separation plant where they are dissolved in nitric acid. The resulting liquor is then subjected to a primary separation process which results in the emergence of two separate streams. One contains mixed uranium and plutonium and the other contains the mixed fission products plus some transuranic elements representing what is known as *high-level nuclear waste*. The stream containing the mixed uranium and plutonium is sub-jected to a further separation process to produce two further streams, one in the form of plutonium nitrate (the separated plutonium) and the other in the form of uranyl nitrate (this is the recovered uranium). The overall process is shown diagramatically in Figure 11.2.

After passing through a purification stage the uranyl nitrate stream has its nitrate content removed and it is then converted to uranium trioxide (UO_3) powder which is later used as feedstock for the manufacture of fresh fuel elements at the Springfields fuel manufacturing plant. The plutonium nitrate stream is similarly passed through a purification stage and later used at Sellafield as feedstock for the manufacture of mixed oxide (PuO_2/UO_2) fuel for the PFR at Dounreay.

The separation of uranium and plutonium from the spent fuel is carried out by a solvent extraction process using mixer–settler stages and a powerful solvent known as tributyl phosphate in odourless kerosene (TBP/OK). Well in excess of 99 per cent of all the fission products present in the spent fuel is contained in the fission product solution derived from the primary separation process.

To give some idea of the relative amounts of the various elements produced during reprocessing, a spent Magnox fuel element typically contains about 99.5 per cent unburnt natural uranium, 0.3 per cent fission product waste and 0.2 per cent plutonium. Corresponding figures for an AGR-type fuel element, which contains enriched fuel, would typically be unburnt uranium, containing about 1 per cent unburnt U-235, about 3 per cent fission products and actinide elements of about 1.5 per cent – principally plutonium.

Magnox fuel reprocessing yields approximately 5 m³ of highly radio-active liquid for each tonne of fuel processed, which is reduced by evapor-ation to approximately 0.05 m³ of high-level radioactive waste per tonne of fuel reprocessed. This waste contains about 3 TBq of beta–gamma fission product radioactivity per litre of separated liquor before concentration by evaporation and is stored in high-integrity tanks with continuous heat

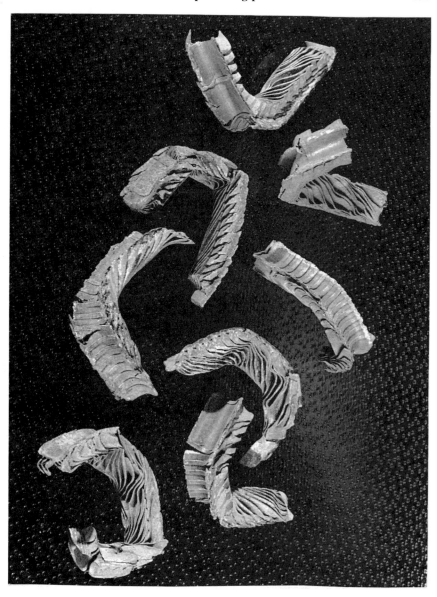

Figure 11.5   Debris from Magnox de-cladding operation

removal. The liquor also contains small quantities of metal salts originating from the dissolution of minor constituents of the fuel and some entrained and dissolved solvents which are later removed by steam distillation.

Although nearly identical in isotopic composition, the reprocessing of oxide fuels yields liquids which, prior to evaporation, contain much higher levels of activity. This is because of the greater quantities of fissile material and the much higher burn-up ratings. However, after evaporation the oxide fuel yields approximately 0.25 m^3 of high-level radioactive waste per tonne of fuel reprocessed and the activity level is similar to that from Magnox reprocessing.

12 Radioactive waste: its nature, storage, treatment and disposal

Any industrialized country creates enormous quantities of waste materials, some of which are highly toxic or dangerous to handle, or at least extremely unpleasant. The traditional way of disposing of such waste has been to discharge it into streams, rivers or the sea, to burn it, bury it or discharge it directly to the atmosphere through car exhaust pipes, household chimneys, factory chimneys, ships' funnels and sewage ventilator pipes.

Domestic and industrial waste takes many forms and is of widely different origins. The metals industry, for example, uses large quantities of deadly arsenic in the manufacture of certain alloys and, in consequence, is responsible for producing large quantities of arsenicous wastes. Arsenic is also used in the manufacture of certain insecticides, sheep dips, rat poisons and fly papers and in the semiconductor industry for manufacturing transistors and silicon chips, etc. Arsenic, being an element, is not degradable with time and lasts for ever; it cannot be destroyed (except in a nuclear reactor or particle accelerator) and only safe storage or dispersion will reduce its danger to life.

Mercury, another highly toxic material, is much used in the production of street lighting lamps and in the electrical industry in mercury arc rectifiers, high-current switches and in Mallory batteries. Mercury, like arsenic, is an indestructible element.

Cyanide, and potassium cyanide – one of the most powerful of all poisons – is used in the mining industry for separating gold and silver from

233

their ores and for case hardening of steel; it is also used in certain fumigants and in photographic processes.

In the electricity generating industry a 2000 MWe coal-fired power station when operating continuously burns approximately 22 000 tonnes of coal every 24 hours (nearly five 1-cwt bags every second!), during which time it discharges directly to the atmosphere about 630 tonnes of sulphur dioxide (SO_2), some of which becomes sulphur trioxide (SO_3) and combines with water vapour to become sulphuric acid (H_2SO_4) – the main constituent of so-called acid rain. It also discharges to the atmosphere each day more than 10 tonnes of fly ash which contains small quantities of cadmium, mercury, chlorine, sodium, arsenic, lead and many other potentially dangerous elements, some of which are radioactive.

There is nothing unusual about the source and disposal of these wastes; it has been going on for many decades and is the price we pay for the relatively high standard of living enjoyed and taken for granted by millions of people throughout the world. To be fair, however, there has been much greater awareness in recent years of the extent to which the environment is being polluted and much greater care is now being taken in disposing of dangerous and environmentally unacceptable waste materials.

NUCLEAR WASTE

The nuclear power industry is also responsible for producing waste materials, most of which are potentially dangerous because of their radioactivity. However, because of the very high energy density of nuclear fuels the quantity of waste produced per unit of electricity generated is considerably less than that produced by the coal- and oil-fired power stations. Furthermore, because the nuclear industry is a relatively young one it has been subject from its inception to strict national and international regulations governing all stages of its construction and operation and in the handling, storage and disposal of its radioactive waste.

Radioactive waste is produced in varying quantities and levels of importance at virtually every stage in the nuclear fuel cycle ranging from the mining and processing of the uranium ore and the use of the fuel elements at the power station, to the reprocessing of the spent fuel. The utilization of radioisotopes in industry, agriculture, education and medicine also gives rise to substantial quantities of waste.

CATEGORIZATION

In the UK, nuclear wastes are categorized into three types defined as *low level*, *intermediate level* and *high level* and may appear in gaseous, solid or liquid forms. The category assigned to a particular waste material is determined by the type of radiation emitted (alpha, beta or gamma) and the concentration of the radioactivity. These two factors essentially define the potential of the material for causing harm to animal life – and human life in particular – and to the environment. Based on advice arising from the Radioactive Waste Management Advisory Committee, the three categories of waste are defined as follows:

Low level   These are wastes which contain radioactive materials other than those which are acceptable for domestic dustbin disposal (very low level), but whose radioactive concentrations do not exceed 4 GBq/t of alpha activity or 12 GBq/t of beta/gamma activity.

Intermediate level   These are wastes which contain radioactive materials whose radioactive concentrations exceed those which define low-level wastes but which do not generate significant quantities of heat.

High level (also known as heat-generating wastes)   These are wastes which contain radioactive materials whose radioactive concentrations may give rise to significant quantities of self-generated heat such that it may affect the design of storage and disposal facilities.

Low-level waste

This type of waste comprises laboratory refuse such as protective clothing, rubber gloves, tissues, swabs, plastic bags, glasswear, used syringes, air filters, empty containers, contaminated tools, plus liquid effluents from sinks and drains and radioactive gases present in ventilation systems. Such materials are found in hospitals, teaching establishments, research laboratories, nuclear power stations, isotope production plants and wherever unsealed radioisotopes are employed. They are also found in nuclear fuel reprocessing, manufacturing and enrichment plants.

Although very low in radioactivity, low-level waste is very large in volume because of its bulky nature; Britain alone generates about 20 000 tonnes of such waste every year.

Intermediate-level waste

This consists mainly of cladding material removed from spent fuel elements (sometimes referred to as 'hulls') and, to a lesser extent, of contaminated pipework, control-rod mechanisms and structural materials found in the core of a reactor (or in the target area of a powerful particle accelerator) plus contaminated filters and ion exchange resins from cooling ponds and water-cooled reactors, and sludges from fuel dissolvers and chemical treatment processes.

The total volume of intermediate-level waste stored in Britain at present is about 40 000 m^3.

High-level waste

This type of waste originates entirely from the reprocessing of spent fuel and is liquid: nitric acid, in which are dissolved and suspended fission products which have been extracted from spent fuel elements. Also present in the acid are residual traces of uranium and plutonium, plus small quantities of so-called *higher actinides*, or *transuranics*. These are the elements neptunium (Np), americium (Am) and curium (Cm) which were formed, like the plutonium, by neutron capture events in the fuel. Their atomic numbers are all in excess of 92 (uranium), hence the name *transuranics*.

High-level waste is sometimes described as heat-generating waste because of the heat which is produced by the decay of its highly concentrated radioactivity. The total volume of high-level waste currently stored in Britain is about 1100 m^3. However, only about a quarter of this is actual radioactive waste, the rest is nitric acid. High-level waste is described more fully later in this Chapter.

WASTE MANAGEMENT

Overall responsibility for national policy on the management and disposal of radioactive wastes in Britain is vested in the Department of the Environment (DoE) and with the Scottish and Welsh offices. However, because of the many ways in which such wastes are created, stored, processed, transported and disposed of, a number of other Government Departments are actively involved; in particular the Ministry of Agriculture Fisheries and Food (MAFF) and the Departments of Transport (DTp), Energy (DEn), and Health and Social Security (DHSS); the roles played by these departments are described in the Appendix.

National policy is formulated from an assessment of relevant social, political and economic factors and of scientific and technical advice provided by research organizations and through international discussions and agreements whenever possible or appropriate. Much of the scientific advice is provided by research undertaken by the United Kingdom Atomic Energy Authority (UKAEA) (which is administered by DEn) and by the National Radiological Protection Board (NRPB), which is administered by

DHSS. However, the main source of advice stems from the Radioactive Waste Management Advisory Committee (RWMAC) which provides independent advice to the Secretaries of State for the Environment, Scotland and Wales, and which has, among its many members, scientific representatives from British Nuclear Fuels plc (BNFL) and the Central Electricity Generating Board (CEGB). DoE also maintains three separate advisory committees of its own; these are the Radioactive Waste Liaison Committee (RWLC), the Radioactive Waste Technical Liaison Committee (RWTLC) and the Plutonium Contaminated Wastes Steering Committee (PCWSC). Research undertaken by the UKAEA on plutonium-contaminated wastes is coordinated by the Plutonium Contaminated Materials Working Party (PCMWP) whose members comprise the UKAEA, DoE, BNFL and the Ministry of Defence (MoD).

A comparative newcomer to the scene is the organization known as UK Nirex Limited; the word Nirex is an acronym derived from Nuclear Industry Radioactive Waste Executive. Nirex was originally established in 1982 as a partnership comprising the four organizations BNFL, CEGB, SSEB and the UKAEA and had the responsibility to develop plans for dealing with solid, low- and intermediate-level radioactive wastes which present no problem of heat generation but whose radioactivity is above the very low levels which would otherwise allow them to be disposed of in the same way as domestic refuse.

The original Nirex organization had no corporate legal identity but following comprehensive discussions between the four partners and with Government, a limited company named UK Nirex Limited was incorporated on 20 November 1985. The four original partners are the major shareholders (CEGB 42.5 per cent; BNFL 42.5 per cent; UKAEA 7.5 per cent; SSEB 7.5 per cent) with the Secretary of State for Energy holding a 'special' share on behalf of the Government.

The operating costs of UK Nirex Ltd are shared by the share holding organizations who also provide the senior representatives which make up the directorate. The directorate monitors the activities of, and sets guidelines for, the UK Nirex company which is based at the Harwell Laboratory. This company is made up of scientific, engineering and administrative staff whose job it is to ensure that the work necessary to plan, develop and operate waste disposal routes is carried out either by itself or by appointed contractors.

UK Nirex Ltd holds periodic meetings with the NRPB, various regulatory authorities, shareholders and customers, so as to exchange views and information on UK Nirex Ltd activities and future plans. Figure 12.1 shows the relationship between UK Nirex Ltd and other organizations associated with the disposal and management of nuclear wastes.

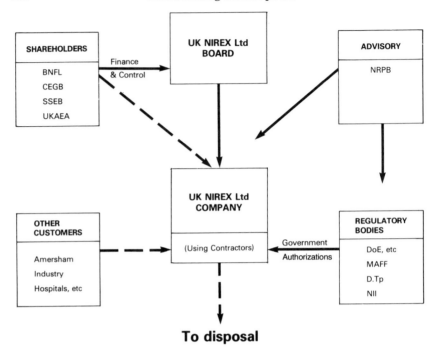

To disposal

Figure 12.1   Relationship between NIREX and other organizations associated with the management and disposal of nuclear wastes

THE ORGANIZATION OF WASTE DISPOSAL

The methods used to store, process, transport and dispose of radioactive waste vary considerably with the type and radioactivity of the waste, the type of radiation emitted, and whether the waste is liquid, solid or gaseous in form. At one extreme the very low levels of radioactivity found in the luminous paints (containing radium) used on the dials of old-fashioned watches and alarm clocks may be disposed of quite safely on the local municipal rubbish tip. At the other extreme the highly radioactive liquids created at the fuel reprocessing plant are at present stored on site where they are created and are likely to remain there for the next 50 years or so although, as explained below, these will be converted into a solid form when the vitrification plant at present under construction at Sellafield is completed.

The methods used for the disposal or discharge of radioactive wastes to land, sea or air are subject to the conditions laid down in the Radioactive Substances Act 1960 and, in England, are under the control of the Secretary of State for the Environment. In the case of disposal of radioactive wastes from or on the premises of the UKAEA and sites licensed under the Nuclear Installations Act 1965 (this includes nuclear power stations and Amersham International) the control is exercised jointly with the Minister of Agriculture Fisheries and Food. In Scotland and Wales control is exercised solely by the Secretaries of State for Scotland and Wales. In Northern Ireland control is exercised by the Department of the Environment for Northern Ireland.

Under the Dumping at Sea Act 1974, the disposal at sea of packaged low-level and some intermediate-level radioactive waste, loaded at an English port, has to be licensed additionally by the Minister of Agriculture Fisheries and Food. If such wastes are loaded at ports in Scotland or Wales then licensing would originate from the Secretaries of State for Scotland and Wales, respectively. For Northern Ireland ports the responsibility for licensing would exist with the Department of the Environment for Northern Ireland.

Disposal arrangements for radioactive waste are also subject to the provisions of the Health and Safety at Work Act 1974, for which the Secretary of State for Employment is responsible in England, Scotland and Wales. In Northern Ireland the legislation applicable is the Health and Safety at Work Order 1978 and the Secretary of State for Northern Ireland is responsible. In addition to this, any dumping site which may be proposed by UK Nirex Ltd will be subject to licensing by the Nuclear Installations Inspectorate (NII), which is part of the Health and Safety Executive.

In England the Secretary of State for the Environment is responsible for initiating and co-ordinating the monitoring of those environmental pathways by which radioactive nuclides can reach people. Similarly, the Minister of Agriculture Fisheries and Food is responsible for initiating and co-ordinating the monitoring of pathways by which radioactive nuclides may reach people by way of food chains (crops, etc.) and the marine environment, for example fish. In Scotland, Wales and Northern Ireland the respective Secretaries of State are responsible for the monitoring of all such pathways.

The responsibility for consideration of the health consequences of discharges and disposals of all kinds of radioactive wastes in England lies with the Secretary of State for Social Services, who is advised by the Committee on the Medical Aspects of Radiation in the Environment (COMARE); in

Scotland, Wales and Northern Ireland this responsibility lies with the respective Secretaries of State.

The responsibility for the management of radioactive wastes arising from Britain's defence programme lies with the Secretary of State for Defence and there exists close liaison between the Ministry of Defence and the relevant civil departments to ensure that the standards which are applied within MoD are as rigorous as those observed by civil organizations and that the disposal of both kinds of waste, taken together, is environmentally acceptable.

There is considerable collaboration between countries on methods used for disposing of radioactive wastes and regular exchanges of views and discussions on the results of related research. Some of the more important organizations in this field are the International Atomic Energy Agency (IAEA), the Nuclear Energy Agency (NEA) of the Organization for Economic Co-operation and Development (OECD), the Commission of the European Communities (CEC), the Council of Mutual Economic Assistance (CMEA), the International Council of Scientific Unions (ICSU) and the International Commission for Radiological Protection (ICRP).

Inspectorates

Those government departments with a responsibility for the disposal and storage of radioactive wastes each appoint their own teams of inspectors to ensure that the relevant legislation is being correctly observed. The Department of Employment, for example, through the Health and Safety Executive (HSE), appoints the Nuclear Installations Inspectorate (NII) to license and monitor all aspects of nuclear installations, including the provisions made at such installations for the processing, storage and disposal of radioactive wastes. The Department of the Environment appoints the Radiochemical Inspectorate (RCI) which, in addition to its role of monitoring all land-based radioactive waste repositories and disposal sites, is also responsible for the authorization and monitoring of small quantities of low-level wastes disposed of at local tips and to sewers.

The monitoring of discharges of radioactive materials to the atmosphere is undertaken by a separate inspectorate from the HSE and known as HM Alkali and Clean Air Inspectorate (ACAI). The Ministry of Agriculture Fisheries and Food also operates an inspectorate team which is set up and called upon as required to monitor discharges of radioactive wastes to rivers, streams, lakes and oceans.

In Scotland the various inspectorates described above are combined in a single organization known as HM Industrial Pollution Inspectorate (IPI).

DISPOSAL METHODS

There are, essentially, two methods adopted for the disposal of radioactive waste; one is direct dispersal to the environment, the other is shielded containment. The dispersal method relies upon massive dilution and slow releases to the atmosphere, rivers, oceans and designated rubbish tips so that the quantities discharged present no significant risk to human populations by any pathway. The shielded containment method of disposal is essentially an isolation method which relies upon secure packaging of the waste materials to prevent leakage and adequate shielding to ensure that there is no likelihood of anyone being irradiated whilst the radioactivity of the waste decays to an insignificant level. Which of the two methods is adopted depends upon the half-life and type of radiation emitted (alpha, beta or gamma), its concentration and toxicity, and whether it is in liquid, solid or gaseous form. As a general rule, low-level solids, liquids or gaseous wastes are disposed of by dispersion to the environment; intermediate-level wastes are currently stored in shielded containers but will eventually be mixed with cement and cast into solid blocks which will be buried on land. High-level wastes are at present stored in liquid form in tanks at the reprocessing plant but will eventually be concentrated and mixed with glass-making materials and cast into blocks of glass. Present government policy is that these blocks will be stored in shielded surface repositories for 50 years or so before being finally disposed of.

Low-level solid wastes

The sort of low-level solid wastes arising from establishments using radioactive materials are carefully monitored for their radioactive content and if found to contain less than 740 MBq/m^3 of alpha activity, less than 2220 MBq/m^3 of beta activity, and emitting a surface radiation of less than 7.5 mSv/h, are transported to a 120 hectare (300 acre) fenced, controlled and licensed disposal site at Drigg, near Sellafield in Cumbria, which is owned and operated by BNFL. Here they are placed in shallow trenches which are isolated from the underlying sandstone by a natural layer of clay and then covered by at least 1 m of soil. A similar site is owned and operated by the UKAEA at its Dounreay site in Caithness. The safety of both sites has been endorsed by the Department of the Environment following detailed surveys to ensure that any radioactivity leached by groundwater cannot enter the drinking water supplies.

Although the Dounreay and Drigg sites have sufficient capacity for many years to come it is accepted that a third site for such wastes will be required sometime around 1990; the finding and operating of such a site is the responsibility of UK Nirex Ltd.

Some of the solid wastes which possess slightly higher radioactivity than that accepted by the Drigg and Dounreay sites were, from 1949 until 1982, packed into small steel drums which were themselves each placed centrally within larger steel drums and the intervening space between them filled with concrete; the physical arrangement of such a drum is illustrated in Figure 12.2. Such drums were then dumped at sea each year at a location nearly 800 km (500 miles) south west of Lands End within a rectangular area bounded by 27 km × 115 km and where the water is about 4 km (2.5 miles) deep. The site, which is well clear of shipping lanes and submarine cables, where little fishing is done and where fish density is very low, was approved by the OECD following a detailed environmental assessment undertaken by a group of international experts.

The practice of deep sea disposal by the UK was also endorsed in 1980

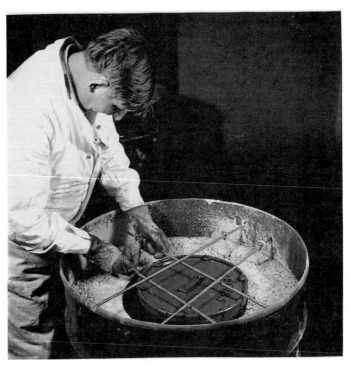

Figure 12.2 Preparing containers for the disposal of radioactive waste at sea

by the independent RWMAC which pointed out in its report of that year that the radioactivity of the material disposed of in the sea is still only a very small fraction of the upper levels which have been internationally assessed as environmentally acceptable and that the added radioactivity represents an infinitesimal addition to the natural radioactivity present in the sea.

The amount of processed radioactive waste which was to have been dumped at sea in 1983, but which was prevented by the action of the British National Union of Seamen, amounted to about 4000 tonnes almost 90 per cent of which was non-radioactive concrete and steel packaging; this figure represents about 4 per cent of the annual upper tonnage limit set for control purposes at any one dumping site. The alpha-emitting content of this waste represented about 2 per cent of the permissible upper limit whilst the beta–gamma-emitting content represented about 1 per cent.

On recent British sea dumping operations there has been an on-board observer from another country representing the Nuclear Energy Agency of the OECD, and also a representative from MAFF who is there to ensure that the dump is carried out in the designated area according to internationally-agreed procedures and that it complies with the terms of the licence granted by MAFF. Furthermore, inspectors from DoE and MAFF examine the waste packaging processes and the packages themselves before they are placed on the ship to ensure that they comply with national and international requirements.

Nevertheless, in spite of all these precautions the very act of dumping radioactive waste in the sea is seen by many people in various parts of the world as being both irresponsible and unnecessary. The controversy was inflamed by the disagreement which occurred between delegates attending the seventh consultative meeting of the so-called London Dumping Convention (more accurately described as 'The 1972 Convention on the Prevention of Marine Pollution by Dumping of Wastes and Other Matters') which was held in February 1983. There were 33 countries represented at this meeting out of the 52 which ratified the recommendations of the first convention held in London in 1972. The disagreement arose when delegates from two Pacific island nations proposed an amendment to the Annexes of the 1972 Convention which would prohibit further dumping of all radioactive wastes at sea. The reason for their proposal was based largely on the claims made in a controversial scientific paper prepared by Professor Jackson Davies of University of California which was tabled to support the proposal, as is required by the rules of the Convention. In essence this paper questioned the validity of the assessments of safety made earlier by scientists from the IAEA and NEA. However, several of the delegates at the convention questioned the conclusions presented in the paper by Prof. Jackson Davies, as did one of the authors cited in the paper who publicly criticized the way in which his own findings had been used.

In order to restore agreement at the Convention the UK delegate proposed that a highly qualified group of experts should be set up to re-examine the scientific factors for and against dumping at sea and that no changes should meanwhile be made to the existing recommendations until delegates had had an opportunity to study the findings of the group; the proposal specified that these should be presented to the Convention within two years since this would also enable the group to examine the results of current IAEA and NEA studies on the scientific basis of sea dumping which were due for completion within the same timescale. However, although there was general agreement to the UK proposal it was rejected in favour of a resolution introduced by the Spanish delegation which recommended the suspension of all sea-dumping pending the findings of the expert group's report; this was supported by 19 countries with six against and five abstentions.

Delegates at the Convention recognized that the vote in favour of the Spanish motion was not binding on members for the important reason that it was not supported by the scientific evidence required to amend the Convention Annexes.

Subsequent to this convention a draft report was issued, prepared by the United Nations Environmental Programme Technical Group, in which most of the claims made in Professor Jackson Davies' paper were considered to be grossly inaccurate. Also, at a subsequent meeting held in May 1983, which was organized by the NEA at the request of the Spanish delegation, the technical case for the safety of the proposed 1983 dump was again examined and in the presence of other contracting parties the Spanish delegates announced that they were satisfied with the scientific basis on which the 1983 dump was to be conducted and there and then withdrew their objections. These events appeared to have little effect on the attitude of the National Union of Seamen which refused to lift the ban it had earlier imposed on the dumping at sea of nuclear waste, thereby forcing the nuclear industry to abandon its proposed 1983 dump.

In April 1984 a review body was set up jointly by the TUC and the Secretary of State for the Environment under the leadership of Professor Fred Holliday, Vice Chancellor of Durham University. The aims of the review were to examine all aspects associated with Britain's practice of dumping radioactive waste at sea and to make recommendations. The report (*Report of the Independent Review of Disposal of Radioactive Waste in the North-East Atlantic* (Chairman Professor F. G. T. Holliday), London: HMSO, December 1984) prepared by the review body was published in December 1984, the gist of which were the recommendations that sea dumping should not be resumed until other reviews had been completed and that the Government should urgently publish a comparative assessment of all disposal and storage options, as called for in the Fifth and Tenth Reports of the Royal Commission on Environmental Pollution. It further recommended that no persistent plastics or other persistent synthetic materials which may float should be included in any future sea dumping operations.

The Government has accepted the recommendations of the Holliday Report, with the exception of that which recommended a marine monitoring programme, which it intends to discuss with other member countries of NEA.

Following the recommendations of the Convention's Seventh Consultative Meeting held in 1983, a panel of 22 international experts, nominated by the IAEA and the International Council of Scientific Unions, prepared a basic document for consideration by an expanded panel including representatives of governments and international organizations. Unfortunately, the experts were unable to reach unanimous agreement on what recommendations should be submitted to the Ninth Consultative Meeting due to be held in the following September, which meant that when the meeting was held the members present would have to reconsider the existing ban on sea disposal of low-level wastes without clear recommendations from its own scientific advisors.

The Ninth meeting went ahead as planned in September 1985 and a resolution proposed by Spain to continue the ban on sea dumping was carried by 25 votes to six with seven abstentions. No time limit was set for the duration of the ban but it was agreed that at future conventions it would be up to those countries who were in favour of sea dumping to provide proof that it would cause no harm to human health or to the environment.

Low-level liquid wastes

This form of waste arises from virtually all establishments handling radioactive materials and takes the form of liquids – usually water – which have become mildly contaminated after having been used, for example, in wash-basins and showers in washrooms and changing rooms or, for example, for flushing out storage tanks, the washing of laboratory glassware and the laundering of protective clothing. The radionuclide content of such liquids will, of course, depend upon the type of work being done in a particular laboratory and can therefore, in theory, contain virtually all known radionuclides, albeit in extremely low concentrations.

Liquid wastes of this type are first treated to remove as much as possible of their radioactive content, after which they are heavily diluted and discharged to the seas, rivers and estuaries or, in the unique case of the Trawsfyndd nuclear power station in North Wales, into a man-made fresh-water lake created specially to provide a source of cooling water for the station's turbine steam condensers. The treatment varies according to what is required and includes such techniques as ion exchange columns, fine linen filters, reverse osmosis and flocculation. Reverse osmosis is a filtration technique widely used in desalination plants producing drinking water from the sea and relies upon the diffusion of a liquid through an ultra-fine membrane. Flocculation is a technique widely used in drinking water treatment plants and involves the removal of solutes by precipitation. In this process finely-divided particles in the liquid are made to coalesce, or coagulate, into groups of larger particles which may then more easily be filtered out of the liquid.

The quantities of low-level liquids which may be discharged to a particular site are closely controlled by the Department of the Environment and the MAFF and vary according to the location and method of disposal. Most nuclear power stations, for example, discharge their low-level liquid wastes via pipelines into the sea or estuaries, whereas those arising from the Harwell Laboratory are discharged into the River Thames; similar wastes from BNFL's fuel manufacturing plant at Springfields are discharged into the tidal reaches of the Ribble estuary. Low-level liquids from the UKAEA's Winfrith and Dounreay sites are discharged via pipelines to the sea, as are those from BNFL's reprocessing plant at Sellafield; those from Sellafield are discharged into the Irish Sea via a 2 km-long pipeline.

The discharge levels permitted by the authorizing Ministries are specified only after careful assessment of all potential mechanisms which might bring about reconcentration of the most hazardous radionuclides in the waste and the routes via which these may then find their way back to human life. Also considered is the radiation dose which might be received by those people most likely to be affected by such an occurrence, that is, the so-called *critical group*. Discharge authorizations are under continuous review by the Ministries and reviewed from time to time.

Since no two discharge sites are the same in all respects it is customary for the authorized discharge levels to be set uniquely for each particular site. Low-level liquid wastes discharged to the River Thames from the Harwell Laboratory, for example, must comply with the following formula:

$$2500[Ra] + 420[\alpha] + 50[Ca + Sr] + [\beta]$$

shall not exceed 740 in any calendar month

Where: [Ra] represents the number of GBq of Radium discharged;

[α] represents the number of GBq of all other α-emitting materials discharged;

[Ca + Sr] represents the number of GBq of radiocalcium and radiostrontium discharged;

[β] represents the number of GBq of all other β-emitting materials discharged, except tritium.

Each term in the formula may be regarded as 'equivalent gigabecquerels' (GBq).

Authorization also permits in addition to the above a monthly maximum discharge of 740 GBq of tritium.

To give some idea of how the amounts of liquid wastes actually discharged to the Thames from the Harwell Laboratory compare with those legally permitted, the figures for the *whole* of 1984 were 703 GBq of ($\alpha + \beta$)-emitting material (excluding tritium), plus 1739 GBq of tritium. These levels represent 8 and 20 per cent, respectively, of the authorized limits.

Dounreay

Discharges of liquid wastes to the sea (via a pipeline) from the UKAEA's Dounreay establishment are authorized by the Scottish Development Department on behalf of the Secretary of State for Scotland and are limited to a total of 222 TBq of ($\alpha + \beta$)- emitting material in any period of three consecutive calendar months, of which not more than 22.2 TBq may be strontium-90 and not more than 2.22 TBq may be α-emitting material.

For comparison, the total amount of liquid waste discharged to the sea from the Dounreay site during the *whole* of 1984 was 52.3 TBq of ($\alpha + \beta$)-emitting material, of which 0.62 TBq were α-active materials and about 13 TBq of Sr-90. The figures corresponding to the largest quantities actually discharged in any three-month period during the year represented 13.4, 15 and 28.2 per cent, respectively, of the authorized limits for $\alpha + \beta$, α-only and Sr-90 materials.

Winfrith

The UKAEA's Winfrith establishment, like that at Dounreay, is a coastal site which also discharges its low-level liquid wastes direct to sea (in this case the English Channel) via a pipeline. However, because of the nature of the sea and local topography the authorized discharge levels differ from those laid down for Dounreay.

For the Winfrith site, authorization permits the discharge to sea of up to 92.5 TBq of radioactive material in each calendar month, provided that the individual contributions to this amount do not exceed 27.75 TBq of ruthenium-106 (Ru-106), 3.7 TBq of Sr-90 and 3.7 TBq of α-emitting material. For comparison, the highest monthly discharge figures for the Winfrith site throughout 1984 represented 18, 1.05, 4.5 and 0.8 per cent, respectively, of the authorized discharge limits for total activity, Ru-106, Sr-90 and α-emitting materials. The total annual discharge for 1984 represented about 10 per cent of the authorized limit, 89 per cent of which was in the form of tritium oxide.

Sellafield

BNFL's Sellafield site in Cumbria discharges its low-level liquid wastes into the Irish Sea, along with much smaller quantities of similar types of waste originating at BNFL's own nuclear power station at Calder Hall and at the UKAEA's nearby Nuclear Power Development Laboratories (NPDL) at Windscale. From a geographical point of view, the Sellafield, Windscale and Calder Hall installations are effectively on the same site.

The liquid wastes are collected from all over the site and after suitable treatment and monitoring are discharged to the sea via a pipeline. The treatment routes include the Site Ion Exchange Effluent Treatment Plant (SIXEP) and the Salt Evaporator Plant which, together with other plant improvements, have significantly reduced the levels of radioactivity in the discharges to sea. All discharges from the Sellafield site are regulated in accordance with limits laid down jointly by DoE and MAFF who also issue the authorizing certificate and carry out monitoring of the local environment independently of that undertaken by BNFL.

The most recent (1986) sea-discharge limits authorized for the Sellafield site came into effect on 1 July 1986 and are considerably lower than those authorized for preceeding years. The authorization document is detailed and lengthy and its full inclusion here is beyond the scope of this book. However, in addition to specifying quarterly and annual discharge limits for 15 separate nuclides, as well as grouped radionuclides, it makes other stipulations including:

1 BNFL shall not, in any two consecutive days, dispose of relevant wastes in which the alpha activity of alpha-emitting radionuclides exceeds 0.3 TBq, or the beta activity of beta-emitting radionuclides exceeds 60 TBq.

2 BNFL shall use the best practical means to limit the activity of relevant waste discharged.

3 BNFL shall notify the Secretary of State for the Environment, and the Minister of Agriculture Fisheries and Food, within a period of 14 days following the end of any calendar year in which the activity of radioactive discharges (arising from 15 specified nuclides) exceeds 70 per cent of the authorized limits and what means the Company has introduced to limit the activity of future discharges.

For comparison, the actual discharges from the Sellafield site of liquid wastes subject to authorization for the whole of 1985 were as follows; the percentages shown alongside each figure represent the corresponding fraction of the authorized limits in force at that time:

		[%]
Strontium-90 (Sr-90)	52 TBq	7
Ruthenium-106 (Ru-106)	81 TBq	5.5
Total beta (β)	587 TBq	8
Total alpha (α)	6 TBq	27

The 1985 discharges were all lower than the new authorized limits which came into force on 1 July 1986.

The total-α and total-β discharges for 1985 were less than half those for 1984 because of improved filtration procedures and modifications to operational management. In fact, the 1985 total-α discharges represented only one-thirtieth, and the total-β one-fifteenth of the peak discharge levels which took place during the 1970s.

In June 1984, BNFL launched a top-priority study into how the company could further reduce its discharge levels of radioactivity to the Irish Sea and is likely to spend £500 million over the next decade in doing so. A good example of the company's commitment to reducing radioactive emissions is the £120 million Site Ion Exchange Plant (SIXEP) which came on-stream in 1985. The plant is expected to reduce annual discharges to sea of caesium to less than 185 TBq and of strontium to less than 18.5 TBq.

The design of the SIXEP plant is based on an ion exchange process in which positively charged ions of radioactive caesium and strontium are extracted from sea-discharge effluent streams and incorporated into the crystal lattice structure of a zeolite material known as clinoptilolite; a rare, naturally-occurring alumino-silicate mineral found mainly in the Mojave desert region of California. In operation, contaminated water from the spent fuel storage ponds is pumped under pressure through a number of sand filters to remove particulate matter. From here the filtered water is passed through what is called a *carbonating tower sump* in which carbon dioxide is arranged to bubble through the water. Doing this neutralizes the alkalinity of the water which would otherwise cause rapid deterioration of the clinoptilolite contained in the next stage of the filtration process.

On leaving the carbonating tower sump the contaminated water is pumped under pressure through two series-connected vessels, each of which contains a bed of clinoptilolite. From here the filtered water is passed, via a proportional sampler, to the final storage tank ready for discharge to sea.

All storage vessels and pipework are constructed from stainless steel and located behind thick concrete walls to provide radiation shielding for the plant operating staff.

Low-level gaseous wastes

Gaseous wastes from nuclear establishments arise from fuel dissolving processes, the cooling of radioactive materials, for example spent fuel stores, and from the ventilation of laboratories and fume cupboards containing volatile radioactive materials.

The most common radionuclides found in gaseous wastes are krypton-85 (^{85}Kr), argon-41 (^{41}Ar), tritium (^{3}H), iodine-129 (^{129}I) and carbon-14 (^{14}C). Other radionuclides such as xenon-133 (^{133}Xe) and iodine-131 (^{131}I) may usually be ignored because of their very short half-lives (3.5 days and 8 days respectively), bearing in mind the relatively long storage times for spent fuel, which is the greatest source of such materials.

All gaseous wastes are filtered and then discharged to the atmosphere via tall stacks at carefully controlled rates which are kept well below the maximum permissible rates specified by the Department of the Environment for such discharges.

The main sources of argon-41 are the very early types of Magnox nuclear power stations in which air is used to cool the steel pressure vessel containing the reactor core. Neutron activation of the non-radioactive argon-40 content of the air at this location gives rise to the radioisotope argon-41 which is then released to the atmosphere. Fortunately, the very short half-life of argon-41 (1.8 hours), plus the fact that there are very few reactors of this type in operation means that this particular radionuclide presents no significant problem as far as environmental pollution is concerned. Later types of Magnox power stations use reactors having concrete pressure vessels which are cooled with water or, as in the AGR, with carbon dioxide (CO_2) gas in a closed circuit.

Most of the radioactivity discharged to the atmosphere in the form of gaseous wastes originates at the reprocessing plant after the spent fuel has been separated from its cladding and dissolved in nitric acid. These operations release the fission product gas krypton-85, along with much smaller quantities of carbon-14, tritium (hydrogen-3) and volatile iodine-129.

Carbon-14 is created from various neutron capture reactions with the oxygen- and nitrogen-impurity contents of the fuel, its cladding and coolant and, in gas-cooled reactors, with the carbon content of the graphite moderator and CO_2 coolant gas. Some carbon-14, in the form of CO_2, also escapes to the atmosphere at the nuclear power station site due to controlled discharges and minor leaks. Most CO_2 released in this way is composed of the naturally-occurring non-radioactive carbon-12 and carbon-13 isotopes but, due to neutron activation in the reactor core, some of these will be transformed into the radioactive carbon-14 isotope having a half-life of 5730 years.

The distribution of krypton-85 (Kr-85) throughout the world due to atmospheric discharges from nuclear plant does not, at present discharge rates, present an environmental problem. The fact that this is so is helped by the relatively short half-life of Kr-85 (10.6 years) which prevents a substantial build-up. However, if the world's nuclear power programme continues to expand at its present rate then it may be necessary, during the early part of the next century, to restrict further discharges and use other means of disposing of this particular nuclide. In anticipation of this requirement many laboratories throughout the world have been researching several methods in which Kr-85 may safely be stored for between 50 and 100 years, until its natural decay renders it harmless.

Two methods of retaining Kr-85 have been studied by the UKAEA, the first being the simple one of storing the gas under pressure in steel containers. The second method is novel and was first developed at the Harwell Laboratory; it involves implanting successive layers of micro-bubbles of Krypton gas into the surface of a copper electrode. Many hundreds of litres of krypton may be stored in this way, safely locked into the atomic structure of the host metal for as long as is required.

The very small quantities of tritium and iodine-129 (I-129) at present produced by the nuclear industry means that these particular nuclides are unlikely to present a serious problem in the foreseeable future; this is especially true for tritium with its relatively short half-life of 12.3 years. In the case of I-129, however, its very much longer half-life of 17 million years means that its releases demand constant scrutiny.

Intermediate-level wastes

Items of solid waste which fall into this category include spent fuel cladding debris (like that shown in Figure 11.5), transport containers, and mechanisms and components which have been contaminated with radioactive material or which have been made radioactive by the process of neutron activation in a reactor core or accelerator target area. Other solids include ion-exchange resins used for the clean-up of reactor coolants, flushing liquids and cooling pond sludges.

Filters and contaminated materials will contain considerable fission product activity whereas those made radioactive by neutron activation will generally contain only radioisotopes of their constituent elements. Neutron activation of the alloy stainless steel, for example, will contain radioisotopes of the elements iron, chromium, cobalt, nickel, carbon and niobium, depending upon its composition. The main problem associated with intermediate-level solid wastes is not so much the level of activity as their very large bulk.

Intermediate-level liquid wastes arise mainly from the fuel dissolving stages at the reprocessing plant and it is here where most intermediate-level wastes of all types are produced. The bulk of these liquid wastes can, after suitable treatment, be classed as low-level liquids and discharged to the environment in the authorized manner. The remaining concentrates may then be stored in high-integrity tanks or incorporated into cements or appropriate matrices. Considerable research is currently devoted in many parts of the world to the immobilization and safe storage of liquid wastes of this type and many methods are being studied including thermosetting resins and polymer-impregnated cements.

Cement is an attractive material for containment since it is strong, inexpensive, non-combustible and unaffected by radiation. Its resistance to leaching by groundwater, although quite high for the actinide elements, is relatively poor for some radionuclides, especially the important element caesium, but this can be overcome by surrounding the cement with a metal drum which will contain these elements for extended periods of time.

When cement is used to incorporate intermediate-level wastes it has been found that some materials affect not only its setting characteristics but also the physical and chemical stability of the solidified waste form. A particularly troublesome type of solid waste is the fuel element cladding debris originating from Magnox, Zircaloy and stainless steel-clad fuels. Such debris is highly radioactive due to tiny fragments of uranium, plutonium and fission products sticking to the inner surfaces of the cladding when it is stripped from the fuel, and also to neutron activation of its alloy constituent elements such as nickel-63 (Ni-63), iron-59 (Fe-59) and cobalt-60 (Co-60). Representative of the levels of radioactivity involved is that arising from spent fuel originating from a thermal reactor and stored for one year. Neutron activation products in the cladding will give rise to about 370 TBq of activity per tonne of fuel processed, which will be accompanied by about 74 TBq of fission product activity; this example assumes that 0.2 per cent of the irradiated spent fuel will adhere to the walls of the stripped cladding. Apart from the damaging effect this radiation might have on the stability of any polymers present in the cement, some of the materials themselves also introduce problems unconnected with radioactivity. Magnox cladding, for example, corrodes in cement and releases hydrogen which can affect the chemical and physical stability of the cement.

An alternative to cement for incorporating intermediate-level wastes is the material bitumen. This has been used for such purposes in Belgium, France and Germany and has a greater resistance to groundwater leaching than has cement. The waste to be incorporated is mixed with hot fluid bitumen which is then allowed to cool and set into the required shape. Its other advantages over cement are its larger capacity for incorporating waste – resulting in physically smaller blocks for a given quantity of waste – plus its lighter weight making for ease of transportation. It is also more able to incorporate 'difficult' materials such as rubber gloves and ion exchange resins which do not readily bond with cement. Its main disadvantages, when compared with cement, are its much reduced radiation shielding ability and its readiness to flow and even burst into flames when hot. For these and other reasons bitumen is not being considered as a suitable material for use in the UK.

The nuclear industry both in Britain and worldwide is devoting considerable effort into researching different materials and various ways of immobilizing intermediate-level wastes for long-term storage and also into developing more efficient processes for the removal of radioactive surface contaminants.

High-level waste

Virtually all high-level (heat-generating) waste is in liquid form and, in Britain, originates at the Sellafield reprocessing plant following the spent fuel dissolution and liquid separation stages. Such waste consists of about

Figure 12.3 Looking inside one of the high-level liquid waste storage tanks under construction at BNFL's reprocessing plant at Sellafield

1100 m^3 of nitric acid impregnated mostly with highly active fission products accompanied by relatively tiny quantities of uranium, plutonium and other transuranic elements such as neptunium, americium and curium. The liquid is reddish in colour with tiny fawn-coloured particles suspended in it and stored in 16 high-integrity stainless steel tanks.

The latest type of storage tank, illustrated in Figure 12.3, has the shape of an upright closed cylinder 6.1 m (20 feet) in height and the same in diameter. Surrounding the bottom and most of the sides of this tank is a second tank, also made of stainless steel, which functions as a water jacket and secondary containment. Water circulated between the two tanks assists in removing heat generated by the highly active liquid stored in the inner tank although most of the heat is removed by a complex assembly of independent cooling coils located within the inner tank and supported by its lid. Figure 12.4 shows this assembly being lowered into position in the storage tank during the construction phase. These two cooling circuits limit the temperature of the highly active liquor to 60°C and so minimize corrosion of the tank and its internal components.

Agitation of the liquid to assist cooling and prevent settling of the radioactive particles is provided by seven jet ballast tanks which force compressed air up through the liquid forcing it to behave as though it were boiling. Six of these tanks are located around the periphery of the storage tank and can be seen in Figure 12.3. The seventh is suspended centrally from the lid of the storage tank and can be seen amid the cooling coil assembly in Figure 12.4.

Both tanks are surrounded by a 2 m (6.5 feet) thick concrete biological shield which is itself lined with stainless steel to provide emergency containment should both tanks develop leaks. A cross-sectional diagram of the storage tanks and shield is illustrated in Figure 12.5. In the unlikely event of a tank developing a leak, its entire contents can quickly be transferred to one which is always kept empty for this purpose, thereby allowing repair work to be undertaken on the defective tank without delay.

Vitrification

High-level liquid waste, now amounting to about 1100 m^3, has been safely stored at Sellafield since reprocessing first started there (when it was known as Windscale) more than 30 years ago. Much smaller quantities arising from the reprocessing of fuel from the PFR and the DIDO and PLUTO reactors at Harwell are similarly stored at Dounreay, which has its own reprocessing plant. Although there is no immediate need to change the present method of storing such waste, it nevertheless requires constant supervision and the eventual replacement of all the tanks and their

Figure 12.4 Cooling coil assembly about to be lowered into a newly constructed high-level liquid waste storage

256

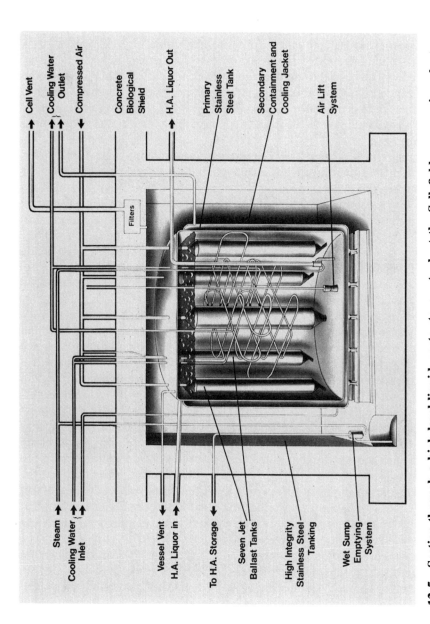

Cell Vent

Cooling Water Outlet

Compressed Air

Concrete Biological Shield

H.A. Liquor Out

Primary Stainless Steel Tank

Secondary Containment and Cooling Jacket

Air Lift System

Filters

Steam

Cooling Water Inlet

Vessel Vent

H.A. Liquor in

To H.A. Storage

Seven Jet Ballast Tanks

High Integrity Stainless Steel Tanking

Wet Sump Emptying System

Figure 12.5   Section through a high-level liquid waste storage tank at the Sellafield reprocessing plant

257

internals. It is for this reason that BNFL and the UKAEA look upon tank storage as no more than an interim measure and that one day all liquid waste will be converted into a solid form which will be more suitable for permanent disposal and will require no supervision.

Considerable research has been devoted in Britain, France and many other countries into the solidification and encapsulation of high-level liquid waste and also into identifying suitable repositories for its eventual storage. The lines of research have been similar everywhere and have concentrated on the conversion of the liquid into an insoluble glass; a process known as *vitrification*.

In the British process a watery slurry of glass-making materials is mixed with a smaller quantity of concentrated liquid waste in a large stainless steel container; the whole contents are then heated to about 1000°C. On cooling, the container holds a solid block of glass weighing about 1.5 tonnes, of which about 25 per cent is nuclear waste. A typical glass composition would be 25 per cent concentrated waste, 42.5 per cent silicon dioxide (SiO_2), 3.5 per cent lithium oxide (Li_2O), 7.2 per cent sodium oxide (Na_2O) and 21.8 per cent borax (B_2O_3). The container, illustrated in Figure 12.6, is approximately 3 m in length, 0.5 m in diameter and has a wall thickness of 12.5 mm.

The British vitrification programme began at Harwell in the late 1950s with the so-called FINGAL (fixation in glass of active liquids) process but was discontinued in 1966 when it was thought that indefinite tank storage would provide a more suitable alternative. The programme was restarted in 1973 under the title HARVEST (highly active residues vitrification engineering study) but which was itself discontinued in 1982 when BNFL decided to opt for a process based on the French vitrification process developed by CEA (the French equivalent of the UKAEA) and known as AVM (atelier de vitrification à Marcoule). The main reasons for adopting the French process are that it had been successfully demonstrated for more than two years using representative quantities of actual waste and that the capital costs of the plant were likely to be significantly lower than those of the Harvest equivalent. A vitrification plant based on the AVM type is at present being constructed at Sellafield and is expected to be operating before 1990.

The essential difference between the Harvest and AVM systems is that Harvest is a batch process whereas AVM is a continuous one. In the Harvest system the waste and the glass-making materials are mixed together as required and poured into a vessel which is also the final containment. The mixing, pouring, evaporation, vitrification and containment is a single-step batch process and is done in large (1.5 tonne)

Figure 12.6   The stainless steel container used in the Harvest experimental vitrification process at the Harwell laboratory

quantities. In the AVM process the liquid waste is fed continuously into a tilted rotary calciner which calcines the liquid, that is converts it into its oxides, and from it produces at its output a continuous stream of brown coloured granules. These are fed directly into an induction-heated pot melter where they are mixed with specially prepared glass and melted to produce a high quality borosilicate glass similar to that produced by the Harvest process. The molten glass undergoes a continuous refining process and is periodically cast into stainless steel containers which hold about 140 kg (0.15 m^3) of glass and which measure 1 m in length and 0.5 m in diameter. The filled containers then have their lids welded in position, after which they are decontaminated and moved out for storage. The whole process is continuous and is illustrated by the diagram illustrated in Figure 12.7.

The calciner consists of a rotating tube located inside a heated furnace. As the liquid waste trickles down the sloping tube the liquid content is evaporated and the radioactive solids appear at the bottom in the form of a dry powder known as 'the frit'. This, along with the glass-making frit, is fed into the melting pot which is heated to about 1150°C. The calcine and the frit fuse together to form vitrified waste. When the molten glass reaches a predetermined level in the melter it is poured into the stainless steel container and the lid is fuse-welded into position.

After the filling stage the containers are washed and then automatically transferred to a storage area where they are stacked ten-deep in a matrix of stainless steel tubes arranged inside a thick-walled concrete cave. The tubes are closed at the bottom and their tops covered by removable shielding plugs which allow access to each tube. Heat generated by the vitrified waste is conveyed to the storage tubes where it is removed by air circulated over the outer surfaces by natural convection. Since this air does not come in contact with the containers which hold the vitrified waste it may safely be discharged directly to the atmosphere without filtering.

The entire AVM vitrification process is designed to reduce as far as possible the need for human intervention, thereby ensuring that operator exposure to radiation is kept to a negligible level.

The AVM process began, as the name suggests, at the French nuclear plant at Marcoule and was itself a development of a full-scale vitrification plant, also at Marcoule, known as PIVER. The PIVER design was based on a batch process very similar to the Harwell-developed Fingal and Harvest processes. During its operating life from 1969 to 1973 the PIVER plant successfully vitrified 12 tonnes of high-level liquid waste and in so doing incorporated 37 000 TBq of radioactivity in 164 containers; these containers are currently stored in air-cooled caves at Marcoule.

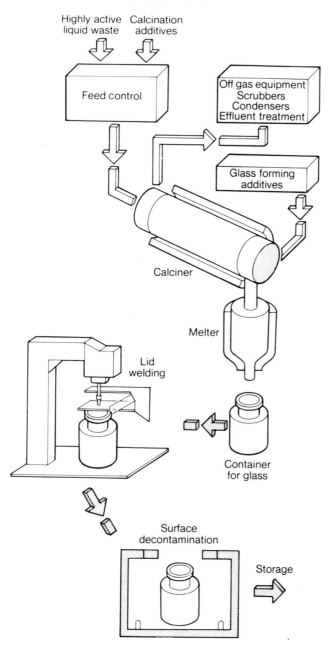

Figure 12.7 Operating principles of the AVM vitrification process

Construction of a full-scale AVM demonstration plant was started at Marcoule in 1971 and experiments using actual radioactive materials began in 1978. Initial experiments were carried out 'cold' over a period of 18 months, during which time every component in the plant was deliberately removed and replaced so as to gain operating experience; for ease of handling the design ensures that no component weighs more than 500 kg.

Reliability of the calciner was verified by operating it for 6000 hours whilst being fed with simulated radioactive liquor. The purpose of the test was to ensure that the calciner could handle the required 40 litres of liquor per hour without 'choking' on its product before it was fed with real radioactive liquor.

A second, much larger, vitrification plant with a capacity of 100 tonnes per year has been built to the AVM design at Cap la Hague, near Cherbourg, the main fuel reprocessing plant in France, and plans exist for extending its capacity in the future.

In November 1980, BNFL announced its intention of constructing a full-scale vitrification plant based on the AVM process and that it would be built at Sellafield under a licence granted by Société Générale pour les Techniques Nouvelles (SGN), the French company which designed the AVM plants at Marcoule and Cap la Hague. The Sellafield plant is at present under construction and is due for completion before the end of this decade.

When completed, the Sellafield plant will have two separate vitrification lines and will be able to vitrify all the liquid waste arising from the reprocessing of up to 2500 tonnes of Magnox fuel, or 800 tonnes of AGR fuel, or 500 tonnes of LWR fuel each year. A full-scale pilot plant is already operating at Sellafield but is being fed with non-radioactive liquor and used to train operating staff and develop new handling techniques.

To give some idea of the rate at which high-level waste is produced, a 1000 MWe nuclear power station operating continuously for one year is responsible for producing about 4 cubic metres of vitrified high-level waste, which is enough to fill 27 AVM containers. If a lifetime of 30 years is assumed for such a station then it would, at the end of its life, if operated continuously (which it never would, of course), give rise to 810 containers of vitrified high-level waste. Demonstrated another way, the photograph shown in Figure 12.8 illustrates how much vitrified high-level waste would be produced by one person if that person's *lifetime consumption* of electricity (domestic, industrial and other) were to be generated *entirely* by nuclear power. The glass block weighs about 0.45 kg (1 pound), of which only 25 per cent is nuclear waste. This example assumes an average life expectancy of 72 years and an average electricity consumption per caput of about 4.4 MWh per year.

Figure 12.8   This block of glass represents the amount of vitrified high-level waste resulting from one person's lifetime use of electricity if all that electricity were to be generated by nuclear power

How good is glass?   Glass is one of nature's most stable materials. It is virtually impermeable to most liquids and has been used by man for thousands of years in the making of drinking vessels, storage containers, windows and decorative ornaments. Naturally-occurring glasses, such as those known as 'obsidians' (volcanic glass of granite composition), have survived millions of years without appreciable deterioration and without protection from the environment. Man-made decorative glassware over 3000 years old can be seen in nearly mint condition in museums throughout the world, still displaying the original colours and patterns. Small wonder then that the nuclear industry should instinctively look towards the use of glass for incorporating its high-level waste, bearing in mind that 300 years or so is about the longest period required for its safe storage.

The most important effects to be considered when glass is intended for the storage of nuclear waste are its leachability and its devitrification sensitivity to high temperatures and nuclear radiation. Leachability is important because of the possible intrusion of flowing water in what was originally a water free underground repository, and the likelihood of this conveying radioactivity to other areas and drinking water supplies.

Sensitivity to high temperatures is important because of the considerable decay heat generated by the fission products incorporated in the glass. Similarly, sensitivity to nuclear radiation is important because of the considerable radiation emitted by the same fission products. Certain types of glass are known to crystallize, or 'devitrify', when subjected to high temperatures, an effect which results in a reduction in the overall stability of the glass.

The effects of nuclear radiation are more complex than that of high temperature. The absorption of alpha particle radiation, for example, gives rise to the creation of helium gas bubbles (remember that an alpha particle is a helium nucleus without its orbital electrons) which, in turn, can cause localized swelling of the glass block and possible damage to its container. You will recall that a similar thing happens to fuel and its cladding when it is subjected to intense radiation in the core of a nuclear reactor. Similarly, the absorption of radiation can cause the build-up of stored energy which, if suddenly released, could cause the glass block to fracture. Radiation can also cause the molecular structure of the glass to be upset by the displacement of atoms when struck by energetic particles.

Needless to say the nuclear industry worldwide has devoted considerable time to researching the above effects and the results so far obtained are most encouraging. In the study of leachability, for example, it has been found that the higher the formation temperature of the glass (usually around 1150°C), the lower is the leaching rate and that results obtained using distilled water, sea water and river water were virtually the same. Experiments have shown that water at 40°C flowing over the type of glass to be used for vitrification of nuclear waste would take about 100 years to leach away 1 mm of the surface – and this ignores the protective presence of the thick-walled stainless steel container which would normally be present and which would have to be leached away first before the glass were exposed to the water. On the debit side it has been found that leaching rate is strongly dependent on the temperature of the flowing water. In fact, between 5°C and 100°C the leaching rate increases by as much as one thousandfold. In one of the many leaching tests carried out at the Harwell Laboratory, boiling water was arranged to flow continuously over a sample of locally-made glass and its loss of weight was measured at weekly intervals to determine the degree of leaching. Typical results from this test indicate that the leaching rate is in the region of 200 μg/cm^2 of surface area per day for the glass, a figure which compares well with the corresponding figure for granite which is only about ten times less.

Sensitivity of the glass to radiation damage caused by beta radiation was measured at Harwell by irradiating a specimen at a single point on its surface with a continuous beam of electrons at 0.5 MeV energy derived from a Van de Graaff particle accelerator. In a period of only one week the specimen received a dose of beta radiation equivalent to that likely to be received by a block of vitrified waste in over 300 years. At the end of the test no observable changes could be detected in the density of the glass or to its leach rate.

Sensitivity of the glass to alpha radiation, much of which is likely to arise from the americium and curium content of vitrified waste, was determined by doping a sample of pure glass with a relatively large amount of the alpha-emitting nuclide plutonium-238. This particular nuclide has a comparatively short half-life of 90 years, which means that it is highly radioactive. After a period of storage of one year, during which time the doped glass received an alpha radiation dose equivalent to that which would be received by a sample containing vitrified waste in about 1000 years, no detectable change was observed in its density and leach rate and almost all of the helium produced was retained by the sample. After a further year of storage – equivalent to 20 000 years of vitrified waste storage – the leach rate of the doped glass had increased by about 50 per cent above its pre-doped value, a value of little significance.

How long must it be stored? When nuclear waste is first separated from the spent fuel it is highly radioactive and generating considerable heat because of the decay of the short-lived fission products; hence the need for cooling of the high-level liquid waste storage tanks at Sellafield. If such waste were to be vitrified shortly after separation then the resulting glass blocks would have to be cooled either by water or by natural or forced-air circulation; this is what is done at the Marcoule vitrification plant.

If blocks of freshly vitrified waste were to be stored, for example, in a deep cavern in granite rock, then considerable overheating would quickly result because of the poor thermal conductivity of the surrounding rock, leading to damage of the stainless steel containers. It is to avoid this sort of thing happening that BNFL has decided to store freshly vitrified waste on the surface for 50 years or so before it is finally consigned to a permanent underground repository. After such a period of storage most of the short-lived fission products will have nearly decayed away, leaving only the radionuclides strontium-90 (Sr-90) and caesium-137 (Cs-137) as the predominant source of heat. Sr-90 is a β-emitter with a half-life of 28 years; Cs-137 is a β,γ-emitter with a half-life of 30 years; the initial radioactivity of the vitrified waste due to these two nuclides will have decayed by 99.9

per cent after about 300 years, so the only radioactivity remaining in the vitrified waste will be that due to the tiny quantities of Cs-137 and Sr-90 still present, plus that due to the small quantities of the long-lived transuranic radionuclides such as americium, neptunium and curium, and traces of uranium and plutonium which were not completely separated when the spent fuel was reprocessed; the total residual radioactivity being little more than that due to the uranium content of the surrounding granite.

Where will it be stored? Present UK policy on the disposal of high-level nuclear waste is that it should first be vitrified (when the BNFL vitrification plant at Sellafield is completed) and the glass blocks then stored on, or close to, the surface for about 50 years or so. What happens after this will depend upon the results obtained from various research programmes going on in various parts of the world.

Some of the many proposals suggested for final disposal include ejection by rocket into space, transmutation of the long-lived alpha-emitting nuclides into shorter fission products in a nuclear reactor, burial in the polar ice caps, burial on or under the ocean bed or in deep stable geological formations. All have been considered seriously.

The technology for disposal to the sun or outer space already exists but the high cost and energy consumed for each launch, plus the need to package the waste into containers which could survive an accidental re-entry, at present outweigh the undoubted advantages of such a method.

Disposal in the ice caps has many attractions, especially in the Antarctic. The sites are well remote from human activities and the technology for emplacement is quite simple. The gentle self-generated heat from each container would ensure that it melted its way through the ice to bedrock below and the hole above it would automatically re-seal after a few minuts due to re-freezing. However, despite these attractions, uncertainties still exist about the physical form of the ice (or water!) at the base of the ice cap and, more importantly, international agreement through the Antarctic Treaty forbids the introduction of radioactivity into the area. This particular option is therefore, at present, not being seriously considered.

Nuclear transmutation – sometimes referred to as nuclear incineration – of the long-lived radionuclides (the actinides), although theoretically feasible, would first require their chemical separation from the predominant fission product content of the waste, after which they would have to be fabricated into suitable fuel elements and placed in the core of a fast reactor; only a fast neutron flux would (and not very efficiently at that) initiate fission in such nuclides as neptunium, americium and curium. Present technology for chemically separating plutonium and residual uranium from spent fuel is highly developed and widely used. Additional separation of the other actinides on an industrial scale would, however, require very complex processes and considerable development time. The repeated reprocessing of the incinerated waste would also itself increase the radiation dose received by operating staff and give rise to even more low-level and intermediate-level wastes for disposal. At an international meeting held in Italy in 1980 and devoted to nuclear transmutation of the actinides, it was generally agreed that the reduction in risk to people brought about by transmutation of the actinides was only marginal when compared with the risks resulting from their retention in high-level waste which had been safely disposed of in geological formations, and that there was therefore little incentive to pursue this particular option.

The geological option is viewed internationally as being a safe and practical method of disposing of vitrified high-level waste and is the one currently pursued in Britain. British research in this field is concentrated on three particular methods: firstly, burial in deep geological formations on land; secondly, burial in ocean bed sediments and, thirdly, deep burial in the ocean bed itself. The attraction of geological disposal is that many regions of the earth are known to have been stable for many millions of years and are therefore unlikely to become unstable during the relatively short time required for the safe storage of nuclear waste.

Stable geological formations on land which are known to be unfissured and free from groundwater are able to provide complete isolation from the biosphere and make ideal repositories for nuclear waste. The types of formations being considered are salt, sedimentary rocks such as clay, shale and slate, and crystalline hard rocks such as granite. Salt formations which would be suitable for waste repositories are generally of two types; those which occur in nearly horizontal thick layers and those which have been forced upwards by earth movements into dome-shaped layers, some of which are many hundreds of metres thick.

The only salt formations in Britain are of the horizontal layer type and are not thought to be thick enough to support a waste repository. The USA, on the other hand, is very rich in both horizontal layered and domed formations and in Germany and Holland salt domes are numerous. Not surprisingly then, these countries devote considerable research into the possible use of such formations for the storage of nuclear waste.

Salt formations make very good waste repositories. Their very existence testifies to a complete absence of groundwater flow and very low moisture content. Salt also has a relatively high thermal conductivity which means that it is able to absorb and safely dissipate the levels of heat likely to be generated by vitrified waste. Another attractive feature is its self-sealing capability should cracks appear in its structure. One disadvantage of salt as a waste repository is that its moisture content is never truly zero, so precautions would have to be taken to ensure that waste containers would be unaffected by the possible accumulation of highly corrosive brine.

Clay and clay-based rocks, known to the geologist as argillaceous rocks, are sedimentary rocks which range in texture from the soft unconsolidated types used for pottery and brick making, to the hard shales and slates which have been formed over long periods at high pressure. The softer clays are virtually impermeable to water flow – hence their use in lining canals many years ago – and, like salt, are self-sealing because of their plasticity. The much harder shales and slates, however, are very easily fissured and cracked by earth movements and are not self-sealing.

Although the rate at which water is able to flow through argillaceous rocks is very low, the residual water content of such rocks is often very high, particularly in clays. It is the presence of this water and what might happen to it if it were subjected to heat generated by high-level nuclear waste which is the subject of current research programmes in Britain. Argillaceous rock formations worthy of study in Britain for potential waste repositories occur in southern Scotland, Cumbria, Gwynedd-Powys, Leicestershire, Nottinghamshire and Somerset.

Rock formations which are receiving considerable attention from the nuclear industry in Britain are the so-called crystalline types represented most familiarly by granite. Apart from its very attractive appearance, granite is a very hard, strong rock and when unfissured is impermeable to water flow; it is also much less affected by heat than are the clay-based rocks. On the debit side it is brittle and may be fractured by earth movements or by expansion caused by excessive heating of large volumes. Its thermal conductivity is also not as high as that of salt, for example, so if it were used as a nuclear waste repository it would conduct away the generated heat only very slowly.

The most promising areas of Britain containing the right type of granite are in Scotland and Northern England. The granites of Devon and Cornwall have so far been found to be unsuitable. Surveys carried out for the nuclear industry by the Institute of Geological Sciences have identified over 100 potentially suitable sites ranging in area from about 5 km^2 to 6000 km^2. Bearing in mind that these areas represent about 16 per cent of the UK land mass and that only about 0.1 km^2 would be required to accommodate all the high-level waste likely to be generated in Britain by the turn of the century, there seems little doubt of a suitable site being found.

A conceptual impression of a high-level waste repository is illustrated in

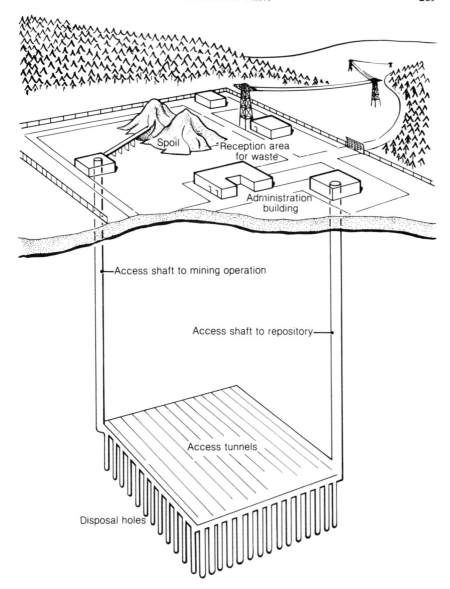

Figure 12.9   Conceptual design of a repository for high-level waste

Figure 12.9. The actual design will depend upon the geological nature of the rock formation, its thermal conductivity and the expected rise in temperature created by the waste being stored. The repository would be placed at least 300 m below the surface so as to be well clear of those rocks closer to the surface which have been fissured during ice ages.

Although no designs have yet been finalized, an upper temperature limit of 100°C has been assumed for the repository so that calculations can be done on the likely spacing of waste containers and the amount of forced air cooling required, if at all. Tentative calculations based upon this criterion show that the heat output from waste containers must not be allowed to exceed 1 kW at the time of admission to the repository and that if waste containers were arranged in a 3-dimensional matrix like that illustrated in Figure 12.9, they would have to be spaced from each other by about 15 m unless forced-air cooling were to be used, in which case the spacing could be considerably less.

The conceptual design of Figure 12.9 consists of a system of parallel tunnels spaced about 20 m apart, each of which gives access to boreholes 300 m deep drilled into the tunnel floors and into which the waste containers would be stacked, one on top of the other. Each container would have packed around it a clay-like material chosen for its ability to resist the ingress of water to the borehole and to adsorb any dissolved radioactive material which might have seeped out of a damaged container. As the boreholes became full so the tunnel would be progressively back-filled with rock spoil produced during excavation until eventually the whole repository would be sealed-off and the surface landscaped.

Deep ocean disposal   Under the terms laid down by the so-called London Dumping Convention it is forbidden for any of the signatory states to dispose of high-level nuclear waste to the oceans of the world. On the other hand, there is considerable opinion among the nuclear power states that the deep ocean could prove to be an even better repository than many land-based alternatives and that, if used as such, it would pose negligible risk to man and to the environment. Such a repository therefore, although unavailable at present, is worthy of detailed scientific investigation, if only to prove its unsuitability.

There are essentially two methods of ocean bed disposal for nuclear waste; one involves the emplacement of high-integrity waste containers deep in the ocean bed sediments, the other involves the emplacement of the containers on the top surface of the sediments, as was done when sea-dumping of low-level waste was practised. Deep sediment disposal is, in many ways, similar to land-based geological disposal with the main difference being that the ocean sediments are permanently water saturated. In this case the first barrier to the release of radioactive material is the container itself and its ability to remain intact for 300 years or so in such an environment. Since this depends upon the thickness and type of material used for the container, plus the integrity of its hermetic seal, it is well within the bounds of present technology to design a suitable container which would remain intact for at least 1000 years in water-saturated sediments. However, since the thermal conductivity of the surrounding sediments is less than ideal it will almost certainly be necessary to assign the waste containers to a term of intermediate storage before they are emplaced in the sediment so as to reduce the possibility of excessive temperature rise, which may damage the containers; the alternative would be to reduce the amount of radioactivity in each container. The other consideration is the ability of the container to withstand continuous nuclear radiation. However, since radiation and the heat it creates are inseparable, effecting a cure for one, for example interim storage or reduced radioactivity content will, in most cases, automatically effect a cure for the other.

The second barrier in deep sediment disposal is the physical form of the waste itself. As discussed earlier in this chapter, high-quality boro-silicate glass is highly resistant to leaching by water – especially low-temperature water – and if the waste is in this form then it should provide a most reliable barrier to the sudden release of radioactive material from a damaged container. In fact, calculations based on laboratory experiments show that vitrified waste in a damaged container in deep sediment would take between 1000 and 5000 years to dissolve fully into the surrounding sediment. The third barrier is the sediment itself. This would act as a very efficient filter to many of the dissolved radionuclides because of its high adsorption properties. The fourth and final barrier is, of course, the ocean since this would cause massive dilution and dispersal of those radionuclides which manage to seep out of the sediment.

Surface sediment disposal   This method of disposal is much simpler and cheaper than the deep sediment method and it has the added advantage of good thermal conductivity, thereby obviating the need for interim storage or a reduction in the radioactive contents of the waste containers. The outstanding disadvantage, of course, is the absence of the very efficient barrier which was provided by the sediment in the deep sediment disposal method.

Where in the ocean?   Should the oceans of the world ever be used for the disposal of high-level nuclear waste, it is natural to ask the question 'where in the oceans would it be placed?' The most obvious places where *not* to place nuclear waste would be those areas which form the boundaries of the seven tectonic plates which make up the earth's crust; these are the most unstable places on earth and are usually associated with earthquakes. From this it is equally obvious that the best sites on the ocean bed would be those which were as close as possible to the central regions of the tectonic plates since these would be the most stable places.

Other areas of the ocean bed which should be avoided are the mid-ocean ridges since these are associated with steep mountain ranges with very little depth of sediment. Similar reasoning applies to the steeply sloping sediments found at the continental edges because these are often subjected to strong turbulent currents which cause underwater avalanches and considerable disturbance to the sediment.

Less obvious areas of the ocean to be avoided are those in the extreme northern and southern latitudes because of the presence of unstable glacier-borne boulders on the ocean bed and the rapid vertical mixing of ocean waters.

Current research aimed at determining the feasibility of using the ocean for high-level waste disposal is being undertaken as an international collaborative venture by Canada, France, Japan, the UK and the USA, under the auspices of the Nuclear Energy Agency of the OECD. The research includes the identification of suitable areas and the long-term safety of any potential repository.

What are others doing?

Although vitrification is currently the most favoured method of immobilizing high-level waste, research is being undertaken in various parts of the world on possible alternatives. These include, for example, work being done in Belgium, Germany and the USA on incorporating beads of glass (or ceramic oxide) containing the waste in a block of metallic lead. The lead, being a metal, has a good thermal conductivity and is therefore able to dissipate the heat generated by the waste; it would also function as a very good leach-resistant barrier.

Work is also being done in the USA on the manufacture of porous glass beads which have an abnormally high silica content. The beads are soaked in high-level liquid nuclear waste for some considerable time, after which they are fired in an oven to seal in the absorbed waste. The beads are then coated with an inactive layer of silica and re-fired to form an impermeable surface layer. The resulting product is found to have a very high leach resistance and shows great promise.

Another very interesting possible alternative to vitrification is the conversion of high-level nuclear waste into a ceramic form composed of highly insoluble substances. Pioneered at the Pennsylvania State University (USA) the idea has been taken up by Professor Ringwood, a geochemist at Melbourne University and Director of the Research School of Earth Sciences there. In his method, which he calls SYNROC (an acronym derived from *syn*thetic *rock*) high-level nuclear waste is mixed with molten minerals of types which are known to be the basic constituents of igneous rock systems. The mixture is then allowed to cool and solidify and the result is a mass of artificial crystalline rock which possesses all the characteristics of naturally-occurring rock.

The individual nuclides which make up the nuclear waste are evenly distributed throughout the SYNROC and are locked into its highly stable crystal lattice structure. The amount of waste contained in SYNROC is typically between 5 per cent and 10 per cent.

The types of rock produced by the SYNROC process are identical to those naturally-occurring ones which are known to have been stable for periods of up to 20 million years in a wide range of geological and geochemical environments. SYNROC therefore appears to be an ideal containment for high-level nuclear waste and if cast into blocks and buried in deep stable rock formations would provide a nearly perfect solution to the problem of high-level waste disposal. Unfortunately, SYNROC research has yet to be taken beyond the laboratory stage and full-size samples containing real – not simulated – radionuclides will have to be produced and subjected to rigorous testing before any conclusions about its suitability can be made. At present there are four types of SYNROC under investigation, all in the laboratory. Typical SYNROC is composed of the minerals perovskite ($Ca.Ti.O_3$), barium felspar ($Ba.Al_2.Si_2.O_8$), hollandite [$(K.Ba_{\frac{1}{2}})Al.Ti_3.O_8$], leucite ($K.Al.Si_2.O_6$) and zirconia ($Zr.O_2$). Each of these belong to a class of minerals which is known to be able to accept radionuclides from high-level waste into their crystal lattices and to retain them tenaciously for many millions of years when subjected to a wide range of geological and geochemical conditions.

HOW MUCH WASTE WILL THERE BE?

The total installed capacity of nuclear generating plant in the UK in 1986 was approximately 10 000 MWe. Since each 1000 MWe nuclear power station is responsible for creating about 4 m^3 of vitrified high-level waste each year of operation, the nuclear generating capacity is currently creating the equivalent of 40 m^3 of vitrified high-level waste per year. If it is (optimistically) assumed that *all* of these stations will continue to operate for a further 14 years from 1986, by the year 2000 they will have created a total of $14 \times 40 = 560$ m^3 of vitrified high-level waste. If it is further assumed that all of this is to be stored in AVM-type containers, remembering that each has a storage capacity of 0.15 m^3, a total of $560 \div 0.15 = 3733$ containers will be required.

To this figure must be added a further 7333 AVM containers to hold the (eventual) mass of vitrified high-level waste arising from the stock of 1100 m^3 (in 1986) of high-level liquid waste at Sellafield, bringing the total to 11 066 containers overall.

These calculations have deliberately ignored the construction of new nuclear stations, two of which are nearing completion, and the possibility of a substantial increase in the nuclear generating capacity by the year 2000. Also ignored is the important fact that many of the present nuclear power stations will be nearing the end of their planned lives by 2000; some were built as long ago as 1956 and are already over 30 years old.

When estimating future nuclear generating capacity, and the ensuing waste which will be created, it is worth remembering that it takes between 8 to 10 years to construct and complete the commissioning trials of a nuclear power station and that it would be most unusual for more than two or three nuclear stations to be under construction at the same time.

The present estimate of there being 11 066 AVM containers by the year 2000 means that if they were stored on their ends, side by side, then the total land area required for their storage would be about 2200 m^2 (each has a diameter of 0.5 m), that is less than half the area of an average sized football pitch! Remember also that these figures represent *all* the high-level waste created in Britain since the first nuclear reactor was built there in 1947 and *all* that expected to be created up to the year 2000 (a 53-year period) from the present nuclear generating capacity.

REFERENCES AND FURTHER READING

Cmnd 884. 'The Control of Radioactive Wastes.' London: HMSO, (1959)

'A Review of Cmnd 884: The Control of Radioactive Wastes.' Department of the Environment, September 1979

'Cmnd 6618. Royal Commission on Environmental Pollution. Sixth Report: Nuclear Power and the Environment.' London: HMSO, (1976)

'The Windscale Inquiry.' Report by Hon. Mr Justice Parker. Presented to the Secretary of State for the Environment on 26 January 1978. London: HMSO

Garner, R. J. 'Environmental Contamination and Grazing Animals.' *Health Physics*, vol. 9, (1963)

'Radioactive Waste Management Advisory Committee. Sixth Annual Report.' London: HMSO, 1985

Hunt, G. J. 'Radioactivity in surface and coastal waters of the British Isles, 1985.' Aquatic Environmental Monitoring Report No. 14. MAFF. Directorate of Fisheries Research. Lowestoft, 1986

BNFL. Health and Safety Directorate. 'Annual Report on Radioactive Discharges and Monitoring of the Environment, 1985.' (Published 1986)

Meggitt, G. C. and Graham, A. C. 'Radioactive Waste Disposal by UKAEA Establishments during 1984 and associated environmental monitoring results.' Safety and Reliability Directorate. UKAEA. Report SRD R388. April 1986

Proceedings of the Seventh Consultative Meeting of the Contracting Parties to the London Dumping Convention. London, 14–18 February 1983, LDC 7/12. (Freely available for inspection from the Library, International Maritime Organisation, 4 Albert Embankment, London SE1 7SR, tel. 01 735-7611.)

13 Decommissioning

There comes a time in the life of any power station, whether it be nuclear, coal or oil-fired, when it becomes technically obsolete or simply worn out and uneconomic to operate any longer. Once this end-of-life stage has been reached the station is shut down and, eventually, the entire plant is taken to pieces; the buildings demolished, the debris disposed of and the site left in a visually acceptable and safe condition. The entire process is known as *decommissioning*.

With the non-nuclear type of power station the decommissioning process is a relatively simple one and could, if thought necessary, be completed in a matter of months. With the nuclear type of station, however, things are not so straightforward because of the residual radioactivity of material in and around the core of the reactor, due to neutron activation and contamination of boilers and other auxilliary plant. For this reason decommissioning of a nuclear station is usually carried out in three distinct stages.

STAGE 1

This involves shutting down the reactor, removing all of its fuel elements and transferring them to the on-site interim storage facility for spent fuel, and thence to Sellafield for reprocessing. The control and shut-down rods are also locked in their in-core positions and their operating mechanisms disconnected to prevent the reactor being started-up again. In this condition the station is completely safe and there are no radiation hazards from the external surfaces of the reactor shield. In fact, with all fuel removed from its core the shut-down reactor is safer than it has been since it first began operating. However, because the on-site interim store will contain spent fuel elements from the decommissioned reactor for up to two years or so, surveillance and routine monitoring of the station is still necessary to ensure normal safety standards.

For example, the Magnox reactor at the Wylfa nuclear power station, when operating at normal power, contains in its core nearly 600 tonnes of uranium fuel operating at surface temperatures in excess of 350°C and with a coolant gas pressure of 400 psi (27 atmospheres). And yet, as illustrated in Figure 8.5, it is safe enough under such conditions for workers to stand on top of the reactor shield during refuelling operations without the need for protective clothing or restrictions on their working time in that location. Clearly, when this same reactor reaches the end of its useful life and is eventually decommissioned to stage 1 there will be no fuel in its core to produce radiation and get hot, nor any pressurized high-temperature coolant gas. *The decommissioned version of the Wylfa reactor will therefore present no risk to anyone going near it or actually standing on its top shield; the same applies to any other type of reactor.*

STAGE 2

This part of the decommissioning process involves the dismantling and removal from site of every structure except the reinforced concrete biological shields of the reactors and the radioactive and contaminated contents of the core areas, for example graphite moderators, core support structures, fuel channel guides, control rods and pipework. Routine monitoring and site surveillance would still be maintained but at a much reduced level. Stage 2 represents about 90 per cent clearance of the site.

STAGE 3

This represents the final stage in the decommissioning process and involves the removal of every structure and piece of material from the site and the return of the site to a 'green field' condition suitable for safe access by the public and for redevelopment. No further monitoring or surveillance is required once stage 3 has been completed.

OPTIONS

The decision of the power station operator to undertake the first of the decommissioning stages will be based upon considerations of safety, cost, technical difficulties and the effect, if any, on the local environment. The completion of stage 1, for example, presents virtually no radiation hazards or technical difficulties but does involve costly long-term monitoring and surveillance and does little to improve the local environment except to reduce the amount of traffic to and from the site. Stage 1, in effect, reduces the station to a care and maintenance rating and requires the presence of a significant number of staff to ensure the well being of the station and the security of its valuable equipment.

The completion of stage 2 represents a major undertaking and involves almost complete clearance of the site. On the other hand, since no radioactive materials are involved, normal demolition practices can be employed and some materials and components can be salvaged for re-use or for sale as scrap. There would be a considerable improvement to the local environment as far as visual appearance is concerned but there would still exist the unattractive structure of the reactor shield and the need for limited surveillance. This would greatly limit the re-sale value of the site for partial redevelopment, especially for domestic purposes. On the other hand the site would be ideal for the construction of a more modern nuclear power station.

Stage 3 could be carried out as an on-going project or deferred for many decades after shut-down to take advantage of the decay of the induced radioactivity of the reactor core materials. Decommissioning, decontamination and disposal of the biological shield and core materials would be a very expensive, time-consuming task involving the employment of skilled staff and enhanced surveillance to ensure the safety of the workforce and members of the general public. Nevertheless, whenever the construction and estimated running costs of a new nuclear power station are being considered, an allowance is usually made for the eventual completion of all three stages of decommissioning. These costs are currently estimated to lie between 10 and 20 per cent of the initial construction cost, depending upon the size and the reactor system.

HOW LONG WILL IT TAKE?

Stage 1 of the decommissioning process starts as soon as the reactor has been shut down and would be completed within 5 to 7 years for a Magnox station. Stage 2 could then follow as soon as the interim store had been emptied; about 1 to three years after shut-down. Stage 3 cannot be attempted for some considerable time after shut-down.

The components and structural materials located in the core of a nuclear reactor are made of mild steel, or stainless steel, and will therefore contain the principle elements iron, cobalt and nickel; cobalt is present as an impurity. These elements, having been subjected to many years of neutron radiation will become intensely radioactive due to the process of neutron activation and many different radioisotopes will be produced. The most prominent of these are iron-55 (Fe-55), cobalt-60 (Co-60) and nickel-63 (Ni-63). Fe-55 has a half-life of 2.6 years and decays by electron capture; the decay is accompanied by the emission of very soft X-rays of about 6 keV. Co-60 has a half-life of 5.27 years and decays by beta emission; the decay is accompanied by beta emission at about 300 keV and two hard gamma rays at 1.17MeV and 1.33 MeV. Ni-63 has a half-life of 93 years and decays by beta emission at about 67 keV.

In Magnox reactors and AGRs, also present in the core will be the radionuclide carbon-14 (C-14) formed by neutron activation of the graphite (carbon) moderator. C-14 has a half-life of 5730 years and decays by the emission of a beta particle of about 160 keV. Finally, in reactors of all types there is the presence of the massive concrete biological shield-cum-pressure vessel. However, since concrete is a very efficient absorber of neutrons, most of the neutron activation products will be restricted to about the first 1 metre of thickness of the shield immediately surrounding the core; most of the shield will therefore be non-radioactive.

From an examination of these sources, it is clear that the overall radioactivity present in the steel components of the core is dominated by Ni-63 with its 93-year half-life. On the other hand, because of its long half-life, its specific activity will be quite low and its very weak beta energy (67 keV) will present little hazard to the workers. The short half-life of the Fe-55 radionuclide means that its specific activity will be quite high immediately after shut-down but the extremely soft energy of its X-ray emissions (6 keV) means that this particular radionuclide poses a negligible safety hazard. By far the most important radionuclide from a safety point of view is Co-60 because of the relative high specific activity from its 5.27-year half-life and its highly penetrating gamma radiation.

Knowing the specific activity of a particular radionuclide and also the type and energy of its radiation, although important, does not give the health physicist enough information to assess the potential hazards of a particular decommissioning operation. What is equally important to know is just how much of that radionuclide is present. The graphite moderator in Magnox reactors and AGRs, for example, is present in vast quantities – over 2000 tonnes of it in each Magnox reactor core – and is nearly pure carbon. On the other hand, since carbon has a very low neutron capture cross-section (about 0.005 barns – which is why it is used as a moderator) very little C-14 will be produced. Conversely, the amount of stainless steel present in the core region of an AGR is considerably less than the moderator and its cobalt impurity content very small. The neutron capture cross-section of cobalt, however, is quite large (about 37 barns) and the Co-60 radionuclide therefore builds up at a much faster rate from this source than does C-14 from the moderator. To the health physicist, therefore, the Co-60 is by far the more important of the two radionuclides. Bearing in mind that 10 half-life periods are required for 99.9 per cent of the initial radioactivity to decay, then at least 50 years must elapse before the Co-60 content of the steel can be considered to have disappeared.

DECOMMISSIONING OF THE WAGR

Since no full-size nuclear power station has yet been fully decommissioned anywhere in the world it is not possible to describe the practicalities of such an operation. It is, however, possible to examine the plans which have been made for full decommissioning of the UKAEA's prototype AGR at Windscale, the so-called Windscale AGR (WAGR). WAGR operated at high load factor for the 18-year period 1963 to 1981, during which time it supplied 33 MWe continually to the National Grid. WAGR provided the experimental data required by the CEGB and the SSEB to build the much larger commercial AGRs.

Full decommissioning of the WAGR with its graphite moderator and gas coolant will provide valuable practical experience for the day when the first of the commercial Magnox stations have to be decommissioned.

Estimates show that at the time when the WAGR was first shut down, and ignoring the fuel content, the sphere contained about 7400 TBq of radioactivity and that this was shared between the mild steel and stainless steel components, the graphite moderator and very small amounts in the concrete and heat exchangers. About 7 years after shut-down, during which time complete de-fuelling is assumed to have taken place, it is estimated that the radioactivity will have dropped to about 2332 TBq and that this would be shared among the various components mentioned above and in the proportions listed in Table 13.1.

A number of interesting and very important points emerge from the table:

1 Although the stainless steel content represents only 5 per cent of the total mass of material present in the sphere, its radioactivity represents almost 75 per cent of the total.

2 About 97 per cent of all the residual radioactivity is concentrated in the steel materials, even though these represent less than half the total mass.

Table 13.1
Estimated distribution of radioactivity in the WAGR 7 years after shut-down

Material	Mass (t)	Radio- activity (TBq)	Specific activity (GBq/t)	% of total mass	% of total activity
Mild steel	761	524	692	40.3	22.5
Stainless steel	93	1739	18 685	5	74.6
Graphite	283	65	222	15	2.8
Concrete	750	2.8	370	39.7	0.1
Totals	1887	2332	—	100.00	100.00

3   Although the concrete content represents nearly 40 per cent of the total mass of material, its contribution to the total radioactivity is only one-tenth of 1 per cent.

4   The highest concentration (specific activity) of radioactivity is in the stainless steel components.

Stainless steel is thus by far the most potentially dangerous material to be handled during the decommissioning procedure.

The total residual radioactivity drops by a factor of about ten in the first ten years after shut-down and by a further factor of four in the next 20 years. After this period, however, it takes a further 300 years to achieve another factor of ten reduction. There is, therefore, little to be gained from waiting much longer than about 14 years before undertaking stage 3 decommissioning. It was after careful consideration of all these factors that the UKAEA decided to embark on a 10-year decommissioning procedure for the WAGR in order to gain the sort of experience which would be required when it became necessary to decommission the very much larger commercial nuclear power stations. Progress to date is as follows:

1   All fuel elements have been removed from the core and the circuit has been purged of the coolant gas (CO_2). The core is now at normal atmospheric pressure.

2   Empty fuel channels are being used as repositories for scrap which is being broken down and manipulated by the refuelling machine.

3   The coolant gas circulating pumps have been removed and the four heat exchangers have been stripped of their asbestos insulation.

4 The turbo-generator set has been removed and the steam drums, de-aerators and pipes have been disconnected in preparation for removal of the four heat exchangers, two of which have already been jacked up ready for decontamination.

5 Demolition of the core itself and the containment sphere will commence in 1990 and a new waste-handling facility, adjacent to the biological shield of one of the heat exchangers, will be used to handle the resulting debris. Demolition of the core will be undertaken using specially-designed remotely operated machinery.

6 Radiobiological aspects of the work are being investigated, including techniques for the radioactive assay of all items of debris, so as to meet the requirements laid down by various regulatory authorities and transport regulations, and to ensure efficient packaging prior to disposal.

Total waste

It is estimated that completion of stage 3 of the decommissioning procedure for the WAGR will give rise to about 16 000 t of waste material, only about 10 per cent of which (1600 t) will be radioactive. This will fall into the low and intermediate levels of radioactivity and will include no high-level wastes.

About 60 per cent of the radioactive waste is expected to fall into the very low category and be suitable therefore for shallow land burial. The remaining 40 per cent will be consolidated into reinforced concrete boxes measuring about 2.3 metre cube and disposed of by UK Nirex Ltd. The boxes will have a wall thickness of 230 mm and will be clad in rust-proof mild steel. The plan is to place the waste inside the boxes and then pour in a specially-formulated concrete which will fill the empty spaces and displace all the air. The box will then be sealed with a concrete lid having the same thickness as the walls and the base. In this way the radioactive waste is safely and firmly embodied in a solid monolithic block weighing about 50 tonnes maximum.

The radioactive content of each box will be carefully controlled so as to ensure that each box contains no more than 3700 GBq of radioactivity. The boxes will be filled in such a way that material having the highest radioactivity will be placed in the centre. This method of packing will reduce the external radiation to which workers handling the boxes will be subjected. Regulations state that the radiation dose rate measured at a distance of 2 m from the block must not exceed 100 μSv per hour.

14 Nuclear fuels and cladding materials

An essential ingredient in any type of nuclear reactor is the fuel in its core since it is from this that all heat is produced by the reactor. It has been explained in earlier chapters that the fuel must contain sufficient fissile material to form a critical mass and that it must be arranged in a suitable geometric configuration for criticality to be initiated and maintained. The type of fuel used in all power-producing reactors throughout the world is either natural (or enriched) uranium, or a mixture of plutonium and uranium. This chapter is devoted to a more detailed examination of these materials and of some others which have been mentioned before only briefly.

IRRADIATION DAMAGE

One of the things which limits the burn-up and hence the useful time spent by the nuclear fuel in the reactor core is its susceptibility to irradiation by neutrons, gamma rays, beta particles and especially the relatively massive fission products, most of which are more than a hundred times heavier than a neutron.

Fission products fly apart from the fissioned nucleus at very high speeds but are quickly brought to rest by multiple collisions with other atoms within the fuel itself and give up their kinetic energy in the form of heat; this is the main source of heat produced in a nuclear reactor, irrespective of its type. Calculations, verified by experiment, have shown that virtually all fission products are absorbed within the mass of fuel in which they are formed and that, on average, they travel less than one-millionth of a metre ($1\ \mu\text{m}$) before being brought to rest. The fuel content of a nuclear fuel element is therefore subjected to the most intense and potentially most damaging form of radiation present at any point within the reactor core.

Before a fission product is brought to rest it collides with many of the atoms within the fuel, some of which – if struck head-on – may be permanently displaced from their normal positions in the atomic structure in the material, leaving behind a void or 'atomic vacancy'. The displaced atoms themselves then go on at high speed, colliding with other atoms as they do so, until they all come to rest. It is estimated that about one million 'knock-on' atoms are produced in this way by each freshly created fission product and that the zone affected by them – usually in the direction of the original fission product – extends to about one-hundredth of a micrometre, that is 10 nanometres (10 nm). This zone respresents a damaged region of the fuel and is known as a 'fission spike' or 'displacement spike'. The 'damage' appears as localized swelling, cracking and creep within the fuel and if allowed to continue beyond an acceptable level will eventually cause structural damage to the fuel cladding, adding to that already being caused by other processes, including the direct absorption of radiation by the cladding itself. Cladding damage was studied in Chapter 10 (page 204). Prolonged irradiation of the fuel can also lead to adverse changes in its ductility, thermal stability and strength.

OTHER CONSIDERATIONS

Important though it is, resistance to irradiation damage is not the only requirement of materials to be used for nuclear fuel. Some other, equally important requirements are:

1   *Good chemical stability*   This is especially important in the event of a cladding failure where the coolant – which may be a gas, water, or liquid sodium – is able to come in direct contact with the fuel. The possibility of a violent chemical reaction resulting from such a contact is clearly an unacceptable risk in a nuclear reactor.

2   *High thermal conductivity*   This allows the achievement of high power densities, that is the magnitude of thermal power per unit of core volume, and high specific powers, that is the magnitude of thermal power per unit of fuel mass, without excessive temperature gradients being created across the core diameter and the attendant possibility of the centre regions of the core reaching temperatures which are close to the melting point of the fuel. Fuel

possessing a high thermal conductivity leads to a core of small physical size and hence a less expensive reactor.

Another factor which is influenced by thermal conductivity is that known as maximum permissible *linear heat rate*. Linear heat rate is expressed in units of kilowatts per metre (kW/m) and is an indication of the rate at which heat is being generated along the length of the fuel rod or pin. The *maximum permissible* linear heat rate is that which can safely be tolerated without the melting of the central region of the fuel rod. Naturally, for safety reasons, fuel elements are designed to operate at linear heat ratings which are well below this figure. Table 14.1 summarizes the linear heat ratings achieved by the AGRs, PWRs, BWRs and PHWRs, all of which use fuel in the form of uranium dioxide (UO_2) pellets. Since UO_2 fuel pellet melting does not occur until heat ratings approach 70 kW/m it is seen that the fuels used in all the reactors listed in the table operate at ratings well below this figure.

Since the heat generated by a fuel element depends upon its position in the core – being maximum in the geometrical centre and minimum at the top, bottom and side extremities – it is usual to specify both nominal peak and average heat ratings, as has been done in Table 14.1.

3 High melting point   This is a rather obvious requirement which allows the core to operate at a high temperature and, in so doing, bring about a good thermodynamic operating efficiency for the reactor and its turbo-generator set. Less obvious is the requirement for no phase transformations to occur below the melting point of the fuel. Such changes are often accompanied by undesirable effects such as changes in density, leading to

Table 14.1
Linear heat ratings (kW/m)

Reactor type	Nominal peak rating	Average rating through-out core
PHWR	46	30
BWR	44	21
PWR	41	18
AGR	35	18

volumetric swelling of the cladding. This is one of the main disadvantages of pure uranium metal as a fuel, as discussed below.

4   Good physical and mechanical properties   These allow ease of fabrication, and a low coefficient of expansion which reduces the need to accommodate changes in physical size when the cladding material is selected.

5   A high concentration of fissile and fertile atoms   (for example U-235, Pu-239 and U-238) and, ideally, the total absence of any impurity atoms which possess a high thermal neutron capture cross-section. Typical undesirable impurities, and their associated thermal neutron capture cross-sections, expressed in barns (b), are hafnium (105 b) and boron (755 b).

These are formidable requirements and, as might be expected, are at present impossible to satisfy with any known material. Compromises are therefore necessary when selecting nuclear fuel.

URANIUM METAL

Natural uranium in its metallic form was, and still is, widely used as fuel in the world's first generation of nuclear reactors, for example the British Magnox, mainly because of its high thermal conductivity, relative ease of fabrication and its very high fissile/fertile atom concentration. Unfortunately, metallic uranium also possess many serious disadvantages, most of which come about because of the way in which it behaves at various operating temperatures.

Metallic uranium occurs in any of three quite different forms, determined by the temperature prevailing at any given time, each form being defined by the crystalline shape of its atomic structure. The three forms are called 'phases' and are known as the alpha (α) phase, the beta (β) phase and the gamma (γ) phase. The α-phase is characterized by an orthorhombic crystal structure, each cell of which comprises six closely packed atoms. Uranium in this phase is soft and ductile and has a density of 19.04 g/cm^3.

Transition to the β-phase takes place quite rapidly as the temperature is raised above 665°C and the crystal cell structure changes to a more complex tetragonal shape comprising 30 atoms. The uranium in this phase is hard and brittle and has a density of 18.11 g/cm^3.

Transition to the γ-phase is also abrupt and takes place at a temperature of 770°C. In this phase the shape of the crystal cell structure is body-centred cubic comprising 9 atoms and the uranium is very soft; its corresponding density is 18.06 g/cm^3. The γ-phase is maintained for operating temperatures up to the melting point which occurs at 1133°C.

The important point to note about the α- to β-phase transition is that it is accompanied by a significant fall in density, which means that the volume increases whenever this transition occurs. It is for this reason that the metallic uranium used in Magnox reactors is restricted to operating temperatures which keep it well within α-phase.

Another factor, not mentioned above, is that the physical boundaries which make up the three axial directions of each crystal cell do not expand symmetrically with changes in temperature; in fact, while two of the boundaries expand, the third contracts. This results in considerable distortion of the crystal shapes when the uranium is subjected to repeated temperature cycling. The effect is most prevalent with uranium which has a preferred grain orientation and is typified by uranium which has been cold-worked during the fabrication stage. Uranium of this type is also susceptible to irradiation growth at relatively low temperatures and, if the grain size is large, it will also develop a rough irregular surface. The effect of physical distortion due to irradiation and temperature cycling can be virtually eliminated by pre-treating the finished uranium so as to produce a random grain orientation. This is done by first raising its temperature so that it is forced into the β-phase and then rapidly quenching it back into the α-phase. Asymmetric expansion of the grains due to temperature cycling now appears only as surface roughening, which can be reduced by using fine grain material.

The addition of small amounts of a suitable metal such as iron, aluminium, chromium or molybdenum to the uranium whilst it is in the molten state produces a uranium alloy with much improved dimensional stability and a greater resistance to corrosion. Uranium–aluminium alloys are widely used as fuel in high-flux low-temperature research reactors and possess a very high thermal conductivity. Uranium–iron–aluminium alloys are used for Magnox fuel to reduce swelling.

URANIUM DIOXIDE

Uranium dioxide (UO_2) in the form of small ceramic pellets is now the most widely used type of fuel in nuclear power reactors throughout the world, including the British AGR. UO_2 does not experience the phase changes of metallic uranium, it is less susceptible to irradiation damage and its high melting point of 2865°C allows operation at much higher temperatures, leading to better thermo-dynamic efficiencies. It is also chemically inert to attack by hot water, a most valuable characteristic in the event of cladding failure in water-cooled reactors.

The high resistance of UO_2 to irradiation damage means that fuel elements can remain for longer periods in the reactor core, leading to much higher burn-up figures (typically more than 4:1) than are achievable with metallic fuels. Coupled with this is the ability of UO_2 to absorb and retain internally a large proportion of the gaseous fission products produced during burn-up, provided its temperature is kept below about 1000°C. Bearing in mind that about 15 per cent of all freshly produced fission products are gases – principally krypton, xenon and iodine – this is another valuable characteristic of UO_2.

The main disadvantage of UO_2 as a nuclear fuel is its low density; little more than half that of metallic uranium. A low density means fewer fissile atoms per unit volume – leading to a larger core for a given critical mass – and, more importantly, a much lower thermal conductivity; about one-third that of metallic uranium. A low thermal conductivity necessitates a fuel element structure in which the fuel is widely distributed so as to prevent the creation of melting point temperatures at its centre. This is why the UO_2-based fuel elements used in the AGR, PWR, BWR and CANDU reactors take the form of bundles of thin-walled small-diameter metal tubes whereas the metallic fuel of the Magnox takes the form of a thick solid rod. This is also why oxide-type fuel elements are more costly to make than their metallic equivalents, bearing in mind the greater complexity of the fabrication process and the very high manufacturing standards demanded for reactor-grade components.

PLUTONIUM

Plutonium-239 (Pu-239) is a fissile material which can be used as nuclear fuel in both thermal and fast reactors. For reasons previously given, however, plutonium is more efficiently used in the 'fast' type of reactor and preferably when surrounded by a breeding blanket of fertile U-238.

At room temperatures plutonium is a solid metal with a density very similar to that of metallic uranium (19.7 g/cm^3). Unlike uranium, however, it has a very low melting point (640°C) and a much inferior thermal conductivity (about 4.2 W/mK). It is also a highly toxic material – potentially more so than arsenic and potassium cyanide – and requires special handling techniques when it is being machined or handled in liquid or powder form.

Like uranium, plutonium also exhibits different properties at different operating temperatures but in this case there are six operating phases, compared with the three associated with uranium, and each corresponds to a different crystalline structure. The six phases are known as α, β, γ, delta (δ), delta-dash (δ') and epsilon (ϵ). Table 14.2 summarizes the transition temperatures of the phases (for increasing temperature) of plutonium and

Table 14.2
Phase-related properties of plutonium

Phase	Transition temperature (°C)	Density (g/cm³)	Polarity of thermal expansion coefficient
α	–	19.7	+
β	120	17.8	+
γ	206	17.2	+
δ	319	15.9	–
δ'	451	16.0	–
ϵ	476	16.5	+

corresponding density values, plus an indication of the polarity of the thermal expansion coefficients.

The most unusual feature to emerge from the table is the reversal in polarity of the thermal expansion coefficients for the two delta phases. In fact, the negative coefficient of the δ'-phase is two and a half times greater than the largest of the positive coefficients possessed by the α-, β-, γ- and ϵ-phases which differ by no more than 27 per cent from one another. Another feature to note is the very large reduction in density which occurs between the α- and δ-phases; a reduction which is much greater than that which occurs between the uranium phases over a much wider temperature range. Note also that having fallen to a minimum value corresponding to the δ-phase, the density then starts increasing again as the operating temperature forces the material into the δ'- and ϵ-phases. *It is these wide variations in density and irregular thermal expansion coefficients which preclude the use of metallic plutonium as a nuclear fuel.*

MIXED OXIDE FUEL

The conversion of metallic plutonium to its dioxide form (PuO_2) brings about a greatly improved increase in its melting point temperature (from 640°C to 2400°C) and obviates the problems associated with phase changes and varying thermal expansion coefficients. When used as the fissile content of fast reactor fuel, however, PuO_2 usually appears as a 20–27 per cent contribution to a homogeneous mixture of PuO_2 and UO_2 (PuO_2/UO_2) and takes the form of small ceramic pellets encased in bundles of

Table 14.3
Some properties of widely-used reactor fuels

Property	U-metal	UO_2	Pu-metal	PuO_2/UO_2
Density (g/cm^3)	19.04	10.96	19.7	10.7
Melting point (°C)	1133	2865	640	2800
Thermal conductivity (W/mK)	25	8.4	4.2	4.1

long, thin-walled stainless steel tubes. It is the relatively poor thermal conductivity of the mixed oxides which necessitates a dispersed type of fuel element, as was the case with the UO_2 fuels used in thermal reactors.

Table 14.3 summarizes some of the properties of the different types of reactor fuels discussed in this chapter. The values specified, although individually correct, should not be used for accurate comparisons since it is not always possible to compare like with like, especially with regard to temperature.

The density value specified for UO_2 (10.96 g/cm^3) is theoretically achievable; however, practical values depend upon the fabrication process and are typically 10.7 g/cm^3.

FUEL CLADDING

Everything which has been said so far about the desirable characteristics of nuclear fuel, with the exception of fissile atom concentration, applies equally well to the materials used for its cladding. To this list, however, must be added a number of other characteristics which are unique to cladding materials.

A very important characteristic of any cladding material is that it should have a very low (preferably zero) neutron capture cross-section so that as many as possible of the neutrons present in the core are able to reach the fuel where they may then contribute to maintaining the all-important chain reaction. A low capture cross-section also means that little neutron-induced radioactivity will be created in the cladding which, in turn, will reduce the amount of intermediate-level waste arising at the reprocessing plant when the cladding is removed.

In a thermal reactor, a low neutron cross-section is of overriding importance for maintaining criticality; in the fast reactor, where the quantity of fissile material is very much greater (typically 25 per cent), it is of less importance.

Another important characteristic of fuel cladding material is that its thermal expansion coefficient should, ideally, match that of the fuel so as to avoid mechanical stress and to maintain good heat transfer by continued physical contact with the fuel.

Cladding material should be ductile, easy to fabricate and weld and should have a high thermal conductivity. A high thermal conductivity ensures good heat transfer from the fuel to the coolant and also avoids thermal stresses in the cladding caused by large temperature differences existing between its inner and outer surfaces. It must also be chemically compatible with both fuel and coolant since it is always in contact with both at the same time.

There are no perfect cladding materials and those being used in present day reactor fuels have been chosen as a best all-round compromise between low neutron cross-section, high melting point, good thermal conductivity, ease of fabrication and good corrosion resistance. Only four possible cladding materials have an adequately low neutron capture cross-section: beryllium, magnesium, zirconium and aluminium.

Beryllium has a neutron cross-section of 9.2 mb, a high melting point (1280°C) and reasonable thermal conductivity (201 W/mK). Unfortunately, it is a very brittle material, not easy to fabricate, and is expensive.

Magnesium has a reasonably low neutron cross-section (63 mb) but its low melting point (650°C) restricts its possible use to low-temperature reactors. Its most important disadvantage, however, is its high oxidation rate in CO_2 coolant gases.

Zirconium has an attractively high melting point (1845°C) but its neutron cross-section (185 mb) is twenty times greater than that of beryllium and its thermal conductivity (23 W/mK) is by far the lowest of the four possible materials. Another disadvantage of zirconium is that its ore always contains the element hafnium (Hf) with a concentration typically between 0.5 to 3 per cent. Hafnium has a very high neutron cross-section (102 b) and considerable care has to be taken during the refining stages to reduce its concentration to around 0.01 per cent; a difficult task because of the close chemical similarity of the two elements.

Aluminium, like magnesium, has a relatively low melting point (660°C) and a neutron cross-section (230 mb) which is even higher than that of zirconium. On the other hand its corrosion resistance is very good, it has a high termal conductivity (237 W/mK) and it is widely used for the cladding of highly-enriched fuel of the type used in low-temperature water-cooled research reactors like DIDO and PLUTO at the Harwell Laboratory. These particular reactors are moderated and cooled by heavy water whose temperature is limited to about 70°C.

Apart from aluminium, none of the four materials described above is used in its elemental form although all are widely used to form alloys with other elements, the most well known being Magnox and Zircaloy. Aluminium is unsuitable at temperatures much above 100°C because its corrosion resistance and mechanical strength begin to deteriorate.

Magnox

The alloy known as Magnox was developed at the Harwell Laboratory as a suitable cladding material for the natural metallic uranium fuel used in all of Britain's first-generation nuclear power reactors. Its elemental composition (by mass) is 98.7 per cent magnesium, 0.8 per cent aluminium and 0.5 per cent beryllium. Magnox has a much better high-temperature corrosion resistance than elemental magnesium and better mechanical properties when operating at reactor core temperatures which typically range from 200°C to 420°C, depending upon the position in the core.

The Magnox cladding used for the type of fuel element illustrated in Figure 6.2 is machined from a single thick-walled hollow extrusion which has four thick ribs running along its entire length and spaced 90° apart. Unwanted material from the ribs is cut away to leave a number (five in the illustration) of channel spacing lugs at various points along its length. The cooling fins which run along the entire length of the cladding are 10 mm in height and are machined by a fly cutter. The minimum cladding thickness at the base of the cooling fins is approximately 2 mm.

After fabrication, one end of the Magnox tube is sealed with a metal cap which is screwed into position and welded tight. The uranium rod is then inserted into the tube, along with a small aluminium oxide (Al_2O_3) insulating disk at each end and the intervening spaces within the tube filled with helium gas. The open end of the tube is then sealed with a metal end cap which is also screwed and welded into position. Finally the exterior of the tube is subjected to a very high pressure so that it is compressed tightly against the surface of the uranium rod; this operation, along with the helium gas filling, ensures good heat transfer between the rod and its cladding.

Zircaloy

Zircaloy is a zirconium alloy which is used almost exclusively in the PWR, BWR and CANDU types of water-cooled nuclear reactors. It is highly resistant to corrosion by hot water and steam and its mechanical strength is considerably greater than that of elemental zirconium. Its strength, however, progressively decreases as its working temperature is increased and it is totally unsuitable for use at the high temperatures associated with gas-cooled reactors. Another disadvantage of Zircaloy is its susceptibility to hydrogen absorption which, in turn, increases its susceptibility to radiation-induced embrittlement.

There are essentially two types of Zircaloy cladding materials; one is known as Zircaloy-2, the other as Zircaloy-4. Zircaloy-2 is used mainly in BWRs and Zircaloy-4 in the PWRs. The difference between the two alloys is quite small: Zircaloy-2 comprises (by mass, per cent) 98.2 zirconium, 1.5 tin, 0.15 iron, 0.1 chromium and 0.05 nickel; Zircaloy-4 comprises 98.2 zirconium, 1.5 tin, 0.2 iron, 0.1 chromium and virtually no nickel. The embrittlement-sensitivity of Zircaloy-4 is significantly less than that of Zircaloy-2, which is the main reason for its development.

The neutron cross-sections of the two Zircaloys are nearly identical at about 193 mb, as are the melting points (about 2090°C) and thermal conductivities (about 15 W/mK).

Zircaloy cladding may be fabricated quite easily using so-called powder-metallurgical techniques but stringent precautions must be taken during the handling of zirconium powder because it is pyrophoric in air, that is it will spontaneously burst into flames if exposed to air.

Stainless steel

Stainless steel is an alloy in which the major constituent is iron; the other constituents vary considerably from one type of stainless steel to another but, in the nuclear power industry, usually consist of one or more of the elements chromium, nickel, carbon, niobium and molybdenum. Stainless steel is widely used in British reactors for the cladding of AGR fuel and worldwide for the cladding of fast-reactor fuels.

Although the type of stainless steel used for cladding has a very high neutron capture cross-section (about 3.2 b), its good mechanical strength, high melting point (about 1690°C), fair thermal conductivity (33 W/mK) and good corrosion resistance makes it a good choice for AGR fuel cladding which has to operate continuously at temperatures in excess of

800°C, that is red heat. On the other hand, it is its high neutron capture cross-section which brings about the need for enriched fuel in the AGR.

The type of stainless steel used for AGR cladding is an austenitic type known as 20/25Nb. This particular alloy comprises 54.5 per cent iron, 20 per cent chromium, 25 per cent nickel and 0.5 per cent niobium. It is produced in the form of high-purity ingots which are forged and extruded into thick-walled hollow tubes. These are then subjected to a number of fabrication processes and end up as thin-walled tubes 1 m long, 15.2 mm in diameter, with a wall thickness of 0.38 mm. A single spiral cooling fin extends the whole length of the tube and has a height of 0.2 mm.

After fabrication the tube is sealed at one end with a welded cap and an Al_2O_3 insulating disk is inserted. This is followed by a stack of enriched UO_2 fuel pellets, followed by another insulating disk. The free spaces within the tube are filled with helium gas and the open end of the tube sealed with a welded end cap. The exterior of the tube is then subjected to high pressure which deforms the tube so that it fits tight against the pellet stack to ensure good heat transfer between fuel and cladding.

Fast-reactor cladding

Stainless steel is also widely used for the cladding of the mixed-oxide (PuO_2/UO_2) type of fuel used in sodium-cooled fast reactors. In the British PFR the cladding takes the form of thin-walled tubes 2.2 m long and (typically) 8 mm in diameter, fabricated from a type of stainless steel known as M316. This particular alloy was developed specially for PFR cladding and comprises 64.8 per cent iron, 17 per cent chromium, 14 per cent nickel, 2.5 per cent molybdenum and 1.7 per cent manganese. It possesses good creep strength and a high resistance to neutron-induced void swelling.

Fast-reactor cladding is particularly prone to void swelling because of the ease with which neutron–proton (n,p) and neutron–alpha (n,α) reactions are initiated in its alloying constituents by the presence of unmoderated high-energy neutrons. Each n,p reaction creates a hydrogen atom which mostly diffuses out of the cladding and is of little consequence. Each n,α reaction, however, creates a helium atom which does not diffuse so readily and which becomes trapped within the cladding in the form of a small bubble, or void, usually at grain boundaries and dislocations in the crystal structure; this phenomenon was described in greater detail in Chapter 10 (page 206). With many millions of n,α reactions taking place during the time spent by the fuel in the core a considerable build-up of helium voids takes place and these give rise to localized swelling of the cladding. Helium

Table 14.4
Some properties of nuclear fuel cladding materials

Material	Thermal neutron cross-section (mb)	Melting point (°C)	Thermal conductivity (W/mK)	Density (g/cm³)
Beryllium	9.2	1280	201	1.85
Magnesium	63	650	156	1.74
Zirconium	185	1845	23	6.5
Aluminium	230	660	237	2.7
Molybdenum	265	2625	138	10.2
Niobium	1150	2470	54	8.6
Magnox	65	647	156	1.75
Zircaloy-2	193	2090	15	6.57
Zircaloy-4	194	2090	15	6.57
Stainless steel				
type 20/25 Nb	320	1690	33	8.0
type M316	290	1690	33	8.0

production also leads to hardening and embrittlement of the cladding and to loss of ductility.

Fast-reactor cladding, like that for AGR fuel is made from ingots which are forged and extruded into thin-walled hollow tubes. Because of the toxicity of PuO_2, however, loading of the mixed-oxide fuel pellets and welding of the final tube seal has to be carried out remotely or in sealed glove boxes.

Table 14.4 summarizes some of the important factors associated with potential and existing nuclear fuel cladding materials. The figures quoted are intended for general interest only since it is not always possible to compare like with like, especially where operating temperatures are concerned.

15 Neutron moderators and absorbers

The function of the moderator in a nuclear reactor is to slow down (moderate the speed of) the high-speed neutrons produced during each fission event so that they can more readily be captured by other fissile nuclei to sustain the chain reaction. The more effective the moderating process is, the less likely the neutrons are to escape from the core and the greater their likelihood is of being captured by fissile nuclei. This leads to good neutron economy and efficient utilization of the fuel.

Neutron moderation is brought about by a succession of multiple collisions between the neutrons and the nuclei of the surrounding moderating material, in a manner not unlike that which takes place between a fast-moving snooker ball and those which are stationary. To ensure a high capture probability in the fissile material the speed of the neutrons should be reduced to about 2200 m/s (equivalent to a kinetic energy of about 0.025 eV) in as short a distance as possible so as to reduce the amount of moderating material required – and hence the cost and physical size of the reactor core. Neutron speeds in the region of 2200 m/s are known as 'thermal' speeds; hence reactors which make use of a moderator (Magnox, AGR, PWR, etc.) are known collectively as 'thermal' reactors.

The potential suitability of any material to be used as a neutron moderator is a combination of many factors, the most important of which is its ability to slow down neutrons by scattering them without actually capturing any. *Every neutron captured by the moderator is lost to the core.* Another important factor is a high density since this ensures that there will be many closely-packed atoms available for the fast moving neutron to collide with, thereby reducing the distance over which it must travel before becoming thermalized.

To understand what makes a good moderator, consider what happens when a fast-moving table tennis ball makes a head-on collision with a snooker ball which, for the purpose of this example, is assumed to be of the same physical size. The snooker ball, being relatively heavy, will hardly move as a result of the impact and will therefore absorb very little energy from the much lighter table tennis ball which will bounce back in the direction from whence it came with very little loss of kinetic energy. In the reverse situation, where a stationary table-tennis ball is struck head-on by a fast-moving snooker ball, the table-tennis ball will be set in motion by the impact and will extract a tiny amount of kinetic energy from the snooker ball. Ignoring frictional and other forms of energy loss, the sum of the kinetic energies possessed by the two moving balls is equal to that originally possessed by the moving snooker ball. For glancing collisions between a stationary ball and one which is moving, the kinetic energy originally possessed by the impacting ball is shared between the two moving balls and in proportions which depend upon their relative masses and the angle of the collision.

Experiments show that maximum energy is transferred between a moving ball and one which is stationary when both balls are of equal mass and when the collision is of the head-on type. This can be demonstrated quite easily on a snooker table by arranging a snooker ball to make a head-on non-skid collision with a similar ball which is stationary. On striking the stationary ball, the moving ball is brought to a dead stop and the stationary ball is driven off in the same direction as the impacting ball and at the same speed. *There has been a complete transfer of energy between the two balls.*

From the snooker ball analogy it is clear that the ideal moderating material for extracting maximum energy from a fast-moving neutron is one whose atomic structure is composed of atoms whose individual mass values are the same as that of the neutron. Remembering that the mass of a neutron is approximately equal to one atomic mass unit (amu), this means that the nucleus of the ideal moderating atom should contain a single nucleon whose mass is also 1 amu. Remembering also that no atomic nucleus can exist as such without the presence of at least one proton, and that the mass of a proton is virtually the same as that of a neutron, it follows that the nucleus of the ideal moderating atom should consist of a single proton only. Only one element fulfils this requirement and that is hydrogen, the lightest of all elements.

HYDROGEN NUCLIDES

Experiments demonstrate that hydrogen does indeed slow down fast-moving neutrons much better than any other type of material and that it does so mainly by forcing the neutrons to make many glancing collisions with the stationary hydrogen atoms. Each collision causes the neutron to be deflected or scattered from its original course and in the process give up some of its kinetic energy. The collisions continue until the neutron has been thermalized (its kinetic energy reduced to about 0.025 eV) where-upon – it is hoped – it will be captured by a fissile nucleus in the nuclear fuel. The ability of the moderator to scatter neutrons in this way is described by its scattering cross-section (σ_s) which, like its absorption cross-section (σ_a), is expressed in barns (b) or millibarns (mb). The greater the magnitude of σ_s, the better is the scattering ability of the material and the more suitable is it likely to be for moderating neutrons.

Hydrogen has an attractively high scattering cross-section of about 38 b. Unfortunately, it also has an unattractively high capture cross-section of about 330 mb. In fact, its capture cross-section is so high that it is physically impossible to sustain a nuclear chain reaction in *any* mass of natural uranium when hydrogen is used as a moderator. In its elemental form hydrogen also has the disadvantage of being a gas which is of low density and highly flammable. In spite of these disadvantages hydrogen is widely used as a moderator when in the form of light water (H_2O) but only with nuclear fuels which have had their fissile concentrations substantially increased to make up for the neutrons lost by the high capture cross-section, for example, BWR and PWR fuels.

On the credit side light water makes a very inexpensive moderating material which, because of its high thermal conductivity, can also function as a very efficient core coolant. Unfortunately, its relatively low boiling point ($100°C$) at normal atmospheric pressure demands that it be subjected to very high pressure in order to suppress boiling and allow it to operate at a temperature which is high enough to bring about an acceptable thermodynamic efficiency.

The next lightest nuclide in the periodic table is heavy hydrogen (^2_1H), more familiarly known as deuterium (D). The deuterium nucleus contains a single proton and a single neutron, giving it an atomic mass equal to twice that of ordinary hydrogen (^1_1H). Although the scattering cross-section of deuterium (about 3.4 b) is very much less than that of hydrogen, its capture cross-section is extremely low, about 0.5 mb, that is nearly seven hundred times less than that of hydrogen. Unfortunately, in its elemental form, deuterium possesses the same disadvantages as ordinary hydrogen in that it is a highly flammable gas of low density. In the form of heavy water, however – more accurately described as deuterium oxide (D_2O) – deuterium is the world's best moderating material.

The very low capture cross-section of heavy water enables a nuclear chain reaction to be sustained in a reactor using natural uranium as fuel, as is done in the highly successful CANDU reactor. The most important disadvantages of heavy water are its very high cost and its relatively low boiling point (about 101°C) which, like light water, brings about the need for pressurization to suppress boiling. A less important disadvantage associated with heavy water is that each time a deuterium atom captures a neutron it is transformed into the hydrogen-3 isotope tritium (^3_1H). Tritium is a radioactive nuclide with a half-life of about 12 years.

Transforming the deuterium content of a heavy water molecule into tritium converts the molecule into tritium oxide or, simply, *tritiated water*. Over a period of time in the reactor core a heavy water moderator becomes increasingly radioactive as its tritium content builds up until it reaches a level which calls for its complete removal and replenishment. The replenishment will be in the form of freshly-made heavy water or heavy water which has been subjected to a de-tritiation process. Fortunately, because of the very low capture cross-section of deuterium, the tritium build-up is quite slow and the need for the removal of the moderator is infrequent.

HELIUM

The next element in the periodic table, in increasing order of mass, is the gas helium (He). Helium is non-flammable, chemically inert and, although its neutron scattering cross-section (about 760 mb) is considerably less than that of hydrogen and deuterium, its capture cross-section is virtually zero. Unfortunately, being a gas of very low density makes it unsuitable as a moderator in its elemental form and, being a noble gas, it cannot be combined with any other element to form a liquid or solid of higher density. It could, of course, be converted to the liquid form to increase its density but this would require its temperature to be maintained at about −270°C; clearly impracticable, remembering that the operating temperature of the core of a nuclear reactor is typically in excess of 600°C!

LITHIUM, BERYLLIUM AND BORON

The next likely contenders in the periodic table, in order of increasing mass, are the elements lithium (Li), beryllium (Be) and boron (B). Lithium has a moderately good scattering cross-section of 1.4 b but a capture cross-section of over 70 b which makes it totally unsuitable for use as a moderator. The same applies to boron with its scattering cross-section of 3.6 b and its massive capture cross-section of 3800 b. Beryllium, on the other hand, is a very good moderating material with a scattering cross-section of 6.1 b and a capture cross-section which, at only 9 mb, is the lowest of all the metals. Its density, at 1.84 g/cm^3, is about three times that of lithium and its melting point is almost 1300°C. Unfortunately, in spite of its many attractive properties, beryllium also has many unattractive ones. It is, for example, very expensive to produce and fabricate and, when in fine powder form, is highly toxic. For these reasons beryllium has not been used in any power-producing reactors although it has been used in some research reactors, usually in the oxide (BeO) form.

CARBON

Moving further up the periodic table, the next element in increasing order of mass is carbon. Carbon is a remarkable material and occurs in many different forms. Graphite and diamond, for example, are two very different forms of the same element carbon and differ only in the shape of their crystalline structure. Diamond is the hardest of all known materials and when cleaned, cut and polished can be formed into a transparent gemstone. Graphite, on the other hand, is a relatively soft jet-black material widely used in pencils and electrical brush gear and as an additive to lubricating oils. It is also extensively used in the nuclear industry as an efficient neutron moderator and reflector and, in some cases, as a load-bearing material.

Although officially classed as a ceramic, graphite possesses many properties akin to a metal. Its electrical and thermal conductivities, for example, are both quite high whereas its ductility is poor and it is porous. Under atmospheric pressure it does not melt but instead sublimes (goes from the solid state directly to a gas) at about 3650°C. Unlike metals it cannot be cast, rolled, forged or welded and yet, unlike ceramics, it can be machined.

The attractive features of graphite as far as nuclear reactors are concerned are its high neutron scattering cross-section (4.8 b) and its low capture cross-section (3.4 mb). These properties, coupled with a moderately high density (about 1.65 g/cm^3), a high thermal conductivity (160 W/mK) and an extremely high sublimation temperature makes graphite a worthy contender to heavy water as a neutron moderator. In fact, graphite and heavy water are the only two moderating materials which are used throughout the world in power-producing reactors fuelled with natural uranium, the British Magnox reactor being one example.

Although graphite occurs in nature, its impurity content in that state is too high for it to be used as a neutron moderator. It is therefore artificially made from petroleum coke, a by-product of petroleum refining. The manufacturing process, known as graphitization, involves mixing ground coke with a coal-tar binder then extruding the mixture into blocks and baking them at temperatures of around 3000°C. The theoretical density of graphite is 2.26 g/cm^3 but in practice it varies from about 1.6 g/cm^3 to 1.7 g/cm^3, depending upon the manufacturing process. This considerable reduction in density is caused by the porous nature of the manufactured product.

Graphite possess some very unusual characteristics. Its tensile strength, for example, actually increases with increasing temperature and reaches a maximum of about twice its normal value in the region of 2500°C. Above this temperature, however, its tensile strength falls steeply. Its thermal expansion coefficient is strongly dependent on the orientation of its individual crystals, a feature which gives it a very high resistance to thermal shock.

A potentially troublesome feature of graphite is that it readily oxidizes in air at temperatures above about 200°C and also with carbon dioxide (CO_2) at temperatures in the region of 600°C to form carbon monoxide (CO). The production of CO in the CO_2 – cooled Magnox and AGR types of reactors is undesirable for two reasons. Firstly, it is being produced at the expense of the load-bearing graphite which is, in effect, being eroded away. Secondly, the intense nuclear radiation in the reactor core causes dissociation of the CO molecules and the creation of elemental carbon and oxygen atoms which are circulated throughout the coolant circuit along with the CO_2. The carbon atoms are deposited en route to form a slowly thickening film on the surfaces of pipework, circulating pumps and structural materials and, more seriously, on the coolant surfaces of the fuel element cladding causing a progressive deterioration in its thermal conductivity and consequent overheating. The oxidation of graphite has been the subject of considerable research in the development of the British AGR programme and has resulted in the addition of inhibitor gases, such as methane, to the CO_2 coolant in an attempt to suppress or greatly reduce the reaction.

The presence of water vapour in a CO_2 coolant gas also introduces problems because of a steam–graphite reaction which results in the creation of CO, CO_2, hydrogen (H_2) and methane (CH_4); all of which are circulated with the coolant throughout the reactor core and heat-exchanger system and give rise to the deposition of elemental carbon.

Other effects of nuclear radiation on graphite depend to a large extent on the associated temperature and involve changes to its physical dimensions and thermal conductivity, and the creation of what is known as *Wigner energy*. At temperatures below about 300°C, irradiation results in simultaneous shrinkage and elongation in two directions. At higher temperatures most types of graphite tend to shrink uniformly. Thermal conductivity of graphite is reduced with both increasing temperature and irradiation, the irradiation having the greatest influence at temperatures below about 100°C. At temperatures in excess of 300°C or so, however, the variations in thermal conductivity caused by irradiation are of little importance.

Wigner energy (named after E. P. Wigner who, in 1942, predicted the phenomenon) is a form of stored energy caused by the build-up of atoms which have been displaced from their normal locations in the crystalline structure of the graphite by prolonged nuclear irradiation. The effect is most pronounced at temperatures below about 100°C and if the energy is not released in a controlled manner it can lead to a sudden release of energy, accompanied by an equally rapid temperature rise which could be as high as 1000°C. Fortunately, the stored energy can be released by an annealing procedure which involves raising the temperature of the graphite until the induced thermal agitation of the atoms is sufficient for them to be able to re-occupy unfilled vacancies in the crystal lattice. The procedure is known as a *Wigner release*.

It is very important when carrying out a Wigner release to ensure that the deliberate temperature rise of the graphite is maintained at a carefully-controlled rate. In the absence of proper control, the heat created by the released energy of the displaced atoms will reinforce that being applied deliberately and cause even more energy to be released, leading to self-accelerated overheating and destruction of the reactor core by fire. In fact it was just such an accident which caused the fire and destruction of the Windscale Pile in Britain in 1957 (see Chapter 18).

Fortunately, the high operating temperature of the Magnox and AGR types of reactor is sufficient to maintain continuous self-annealing of the graphite and prevents the build-up of dangerous levels of Wigner energy.

MODERATOR COMPARISONS

When comparing the suitability of one moderating material with another it is important to consider not only the relative neutron scatter and capture cross-sections but also the number of collisions which must occur within the material before the neutron is finally thermalized, and the distance it travels throughout the material during the moderating process. Other factors which must be taken into account are the amount of kinetic energy lost by the neutron from each collision and the time taken for the moderating process to be completed.

The usual method of expressing moderator usefulness is by use of the factor known as the *moderating ratio*. This is defined by the expression:

$$\xi \frac{\Sigma_s}{\Sigma_a}$$

where ξ is a factor known as the average logarithmic energy decrement per collision; Σ_s and Σ_a are the macroscopic scattering and absorption cross sections, respectively (the microscopic cross-section refers to a single nucleus whereas macroscopic cross section refers to the total cross-section for the number of nuclei contained in a given mass of material).

The number of collisions (Nc) required to reduce the kinetic energy of a neutron from an initial energy of E_i to a final energy of E_f is given by the expression:

$$Nc = \frac{\log_e(^{E_i}/_{E_f})}{\xi}$$

The kinetic energy of neutrons produced by a fission event have an initial energy of about 2 MeV. If thermalization is assumed to be complete at a neutron energy of 0.025 eV, substituting E_i = 2 MeV and E_f = 0.025 eV in the expression for Nc yields:

$$Nc = \frac{18.2}{\xi}$$

The value of ξ for elemental hydrogen is 1; the corresponding value of Nc is therefore 18.2. Thus, using elemental hydrogen as a moderator, a fission neutron will have to make, on average, 18.2 collisions before becoming thermalized. For elemental deuterium the value of ξ is 0.76 and the corresponding value of Nc is 25; this demonstrates the superiority of hydrogen over deuterium as far as moderating ability is concerned. However, for reasons already discussed, deuterium is by far the better moderator of the two elements when capture cross-sections are taken into account.

Table 15.1 summarizes some of the factors discussed for potentially suitable moderating materials. Note that when hydrogen is in the form of

Table 15.1
Properties of the most widely-used neutron moderating materials

Material	Density (g/cm³)	Mass no.	σ_a (b)	σ_s (b)	ξ	Mod. ratio	Nc	L_s (cm)
Light water	1.0	18	0.664	103	0.93	62	19.6	5.2
Heavy water	1.1	20	0.003	13.6	0.51	2100	36	11.4
Beryllium	1.8	9	0.009	6.1	0.207	126	88	10
Graphite	1.6	12	0.0034	4.8	0.158	216	115	19

light water (H_2O) the corresponding value of Nc increases from 18.2 to 19.6. Similarly, with deuterium in the form of heavy water (D_2O) Nc increases from 25 to 36. This is because of the presence of oxygen which has an inferior moderating ability (ξ for oxygen is 0.120). The most obvious point to emerge from the table is the outstanding superiority of heavy water as a moderator when the moderating ratio is considered; nearly 40-times better than light water and almost ten times better than graphite. The values given for Nc assume $E_i = 2$ MeV and $E_f = 0.025$ eV.

The factor L_s is known as the *slowing down length* (expressed in cm) and represents the average distance travelled by a neutron in the moderator between the time of its creation and the time at which it has reached thermal energy (assumed here to be 0.025 eV).

NEUTRON ABSORBERS

In any nuclear reactor it is essential that provision is made for the smooth control of the neutron reactivity – and hence the heat being generated in the core – and for shutting down the reactor in an emergency or during routine operation.

There are essentially three methods of controlling a nuclear reactor; these are: firstly, controlling the rate of neutron generation by temporarily withdrawing or inserting the fuel or the moderator; secondly, controlling the rate at which neutrons are allowed to escape from the core by the adjustment of a removable reflector; thirdly, controlling the rate at which neutrons are absorbed in an absorber inserted in the core. All three methods have been employed in the past but for the large power-producing reactors the third method is the only one which is widely used.

The most obvious requirement of any neutron absorber is that it should possess a high neutron capture cross-section (σ_a). Other important requirements are a melting point which is well above the peak operating temperature of the coolant, and a high resistance to corrosion by the coolant. A less obvious requirement results from what happens to the absorber after prolonged neutron irradiation, and affects its permissible residence time in the core.

Most neutron absorbers (boron-10 being one exception) are transformed, after absorbing a neutron, to the next higher isotopic form in which the atomic mass is increased by one unit. Although this causes no changes to the chemical properties of the element, the capture cross-section of the newly-created isotope may be – and often is – much inferior to that of the parent isotope. This, of course, means that the effectiveness of the absorber is progressively worsening all the time it is being used and that it must eventually be replaced. A good absorber is therefore one in which this form of deterioration is absent, or is of little consequence.

Table 15.2 lists some of the more important properties of the most widely-used materials for neutron absorption in a nuclear reactor. The values given for σ_a are for thermal neutrons at 0.025 eV.

Boron

Boron is widely used as a neutron absorber for reactor control. Its capture cross-section is relatively high (760 b) and its melting point (2300°C) is

Table 15.2
Properties of neutron absorber materials

Material	σ_a (b)	Mean atomic weight	Melting point (°C)
Boron (B)	760	10.81	2300
Cadmium (Cd)	2450	112.40	320
Europium (Eu)	4240	151.96	826
Gadolinium (Gd)	49 000	157.25	1310
Hafnium (Hf)	113	178.49	2220
Indium (In)	195	114.82	157
Silver (Ag)	63	107.87	961

well above any present-day reactor coolant temperatures. It is also an abundant and inexpensive element.

In its elemental form, boron is extremely brittle with a hardness which is comparable with that of diamond. The usual method of fabrication is by use of powder metallurgy and hot pressings at temperatures in the region of 1000°C. Boron is sometimes mixed with carbon to form the refractory compound boron carbide (B_4C) and hot-pressed in graphite moulds to the required shapes, or sintered with an aluminium metal binder to form the compound known as boral.

Many nuclear reactors use control rods formed from boron stainless steel alloys containing between 2 and 4 per cent boron. The control rods used in the British Magnox station at Wylfa, for example, use over 150 boron–steel control rods, each measuring 8 m (26 feet) in length and 76 mm (3 inches) in diameter. A similar method of control is used in the British AGR stations. In the BWR the control rods take the form of cruciform containers filled with stainless steel tubes packed with B_4C powder (see page 165 and Figure 8.25).

Naturally-occurring boron is composed of the two isotopes B-10 and B-11. B-10 has a relative abundance of approximately 20 per cent and a very high thermal neutron capture cross-section of 3840 b. B-11, on the other hand, has the much higher abundance (80 per cent) but a very low capture cross-section of only 0.006 b. It is the overwhelming presence of B-11 which reduces the capture cross-section of naturally-occurring boron to 760 b. It is, of course, possible to enrich the B-10 content of natural boron but this adds to the overall fabrication costs. Nevertheless, boron enriched with the B-10 isotope has been used in the manufacture of boric acid (H_3BO_3) which is used as a chemical shim, and also as a means of auxiliary shut-down by its massive injection into the water coolant.

The term 'chemical shim' describes a method used to provide smooth control over the neutron multiplication in a water-cooled reactor by varying the concentration of boric acid dissolved in the coolant. As the fuel in the core is gradually burnt up, and the neutron reactivity falls, so the concentration of boric acid in the coolant is reduced by passing it through an ion exchange column. The reduction in boron content compensates for the fuel which has been burnt up, thereby allowing the power level of the reactor to be maintained at its preset value. The big advantage of this method of fine control is that it is effective over the whole volume of the core whereas the mechanical equivalent would involve the withdrawal of discrete control rods at relatively few points in the core, the effect of which would be to upset the neutron flux (and consequent power) distribution.

A phenomenon which is peculiar to boron among the most widely used neutron absorbing materials is the creation of helium gas due to a neutron-alpha (n,α) reaction on the B-10 isotope. When B-10 captures a neutron its nucleus ejects an alpha particle (2 protons + 2 neutrons) which, on capturing two free electrons, becomes a helium atom ($^4_2$He). The B-10 isotope in losing two protons and two neutrons from its nucleus is transmuted into lithium-6 ($^6_3$Li), which also has a high capture cross-section of 940 b. However, since the abundance of B-10 in natural boron is only 20 per cent, the problem of helium gas retention becomes significant only in sealed systems containing enriched boron. The B-11 isotope on capturing a neutron undergoes a neutron-gamma (n,γ) reaction and is transformed to the B-12 isotope without nuclear transmutation occurring; this is the reaction experienced by virtually all popular neutron absorbers.

Cadmium

Cadmium is an excellent material for neutron absorption but it possesses a number of serious drawbacks. The first is its relatively low melting temperature (320°C) which is below the coolant temperatures of most power-producing reactors, including the water-cooled types. Nevertheless, cadmium is widely used for the control of low-temperature research reactors such as the DIDO and PLUTO heavy-water reactors at Harwell and for emergency shut-down in the CANDU reactor.

Another draw back of cadmium is its complex isotopic structure. Naturally-occurring cadmium is composed of eight isotopes ranging from Cd-106 to Cd-116. The Cd-113 isotope has the highest neutron capture cross-section ($\sigma_a = 19\ 800$ b) but, unfortunately, its relative abundance is only 12.5 per cent. To make matters worse, Cd-113 is transformed to Cd-114 on capturing a neutron, the capture cross-section of which is considerably less than that of the parent isotope. This can lead to rapid depletion of the main neutron-absorbing isotope and a relatively short residence time in the reactor if it is of the high-flux type.

Other, less important disadvantages of cadmium are its softness and lack of strength which prevent it being used on its own as a reactor control element. However, this is not a serious problem since the cadmium can be fabricated into the required shape and then clad in aluminium or stainless steel to give it the required strength and, at the same time, protect it from possible corrosion. This is the method adopted in the British designed research reactors at Harwell, Lucas Heights (Australia), Risø (Denmark) and Julich (Germany) where the cadmium is fabricated into the shape of a railway signal arm, tapered at one end and then clad in stainless steel. There are six such arms in the Harwell DIDO reactor (Figure 15.1). These are pivoted at their tapered ends; the other ends are allowed to dip into the

Figure 15.1 View inside the core of a DIDO-type research reactor at Harwell
The coarse control arms are seen interleaved between the vertically-mounted fuel elements

311

heavy water between the fuel elements. Since, for safety reasons there are many more arms than are needed for normal control, it is the tips of the arms which do most of the neutron absorption since it is these areas which spend most time in the high-flux central region of the core.

Harwell has extended the useful operating life of the DIDO control arms from the normal one year or so to as much as four years by replacing the tip areas with the rare earth compound europium oxide. Although europium is a very expensive material it has a high neutron cross-section ($\sigma_a = 4240$ b) and the added advantage that its higher-mass isotopes produced by neutron capture also have relatively high cross-sections, thereby reducing the depletion rate of the parent isotope.

The 30 or so shut-down rods used in the CANDU reactor are tubular in shape and fabricated in the form of a stainless steel–cadmium–stainless steel sandwich in which the cadmium is completely encapsulated. Cadmium is also used in four stainless steel-clad control rods in the CANDU reactor. This is permissible not only because of the protective cladding but also because the outlet temperature of the coolant in the CANDU reactor is typically less than 310°C and therefore below the melting temperature of the cadmium.

Indium and silver

Naturally-occurring indium (In) has a capture cross-section of 195 b and is composed of the two isotopes In-115 (95.7 per cent) and In-113 (4.3 per cent). Unfortunately, its low melting temperature of 157°C makes it unsuitable for most types of reactors in its elemental form. It is, however, used for the control of PWR in the form of a silver–indium–cadmium (Ag–In–Cd) alloy, clad in stainless steel, and having a composition which is typically (by weight) 80 per cent Ag, 15 per cent In and 5 per cent Cd. The high silver content of the alloy gives it a melting point of about 750°C, which is considerably higher than that of cadmium and therefore makes it suitable for use at PWR temperatures which are typically 330°C. A modern PWR will contain as much as 3 tonnes of Ag–In–Cd alloy in its control absorbers.

Although the capture cross-section of natural indium is only 195 b, its In-115 isotope (almost 96 per cent abundance) has a capture cross-section of 30 000 b for neutrons whose kinetic energies lie within the so-called epithermal region (close to and a little above 1 eV) and is therefore of great importance in the control of core neutrons. The same applies to silver. Naturally-occurring silver is composed of the two isotopes Ag-107 (52 per cent) and Ag-109 (48 per cent) and has a thermal neutron cross-section of only 63 b. In the epithermal region, however, the Ag-107 isotope has a cross-section of 630 b and the Ag-109 isotope has a cross-section of 12 500 b. The corresponding epithermal cross-section for the Cd-113 isotope is 7300 b.

It is the high values of the epithermal cross-sections of the individual constituents of the Ag–In–Cd alloy which makes it such an attractive control material in water-cooled reactors where a substantial proportion of the core neutrons have energies in the epithermal region.

Hafnium

Hafnium (Hf) is an excellent material for neutron absorption. It has a high melting temperature (2500°C) and adequate mechanical strength and corrosion resistance for it to be used on its own in the core of a nuclear reactor. Unfortunately, its high cost has prevented its widespread adoption in power-producing reactors although it was used for the control of the 60 MWe (later increased to 150 MWe) prototype PWR at Shippingport, USA, which began operation in 1957.

Naturally-occurring hafnium has a moderately high thermal neutron capture cross-section of 113 b and is composed of six isotopes ranging from Hf-174 to Hf-180. The main attraction of hafnium from a neutron absorption viewpoint is the very large capture cross-sections of its isotopes in the epithermal region, making it a suitable material for use in water-cooled reactors. The isotope Hf-177, for example, has an abundance of 18.5 per cent and an epithermal cross-section of 6000 b. Corresponding figures for the Hf-178 isotope are 27 per cent and 10 000 b, respectively; for Hf-179 the figures are 14 per cent and 1100 b.

Yet another attraction of hafnium is the fact that successive neutron captures by Hf-177 leads (by n,γ reactions) to the higher-mass isotopes Hf-178 and Hf-179, both of which have high epithermal capture cross-sections of their own. This chain of events tends to compensate for the progressive loss of Hf-177 and thereby ensures a long useful lifetime in the reactor core.

BURNABLE POISONS

When a newly-commissioned nuclear reactor is first loaded with fresh fuel it is deliberately supplied with a large surplus so that it can operate for long periods between refuelling. The surplus fuel provides what is called *excess reactivity* and can be almost 30 per cent for the PWR which uses enriched fuel. After the first refuelling operation the excess reactivity is never again as high as its initial value because only about one-quarter of the fuel inventory is replaced at each refuelling operation and the reactor is never again filled with all-fresh fuel.

Clearly, on starting up a freshly-fuelled reactor some means must be provided for suppressing the excess reactivity and to allow it to be 'turned on' as required to compensate for that lost by fuel burn-up and fission-product poisoning. Such control could be provided by conventional control rod assemblies but this would demand a very large number of rods, most of which would eventually be withdrawn from the core and remain unused until the reactor was refuelled; their presence would also upset the neutron flux distribution throughout the core. A chemical shim, in the form of boric acid, for example, can provide some means of soaking up excess reactivity but this form of control is only possible in water-cooled and moderated reactors and there are technical reasons associated with the temperature coefficient of the moderator which limits the amount of boric acid which can be tolerated in the water.

The solution to the problem is to add to the freshly-loaded fuel a material whose neutron capture cross-section is initially very high but which deteriorates at a rate which, ideally, matches that of the burn-up. In this way the core reactivity is maintained steady throughout each operating period for a given power output setting.

A material widely used as a burnable poison in PWRs is boron in the form of borosilicate glass tubes distributed among the fuel pins in vacant control rod guide thimbles in the fuel assemblies. The boron-10 isotope is used since it is transmuted to helium as excess neutrons are captured.

An alternative to boron is the rare earth element gadolinium (Gd) in the form of gadolinium oxide (Gd_2O_3), more commonly known as *gadolinia*. Naturally-occurring gadolinium has an extremely high neutron capture cross-section of 49 000 b and a melting temperature of 1310°C. It is composed of seven isotopes ranging from Gd-152 to Gd-160. The two isotopes having the highest cross-sections are Gd-157 (σ_a = 255 000 b) with an abundance of 15.6 per cent, and Gd-155 (σ_a = 61 000 b) with an abundance of 14.5 per cent.

It is the rarity and consequent expense of gadolinium which prevents it being used in bulk quantities for reactor control. It is, however, used as a burnable poison in BWRs and PWRs and in the DIDO and PLUTO research reactors at Harwell. It is also used in the Canadian CANDU reactor as a means of auxiliary shut-down. In this case the gadolinium appears as a liquid in the form of gadolinium nitrate ($Gd(No_3)_3$) which can be injected into the D_2O moderator under high pressure, thereby rendering the moderator virtually ineffective.

The quantity of gadolinium poison added at each fresh fuel loading operation is carefully chosen so that it is completely burnt up at the end of the operating cycle; for BWRs and PWRs this corresponds to a period of about one year.

CONTROL OF THE FAST REACTOR

Much has been said in previous pages about the importance of a high thermal neutron capture cross-section in neutron absorbing materials used for the control of thermal reactors. This is understandable since most of the neutrons present in the core of a thermal reactor have had their kinetic energies reduced to thermal levels (about 0.025 eV) by the presence of the moderator. In a fast reactor, however, no moderator is employed and the core neutrons have energies well in excess of 2 MeV. How then is a fast reactor controlled?

In some early designs of fast reactor, control was achieved by physically moving some of the fuel elements in and out of the core assembly. In the British Dounreay Fast Reactor (DFR), for example, control was exercised by raising and lowering 120 fuel elements arranged in groups of 10 around the perimeter of the core. However, because of the mechanical difficulties associated with this method of control it was superseded in later reactor designs by conventional control rod assemblies. In the British 250 MWe Prototype Fast Reactor (PFR), for example, control is exercised by five control rods formed from tantalum plates clad in stainless steel. These are supplemented by six shut-down sub-assemblies formed from stainless steel tubes filled with boron carbide.

Although the capture cross-section for tantalum is quite low for thermal neutrons ($\sigma_a = 21$ b), its cross-section for fast neutrons ($\sigma_a = 7.3$ b) is considerably higher than that of those materials normally used for the control of thermal reactors; the fast-neutron capture cross-section for boron, for example, is only 2 b. Another attractive feature of tantalum is its high melting temperature of 3000°C, which makes it suitable for operation at fast-reactor coolant temperatures which are usually in excess of 650°C.

Although the neutron cross-sections of the materials used for controlling the fast reactor are considerably lower than those of similar materials used in thermal reactors, this is of little importance because of the small amount of excess reactivity required by a fast reactor. The reasons for this are twofold; firstly, the fission products created during burn-up of the fuel have relatively low capture cross-sections for fast neutrons and therefore have very little effect on core reactivity. Secondly, the burn-up of plutonium created in the U-238 breeder blanket surrounding the core compensates to a large extent for the burn-up of the plutonium originally present in the fuel. Thus, as more plutonium in the fuel is consumed so more plutonium is created in the blanket, much of which is itself consumed *in situ*. The result of this compensation is that the reactivity of the fast reactor core varies relatively slightly between refuelling periods and, correspondingly, only little control of reactivity is required.

16 Reactor coolants

The fundamental purpose of the coolant in any power-producing nuclear reactor is to remove the heat generated in the core and convey it as efficiently as possible to the heat exchanger where it can be used to produce steam for the turbo-generators. In some reactors – the PWR for example – the coolant is also required to function as the moderator.

Some of the many important requirements of a reactor coolant are a high thermal conductivity and high specific heat (large thermal capacity) so that very high flow rates are unnecessary; also a low viscosity so that little energy is demanded from the circulating pumps. In addition to its heat-transfer properties, a coolant should have a very low neutron capture cross-section so that it does not absorb valuable neutrons and, in the process, itself become radioactive. Neutron capture by the coolant is undesirable not only because it lowers the overall neutron economy but also because it raises the radiation levels present in and around the heat exchangers and associated pipework, thereby demanding additional shielding and potentially increasing the radiation dose received by maintenance personnel.

A reactor coolant should also be able to function without the need for pressurization and remain stable (not decompose) under irradiation at high temperatures; it should also be chemically compatible with the materials with which it comes in contact. This latter requirement is very important, not only because of the weakening effects which chemical reactions might have on fuel cladding, control rods and structural members, etc., but also

317

because radioactive corrosion fragments could become dislodged from the reactor core and transported throughout the coolant system and re-located elsewhere; this would increase problems of maintenance and increased radiation dose to personnel. Finally, if the coolant is a liquid or a solid at normal temperatures then it should have a low melting point and a high boiling point. A low melting point is desirable to avoid the possibility of the coolant solidifying when the reactor is shut down and the consequent need for supplementary heating to keep the coolant in the liquid phase. A high boiling point allows the coolant to operate at the high temperatures required to ensure a good thermodynamic efficiency and, most importantly, to do so without the need for pressurization to suppress boiling.

This is a formidable list of requirements and one which is never completely satisfied in any reactor system.

GASEOUS COOLANTS

Gases have many attractive features when used as a reactor coolant. Air, for example, is plentiful and inexpensive and its physical properties are well understood. Air continues to be used in the GLEEP research reactor at Harwell – the first nuclear reactor in Western Europe – and was used in the BEPO reactor at Harwell which operated from 1948 to 1969. Unfortunately, the poor thermal properties of air and its readiness to attack many core materials when used at high temperature has excluded its use in the large power-producing reactors. Air also has a relatively large neutron capture cross-section, due mainly to its nitrogen content (about 78 per cent) which has a cross-section of 1.8 b. This gives rise to a nitrogen-16 (N-16) radioisotope which emits very energetic gamma rays, all in excess of 6 MeV. The much less abundant argon content (about 1 per cent) has a cross-section of 700 mb and gives rise to an argon-41 (Ar-41) radioisotope which emits both beta and gamma radiation in excess of 1 MeV. Although both of these radionuclides have quite short half-lives (7.3 seconds and 1.8 hours, respectively), they could pose a radiation hazard if generated in large quantities and released to the atmosphere.

Hydrogen has very good heat transfer properties – ten times better than air – but its relatively high cross-section (σ_a = 330 mb), coupled with its high cost and potentially explosive nature makes it unsuitable as a coolant. Helium, on the other hand, has virtually zero capture cross-section and thermal properties which, although not as good as hydrogen, are many times better than air; it is also chemically inert and stable at all temperatures. Unfortunately the comparative rarity and hence high cost of helium makes it unsuitable as a coolant.

A gas which is used in all British commercial nuclear reactors and also in many other parts of the world is carbon dioxide (CO_2). Although its thermal properties are inferior to those of hydrogen and helium, but better than air, its capture cross-section (3.4 mb) is considerably lower than hydrogen and air. It is also relatively inexpensive to produce and free from explosive hazards. CO_2 has excellent radiation stability and does not readily react with core materials, provided its operating temperature is kept below about 650°C. At temperatures much in excess of this figure it starts to react with the graphite moderator.

The big disadvantage of all gaseous coolants is their very low density and hence poor heat-transfer capabilities. The overall cooling efficiency of light water, for example, is more than one million times better than any of the gases described here when such factors as density, viscosity, specific heat and thermal conductivity are taken into account. The only way to improve the cooling efficiency of a gas is artificially to increase its density by pressurization, and this is what is done in the British Magnox reactors and AGRs which operate at gas coolant pressures ranging from about 220 psi (15 atmospheres) to 576 psi (40 atmospheres).

LIQUID COOLANTS

Liquid coolants for nuclear reactors have many advantages over the gaseous types, the most important being their much higher densities and better heat-transfer capabilities.

Light water (H_2O) is the most widely-used type of liquid coolant and is used in PWRs and BWRs where it also functions as a moderator. Its popularity derives from its cheapness and abundance in large quantities in many parts of the world and the fact that its heat-transfer properties are well understood and have been used in a variety of applications over many years. Its outstanding disadvantages are its low boiling point (100°C at normal pressure) and a relatively high neutron capture cross-section (664 mb). The low boiling point is overcome by substantial pressurization, either to suppress boiling, as in the PWR, or to raise the boiling temperature to an acceptably high value, as in the BWR. The high capture cross-section is overcome by the use of enriched uranium fuel.

Heavy water (D_2O) possesses virtually the same heat-transfer properties as light water but differs from it in two very important ways. Firstly, its capture cross-section (3 mb) is 220 times less than that of H_2O and, because of this, it can be used with natural (non-enriched) uranium fuel, as is done in the Canadian CANDU reactor. Secondly, unlike light water, D_2O is very expensive.

Although water is only mildly corrosive at moderate temperatures it is significantly corrosive at the level of operating temperatures found in BWRs and PWRs and this fact must be taken into account when materials are being selected for use in the core and heat exchanger regions of a water-cooled reactor. Water is also subjected to radiolytic decomposition under prolonged irradiation giving rise to separated hydrogen (H) in light-water coolants and deuterium (D) in heavy-water coolants. The effect is not serious, however, and the separated elements can be recombined with oxygen on a routine basis whilst the reactor is operating.

Liquid metals

The poor heat-transfer properties of the gaseous coolants and the neutron moderating influence of the water coolants make them unsuitable for use in fast reactors where no moderator is employed and where, because of the absence of a moderator, a very high power density is present in a small-volume core. It is the high power density which presents the biggest problem in selecting a suitable coolant for a fast reactor.

Liquid metals have many advantages over the gaseous and water types of coolant, principally because their thermal conductivities are so much better. The choice, however, is very limited. A metal which might at first be thought suitable is mercury (Hg) since it is in the liquid phase at normal temperatures and its density is more than thirteen times greater than that of water. Unfortunately, its relatively low boiling point (about 357°C) at normal pressure would demand substantial pressurization to suppress boiling, and its capture and scattering cross-sections are unacceptably high.

Another metal which might be thought suitable as a coolant is lead (Pb). This has a boiling point of about 1750°C, which is well above the required operating temperature for a coolant, which means it could be used without pressurization. Unfortunately, its relatively high melting point (about 327°C) means that the reactor coolant would 'go solid' whenever the reactor was shut down for any length of time unless considerable supplementary heating was maintained throughout the entire cooling circuit.

A metal which satisfies virtually all the requirements of a fast-reactor coolant is sodium (Na) and it is this metal which is used in the British PFR and in all major fast-reactor systems throughout the world.

Sodium is a soft silver-white metal with a density of 0.978 g/cm^3 (almost the same as that of water). Its melting point is 97.5°C (slightly below that of water) but its boiling point is 883°C. Its thermal conductivity is 80 W/mK and its specific heat is 1.23 kj/kg.K. These very attractive features, coupled with the fact that its neutron cross-sections are very low, makes liquid sodium a near-perfect coolant. It is not without its disadvantages, however, some of which are quite serious.

The most important disadvantage of sodium is that it reacts violently with water, virtually exploding when coming in contact with it at high speed. Naturally this presents considerable problems when designing coolant-carrying pipework external to the reactor and in the heat exchanger, in particular where the outer surfaces of sodium-carrying pipes are in direct contact with water. Another disadvantage of sodium is that over a period of time some of its atoms are transformed to the Na-24 isotope by the process of neutron capture. This particular isotope is radioactive and emits two very powerful gamma rays each time it decays, thereby adding to the problems of shielding external to the core. Fortunately, since the capture cross-section of the parent isotope is quite low and the half-life of Na-24 is only 15.4 hours the build-up of radioactive sodium is not a serious problem and, in any case, it diminishes quite quickly once the reactor has been shut down.

Another disadvantage of sodium, unconnected with the actual operation of the reactor, is the fact that sodium is optically opaque. This means that in-core maintenance and inspection has to be carried out 'blind' using such techniques as ultrasonics, whereas with water and gaseous coolant, closed circuit TV cameras can be employed.

In spite of these disadvantages, liquid sodium has proved itself over

Table 16.1
Differences between various reactor coolant materials

Material	Density g/cm^3	Melting point °C	Boiling point °C	Thermal conductivity W/mK	Specific heat kJ/kg.K
CO$_2$	0.00198	−57	−78.5	0.0144	0.845
He	0.00018	−272	−268.98	0.143	5.190
H$_2$O	1.00	0	100	0.645	4.181
D$_2$O	1.10	4	101	0.60	4.181
Hg	13.55	−38.87	356.7	8.4	0.138
Pb	11.35	327.4	1750	34.8	0.130
Na	0.978	97.5	883	80	1.23

many years to be a most suitable coolant for the fast reactor. The most important of its disadvantages (sodium–water reaction) has been overcome by the use of two completely separate heat exchangers, as shown in Figure 9.5 and described on page 185.

Table 16.1 lists some of the important parameters of just a few of the many materials which have been considered for use as a reactor coolant. The values presented are intended for comparison and general guidance only since many of them vary considerably with temperature and pressure and, in some cases, it is impossible to compare like with like, for example water and lead at 100°C.

17 Radiation detection and measuring instruments

The detection and measurement of nuclear radiation in the nuclear industry is necessary for two main reasons; the first is to ensure the efficient and safe operation of a particular installation (reactor, reprocessing plant, etc.); the second is to ensure the safety of the nuclear workforce and members of the public who may live nearby, or who may come close to nuclear materials (radioisotopes, etc.) being transported by road, rail, air or sea. Unfortunately, nuclear radiation cannot be detected by any of the human senses and so to detect its presence and measure its effects it is necessary to make use of materials which are sensitive to nuclear radiation and then to display the effects of such radiation in a form capable of human observation.

Some materials are sensitive to virtually all types of nuclear radiation whereas others are affected either partially or predominantly by only one particular type. Since there are many forms of nuclear radiation and many differing requirements it is not surprising to find there exists a wide range of radiation detection devices and monitoring instruments, each designed for a particular application. It is beyond the scope of this book to examine in detail the operation and design of every type of radiation detection device currently used in the nuclear industry; instead brief descriptions will be given of some of the more widely used types.

RADIATION DETECTION

Most radiation-monitoring instruments are used for the monitoring of
individual personnel and of the local environment in which they work. The
types of radiation they are most frequently required to detect and measure
consist of alpha particles (α), beta particles (β), neutrons (n), gamma rays
(γ) and X-rays (X).

It is the differences in the magnitude and polarity of the charges, and the
differences in the individual mass values which enable the different types of
radiation to be detected, identified and measured. These differences were
explained in Chapter 3.

There are many devices capable of detecting nuclear radiation but the
types most widely used in the nuclear industry for this purpose are the
ionization chamber, the Geiger counter, the scintillation counter, the
semiconductor detector, the quartz fibre electroscope and the film badge.

Ionization chambers

An ionization chamber – or ion chamber as it is usually called – consists of a
sealed vessel (the chamber), usually made of metal, containing a quantity
of gas. The gas may be dry air, nitrogen, argon, methane, etc., and it may
be pressurized so as to increase its density. Ion chambers are mostly used
for the detection of high-energy gamma radiation and, when suitably
modified, for the detection of neutrons. An ion chamber, as the name
suggests, detects the presence of radiation passing through it by detecting
the presence of ionization in its gas filling. The ionization is brought about
by multiple collisions between the invading radiation and the orbital
electrons surrounding the atoms of the gas, each collision causing one
electron to become detached.

An atom which has lost one or more of its orbital electrons also loses its
electrical neutrality and acquires a positive charge, the magnitude of which
is numerically equal to the number of negative electrons lost; an ionized
atom thus become a positive ion. By placing a pair of insulated terminals
within the chamber and holding one at a potential which is negative with
respect to the other, the positive ions in the chamber will be attracted to the
negative terminal and the detached negative electrons will be attracted to
the positive terminal. In this way the invading radiation is converted to
electrical signals which may be observed either as individual pulses, with
each pulse representing the detection of one gamma ray, or as a reading on
a meter which represents the total number of gamma rays which have been

detected over a given period of time; this is the so-called integrating mode and is used to indicate total dose received.

In a typical arrangement, one electrode takes the form of a thin wire stretched taut along the axis of the chamber and insulated from it; this is supplied with a positive potential and functions as the anode. The second electrode, the cathode, consists of the metal wall of the chamber itself and is usually connected to earth (zero) potential. In practice, ion chambers are considerably more complex in design than the simple arrangement described here but the principle is the same. Some ion chambers are constructed from materials whose density and radiation 'stopping power' matches closely that of biological tissue. In this way it is possible to estimate the total dose likely to have been absorbed by people working in a given radiation field.

For an ion chamber to detect neutrons it is necessary to fill it with a gas which has a high neutron scattering or neutron capture cross-section. In this way indirect ionization may be induced either from the passage of high speed gas nuclei recoiling from a neutron collision, or from the passage of high speed charged particles emitted following transmutation of the gas atoms. Typical of many different gas fillings are hydrogen, helium-3 and boron-10 in the form of boron trifluoride (BF_3). An alternative to gas filling is to line the inside walls of the chamber with a hydrogen-containing compound such as paraffin wax or polythene. This is often done when the chamber is used for the detection of fast neutrons. For the detection of thermal neutrons the internal electrodes of the chamber may be coated with a thin layer of boron or lithium-6 and the whole chamber completely enclosed within a thick polythene sphere to thermalize the arriving neutrons.

Some ion chambers, operating as *fission chambers*, have their inside walls or electrodes coated with a thin layer of highly enriched (typically 90 per cent) uranium oxide. Such chambers are highly sensitive to thermal neutrons because of the very high fission cross-section of the U-235 content of the uranium coating. Thermalized neutrons on entering the chamber are readily captured by the U-235 atoms which immediately fission and release pairs of relatively massive fission products travelling at high speed. These cause intense ionization along their tracks and give rise to high-amplitude electrical signals at the output terminal of the chamber. Fission chambers are often used to measure low-intensity thermal neutrons in a reactor core in the presence of high-intensity gamma radiation, to which all ionization chambers are sensitive. This is the sort of condition which exists in a reactor immediately after shut-down and during start-up when very few fission neutrons exist but when much gamma radiation is being produced by the fission products in the fuel. The much larger output signals produced from the neutron-initiated fission events makes it a simple matter to select them in preference to the much smaller signals produced by the interfering gamma radiation.

Geiger counters

The Geiger counter is more accurately described as a Geiger–Muller tube, or GM-tube for short, since its name derives from the two German scientists who invented it many years ago. It is one of the simplest, least expensive and most effective of all radiation detection devices and is still widely used for the detection of β-particles, γ-rays and X-rays.

A GM-tube is essentially a gas-filled ion chamber which operates at a voltage which is usually much higher than the conventional ion chamber. It is usually cylindrical in shape and made from metal or glass with metal internal electrodes. Sizes range from typically 20 mm in length $\times$ 6 mm in diameter to 600 mm in length $\times$ 30 mm in diameter.

The high operating voltage of the GM-tube imparts a very high velocity to the ion pairs produced in its gas by ionization, that is the positive ions and negative electrons, so much so in fact that the ions themselves are capable of initiating further ionization by high velocity collisions with other gas molecules. These, in turn, cause even further ionizing events to occur and the process rapidly builds up into what is called an avalanche condition. This results in a massive output signal pulse from each individual event detected and it is this which gives the GM-tube its high sensitivity.

The process whereby additional ionization is initiated from a single ionizing event is known as gas multiplication and is analogous to the amplification which takes place in a transistor or thermionic valve amplifier.

All GM-tubes produce output signals in the form of individual pulses which are of constant amplitude (for a given operating voltage) irrespective of the type of radiation detected. For example, the detection of one β-particle, one γ-ray or one X-ray by the same GM-tube will, for the same operating conditions, result in an identical output pulse from each source of radiation. GM-tubes cannot, therefore, be used to measure or differentiate between different energies of incident radiation, nor be used to identify type. GM-tubes are simply devices used to detect the presence of any type of radiation which is capable of ionizing its gas filling and will produce single output pulses for each event detected.

The output pulses produced by a GM-tube are either averaged and displayed on a calibrated scale of a current-measuring meter to indicate the mean arrival rate of the ionizing events, or they are amplified and used to produce individual clicks in a loudspeaker or to be counted on an electrical impulse counter. It is the ability to count individual pulses in this way which gave rise to the name Geiger counter.

The photograph illlustrated in Figure 17.1 shows a range of miniature GM-tubes manufactured by the Centronics company. They have operating voltages ranging from about 350 V to 1100 V and are widely used in survey instruments, pocket dosimeters, and for health and safety monitoring. They are all metal-walled gas-filled types capable of detecting gamma radiation over the approximate range 20 keV to 1 MeV.

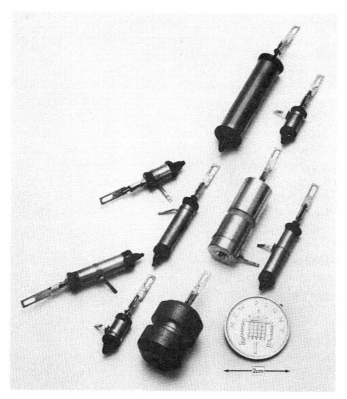

Figure 17.1 A selection from the range of GM-tubes manufactured by the Centronic company (courtesy of Centronic Limited)

The scintillation detector

The ion chamber and GM-tube are gas-filled devices which detect the presence of nuclear radiation from the ionization it produces in the gas filling; such devices therefore convert the radiation directly into equivalent electrical signals. The scintillation detector, on the other hand, is a solid state device which detects the presence of radiation from the tiny flashes of light (scintillations) which are produced whenever a collision occurs between the incident radiation and the counter's atomic structure. The scintillation detector, therefore, is a device which converts nuclear radiation directly into equivalent flashes of light. However, before the very weak flashes of light can be used for serious observation and measurement they must first be converted into equivalent electrical signals and then amplified – often by as much as one million times. The scintillation detector is, in fact, a multi-element device which consists of: firstly, a scintillator for converting the radiation into flashes of light; secondly, a photo-detector which converts the flashes of light into equivalent electrical signals; and, thirdly, an amplifier which raises the amplitude of the electrical signals to a level which can be used for observation on some form of display.

The scintillator itself occurs in many shapes and sizes and is made from a variety of materials to suit the application and the type of radiation it is required to detect. The most widely used material is a crystalline composition of sodium (Na) and iodine (I) which has been 'activated' with thallium (Tl) to produce the desired type of light output. Such a scintillator is known as a sodium iodide crystal and is described by the chemical symbol NaI(Tl).

NaI(Tl) crystals are usually cylindrical in shape and encased in a thin aluminium can which has a glass window at one end; it is through this window that the scintillations are observed by the photo-detector. The inside surfaces of the can are coated with a thin reflecting layer of magnesium oxide so as to improve the overall scintillation efficiency. The most popular sizes of crystal range from about 25 mm to 75 mm in diameter, although much smaller and much larger sizes are made for special applications. Figure 17.2 illustrates a range manufactured by Harshaw.

Sodium iodide crystals have a relatively high density (about 3.7 g/cm^3) and are therefore quite efficient at absorbing energetic X- and gamma radiation; very much better, in fact, than the GM-tube with its low-density gas filling. Although some types of scintillators are used for beta particle detection the majority are used for X- and gamma radiation detection.

Amplification of the light flashes produced by the scintillator takes place in two or more separate stages, the first of which is combined with the

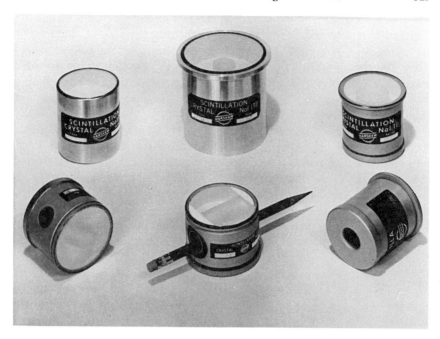

Figure 17.2   A selection of sodium iodide scintillation crystals manufactured by Harshaw (courtesy of Harshaw)

photo-detection process using what is called a photomultiplier tube (usually abbreviated to PM-tube). A PM-tube is essentially a cylindrically-shaped glass envelope – often about the same size as a jam jar – which is closed at one end with a transparent window and at the other end with an arrangement of connecting pins for the internal electrodes.

The inside surface of the window is coated with a thin layer of photosensitive material which emits a short burst of electrons whenever it is exposed to a flash of light; this surface forms what is called the photocathode and is usually connected to earth (zero) potential. The outer surface of the photocathode window is held in firm contact with the transparent window of the scintillator so that the flashes of light which it produces from the incident radiation are 'seen' by the photocathode and converted into equivalent bursts of electrons. The greater the energy of the radiation detected, the greater is the intensity of the flash of light produced and the greater is the quantity of electrons contained within the resulting burst.

The electrons produced by the photocathode are deliberately attracted towards a specially-shaped electrode, known as a dynode, by means of a high positive potential. By the time they reach the dynode the electrons are travelling at a very high speed and possess considerable kinetic energy. They hit the dynode with sufficient force to dislodge many so-called secondary electrons from its specially-prepared surface, the number of which greatly exceeds that of the impacting primary electrons; electron multiplication has therefore taken place.

The secondary electrons released by the dynode are themselves attracted towards a second similarly shaped and carefully-oriented dynode which also carries a high positive potential. The result is that the second dynode releases from its own surface even more secondary electrons due to the impacting secondaries originating from the first dynode, thereby introducing a further stage of electron multiplication.

In a typical PM-tube there are ten dynode multiplying stages, each carrying a higher positive potential than the one immediately below it in the chain. The final electrode is known as the collector, or anode, which is also supplied with a positive potential; its function is to collect the secondary electrons released from the final dynode. The entire multiplication process takes place within a fraction of a microsecond and culminates in a large pulse of current at the collector; this pulse represents the output signal from the PM-tube and its magnitude is representative of the energy of the incident radiation detected by the scintillator. Subsequent amplifying stages, external to the PM-tube, are used to boost the signal still further and to convert it into a form suitable for feeding some form of visual display or an electronic store.

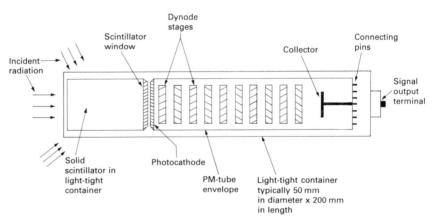

Figure 17.3   Simplified schematic of scintillation detector assembly

Figure 17.3 summarizes the structural features of a typical scintillation detector assembly. For a 10-dynode structure the PM-tube would be supplied with an overall operating potential of about 1500 V. This would be applied directly to the collector electrode, and via a resistive potential divider to the individual dynode stages so as to provide about 100 V to 150 V potential per stage. Such an assembly would yield an overall electron multiplication factor of more than one million.

Figure 17.4 illustrates the range of PM-tubes manufactured by the Thorn EMI company.

Figure 17.4   The range of PM-tubes manufactured by Thorn EMI

The semiconductor detector

A semiconductor detector is, in effect, an ordinary junction diode which has been formed from two small pieces of semiconductor material (usually silicon or germanium). The material used to form one side of the junction will previously have been doped with impurity atoms of a type (for example, arsenic) which causes that material to become what is known as a p-type semiconductor. The material used to form the other side of the junction will previously have been doped with impurity atoms of a type (for example, phosphorus) which cause that material to become an n-type semiconductor. The two pieces of semiconductor material are bonded together by the diffusion process (one is usually diffused on top of the other) to form what is known in the electronics industry as a P-N junction. A semiconductor diode is formed from a single P-N junction, whereas a semiconductor transistor is formed from two P-N junctions.

Semiconductor junction diodes are widely used in the electronic and electrical industries in, for example, radio and TV receivers, electric locomotives and electrical switchgear. The essential difference between the semiconductor diode used in the electronics field and that used as a radiation detector is in the way it is electrically operated. In a television receiver, for example, a semiconductor diode is usually supplied with an electrical operating potential which allows current to flow through the diode, when required to do so. When operated in this way the diode is said to be 'forward-biased' and current flow can be 'turned-on' and 'turned-off' (like water from a tap) by the presence or absence of the forward bias.

A semiconductor diode used as a radiation detector is *always* reverse-biased by an operating potential which prevents the flow of forward current in the diode. The only current which manages to flow under this operating condition is a very tiny current – usually measured in fractions of a millionth of an ampère – known as the reverse current or, sometimes, as the leakage current.

Many years ago it was discovered that when a reverse-biased diode is immersed in a field of nuclear radiation the magnitude of its reverse current increases according to the intensity of the incident radiation. It was further discovered that the increase in current took the form of a large number of individual pulses of current, each corresponding to a single nuclear event, that is the passage through the diode of an alpha or beta particle or a single gamma ray. Even more importantly, it was found that the amplitude of each current pulse was directly related to the energy of the nuclear event which meant that, like the scintillation detector, the semiconductor detector could be used to identify the source of the incident radiation simply by sorting out current pulses of differing amplitude and using them to build up a bar-chart type of image on the screen of a cathode ray tube.

The semiconductor detector, like the scintillation detector, is a solid-state device which, because of its relatively high density, possesses good stopping power for nuclear radiation. Semiconductor detectors are made from either germanium (Ge) or silicon (Si). Because of the relatively low atomic number of silicon ($Z = 14$), silicon detectors are used mainly for the detection of alpha particles, beta particles and low-energy X-radiation. The much higher atomic number of germanium ($Z = 32$), however, makes the germanium detector more suitable for the detection of high-energy gamma radiation, up to about 2 MeV.

Silicon detectors are usually fabricated as physically small devices in a variety of shapes to suit a wide variety of applications. They are used in the medical field as diagnostic probes and for the detection of low-energy X-rays (typically less than 20 keV) encountered in X-ray fluoroscopy. Germanium detectors, on the other hand, are generally fabricated as large-volume devices (typically 250 cm^3) and are widely used for applica-

Figure 17.5   This silicon radiation detector from the Harwell Laboratory is intended for use in alpha-in-air monitoring systems

tions involving high resolution gamma spectrometry. Such detectors would be of cylindrical shape measuring up to about 70 mm × 70 mm and would be cooled by liquid nitrogen to reduce electrical noise and ensure a high resolution. Figure 17.5 illustrates a silicon semiconductor detector from the range of detectors manufactured at the Harwell Laboratory. The detector shown in the photograph has an overall diameter of 33 mm and a sensitive surface area of about 700 mm^2. The diode itself is mounted in a stainless steel container and connected to a 15 V operating potential by way of two electrode connections which protrude out of the back of the container. The detector is used mainly for alpha-in-air monitoring systems.

MEASUREMENT

The quartz fibre electroscope

The quartz fibre electroscope, or QFE as it is more familiarly known, is a radiation monitoring device which has the appearance of a rather fat fountain pen, complete with pocket clip. It is usually worn in the top pocket of a laboratory coat to measure the total radiation dose absorbed by the wearer; for this reason it is also known as a *pocket dosimeter*.

The QFE operates on the same principle as the gold-leaf electroscope – one of the earliest types of radiation measuring devices – and is used exclusively for personnel monitoring. Its principle advantage is that it gives the wearer an immediate indication of the accumulated radiation dose received since the device was last reset to zero.

The QFE functions as a miniature air-filled ionization chamber in which the centre electrode is replaced with a short length of springy quartz fibre which has been coated with a thin film of metal. The metal casing of the QFE forms the outer electrode of the chamber. When the two electrodes of the QFE are connected to a DC voltage of 100 V or so, the capacitance existing between them is charged to the magnitude of the applied voltage and the quartz fibre physically moves to a position which corresponds to zero on a miniature internal scale graduated in units of radiation dose. The movement is caused by mutual repulsion between the quartz fibre and its supporting wire, both of which acquire the same potential during charging. Once the zero position on the scale has been reached the charging voltage is removed.

If the QFE is now subjected to nuclear radiation which is energetic enough to penetrate the outer casing, some ionization of the internal air will occur and some of the electrical charge stored in the internal capacitance will be destroyed. This, in turn, causes the quartz fibre to move away from its fully-charged zero position and to occupy a different position on the scale corresponding to the magnitude of the dose received from the incident radiation. The more prolonged, or more intense, the radiation received, the greater will be the quantity of charge destroyed within the QFE and the greater will be the movement of the fibre across the scale.

Full scale deflection of the fibre across the scale represents complete discharge of the internal capacitance of the QFE and corresponds to an accurately-defined total dose of radiation received. It may take minutes, hours or days to reach full scale deflection, depending upon the intensity of the incident radiation and the total exposure time of the QFE to the radiation.

To read the absorbed dose indicated on the scale the wearer simply holds the QFE towards a source of light and views the scale through a small magnifying lens located at one end (Figure 17.6).

Figure 17.6   The quartz fibre electroscope pocket dosimeter and film badge

The film badge

Although the QFE pocket dosimeter is a valuable monitoring device for providing an up-to-the-minute indication to the wearer of absorbed dose it does not provide a permanent record and its accuracy is affected by humidity, dust and rough handling. For these reasons the QFE is usually worn in conjunction with the so-called *film badge*; a device which, although unable to provide the wearer with an instantaneous reading of absorbed dose, *is* able to provide a permanent record. The QFE and the film badge thus complement one another when used together.

The film badge derives its name from the fact that it contains a tiny piece of photographic film housed in a small plastic container which is pinned to the coat of the wearer in the same way as a badge or identity disk would be worn (Figure 17.6). The film itself is sandwiched between two thin pieces of card to render it light-proof and then slipped between two halves of a plastic container which are hinged at one end and snapped shut at the other.

One face of the container has an open window cut into it which allows unimpeded entry of X-, beta and gamma radiation onto a narrow section of the film, which is virtually unaffected by neutrons arriving via this route. Other 'blind' windows in the form of thin- and thick-walled sections in the plastic container allow beta radiation reaching the film to be segregated into two energy bands which allow the separate determination of beta radiation dose. Two of the 'blind' windows have thin pieces of tin, lead and dural placed under them which preferentially attenuate certain types of radiation and enable the effective X- and gamma dose to be determined. A cadmium–lead filter placed under another 'blind' window, and used in conjunction with the tin–lead filter, is used to determine the dose received from slow neutrons.

The film itself is covered on one side with a sensitive emulsion which will respond to the range of radiation doses expected during normal working conditions and on its other side with a relatively insensitive emulsion which will respond to the much higher doses likely to be received under accident conditions.

The used film is removed from the plastic carrier and replaced with a fresh one at regular intervals (for example monthly) and sent to a central administrative point where it is developed and studied by a health physicist. The measured dose is then entered in the wearer's personal health physics record.

A typical film badge is able to measure absorbed dose over the range 0.2 mSv to 100 mSv using both emulsions, and up to about 10 Sv on the separated less sensitive emulsion.

RADIATION MONITORING

Radiation monitoring can be categorized into three main application areas: fixed installations, portable instruments, and personal monitoring devices. All of these are managed and supervised by a team of professional health physicists whose job it is to ensure a safe working environment for the workforce and the well-being of the local population as far as nuclear radiation is concerned.

Fixed installations

Representative of a fixed installation would be a group of weatherproof instruments used to monitor the gamma radiation levels at the perimeter fence of a nuclear power station. These will be of the ionization chamber type and fitted with audible and visible alarms which will be actuated automatically to warn nearby personnel should the radiation level at any point exceed a preset threshold. The instruments will also be electrically connected to a central control point where chart recordings are maintained of each measurement reading for reference purposes. Other types of fixed-installation instruments will be scattered around the working area of the reactor building itself to monitor background radiation and to detect the presence of neutrons and any radioactive dust or gases in the atmosphere.

Another type of fixed installation is the personnel contamination monitor. This is used to monitor personnel leaving an area in which radioactive materials are used and to check for the presence of contamination on hands, footwear, hair and clothing. Similar instruments are used to detect the unauthorized removal of nuclear materials from a controlled area. A good example of this type of monitor is the IPM7 instrument manufactured by Nuclear Enterprises and illustrated in Figure 17.7. The IPM7 consists of a cubicle, about the size of a telephone box, with radiation detectors mounted under the floor, in the roof, and in the form of a semi-circular array extending from floor to roof against one of the panels. The detectors are ionization chambers of the gas-proportional type which are able to measure beta and gamma radiation of very low energies.

The person to be monitored in the IPM7 stands in the cubicle with his feet aligned above two floor-mounted detectors and his hands placed inside two 'glove compartments'. He leans forward towards the vertical array of detectors and switches on the monitor with his finger tips. After a few seconds a bleeping signal indicates that the first part of the monitoring process has been completed satisfactorily. The person then turns round, leans back towards the vertical array and places his hands on two more

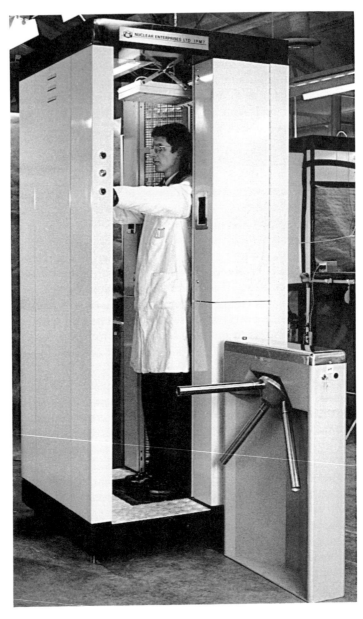

Figure 17.7 The microprocessor-controlled personnel contamination monitor manufactured by Nuclear Enterprises (courtesy of Nuclear Enterprises)

switches to re-activate the monitor; this is the position illustrated in Figure 17.7. After a few seconds a two-tone chime indicates that the monitoring procedure has been completed and that no contamination has been detected. If contamination *is* detected during either of the two monitoring operations then an audible alarm alerts the resident health physicist who consults a body-layout plan on the monitor which indicates precisely where on the person's body the contamination was detected. The two-position monitoring sequence ensures that both sides of the body are examined by the vertical array and that the resulting measurements are almost independent of the size and shape of the person being monitored. The results of the measurement can, if required, be used to control the operation of an exit barrier to prevent anyone disregarding the indication of contamination having been found.

Portable instruments

These are used mainly by health physics staff for the routine monitoring of working areas, clothing, tools, equipment, etc., to ensure that there are no abnormal levels of radiation present and no contaminated materials which the workforce might come in contact with. Representative of the many different types of portable monitoring instruments are the Neutron Dose Equivalent Monitor from the Harwell Laboratory and the Mini-Monitor from Mini Instruments.

The neutron monitor (Figure 17.8) is a battery-powered instrument which is able to measure neutrons whose energies range from thermal up to about 11 MeV. The energy response of the instrument is carefully tailored to match as closely as possible that of the human body so that the indicated dose or dose rate is equivalent to that which would be received by a person standing in the same place. The neutron detector comprises a polythene sphere of 208 mm diameter, inside which is a spherical proportional counter of the helium-3 type surrounded by a perforated cadmium shield. The polythene sphere functions as a moderator for fast neutrons and the cadmium shield serves to equalize the energy response of the detector.

The monitor can be set to measure either neutron dose rate or accumulated dose over a measured period of time and the result of a measurement is displayed on a 4-digit liquid crystal display in corresponding units of millisieverts (mSv) or mSv per hour. The monitor is virtually insensitive to gamma radiation with energies up to about 7 MeV, the gamma-to-neutron rejection ratio being better than 3000:1.

The Mini Monitor illustrated in Figure 17.9 is an inexpensive contamination monitor which is widely used in hospitals, teaching establishments and

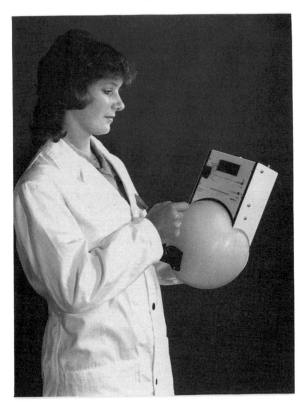

Figure 17.8   The Neutron Dose Equivalent Monitor from the Harwell Laboratory

nuclear laboratories. It uses an end-window type of GM-tube which is sensitive to alpha, beta, and gamma radiation and which is connected to the monitor in the form of a hand-held probe via a short length of flexible cable. An internal ratemeter measures the rate of arrival of incident radiation and displays the results on a front panel meter which is able to indicate up to 2000 counts per second. A small internal speaker may be used to signify the detection of each event by an audible click. This facility is often useful when locating an unknown source of radiation since it allows the user to concentrate on physically positioning the GM-tube whilst seeking to maximize the audible count rate. The Mini Monitor is battery-operated and weighs about 1 kg. It can be used with many different types of detector probes to suit a wide range of applications.

Figure 17.9   The Mini Monitor from Mini Instruments (courtesy of Mini Instruments)

18 Radiological protection

Within the short period of 16 years from 1895 most of the important types of nuclear radiation had been identified, although it took another 21 years before the elusive neutron was discovered; there was widespread use of X-rays and 'radium-rays' for clinical applications ranging from the location of broken bones to the inspection of teeth and the treatment of tumours and other diseases.

It was not long, however, before the harmful effects of X-rays and nuclear radiation became apparent, partly through ignorance and partly because of accidents. Many of the pioneering radiographers, for example, had limbs amputated because of delayed bone cancer and many died from contracting leukaemia. A memorial to these martyrs was erected in Hamburg, Germany, in 1936 listing 178 names! The obvious dangers associated with these types of radiation led to the formation in 1928 of the *International X-Ray and Radium Protection Committee* and the publication of guidance procedures for radiation workers. This organization later became known as the *International Commission for Radiological Protection* (ICRP), since when it has continued to be the much-respected world authority on all aspects of X-rays and nuclear radiation; the so-called *ionizing radiations*.

It was realized at the time of building the early nuclear reactors that such devices represented sources of radiation of enormous magnitude and that special precautions would have to be taken to ensure the long-term safety of the operating staff. It was this consideration which gave rise to the completely new branch of science known as *radiological protection*, more commonly known as *health physics*. The role of the health physicist is to identify and study potential radiation hazards in all places of work where nuclear materials are handled or transported, or where dangerous radiation levels may arise from nuclear devices such as reactors and particle accelerators and to recommend and enforce operating procedures which will ensure the safety and well-being of the workforce and nearby personnel. This responsibility includes awareness of protection measures laid down for members of the public against dangers arising from ionizing radiations.

The working regulations relating to ionizing radiations are formulated both nationally and internationally by teams of specialists and, in the UK, are enforced by numerous Acts of Parliament. These are the regulations which guide the health physicist in his work and which are usually supplemented by local work practices designed to enhance the safety of a particular organization's workforce, over and above that dictated by law.

LEGISLATION

The ICRP is a non-governmental body comprising 13 members of various nationalities. The members are selected every four years by the Commission from its own nominations and from those submitted to it by national delegates to the International Congress. The Commission has established four expert committees, each composed of about 15 persons and chaired by a member of the Commission. Additional experts are invited to participate as required and more than 100 experts from over 20 countries are usually taking part in task-group business for the ICRP at any given time.

Although the ICRP has no legal authority, its recommendations are used as a basis for legislation in Britain and other countries, either directly or through treaties and international conventions. Some of the important Acts and Statutory Instruments involving the British nuclear industry are:

- Radioactive Substances Acts 1948 and 1960
- Factories Act 1961
- Nuclear Installations Acts 1965 and 1969
- Health & Safety at Work Act 1974
- Transport Act 1963
- Radiological Protection Act 1970
- Control of Pollution Act 1974
- Ionizing Radiations Regulations 1985 (SI 1985 No. 1333).

The effect of these Acts and the regulations derived from them is to apply the ICRP recommendations to all activities involving ionizing radiations in the UK.

Other international organizations associated with radiological protection are the International Atomic Energy Agency (IAEA), the United Nations Scientific Committee on the Effects of Atomic Radiation (UNSCEAR) and the Committee on the Biological Effects of Ionizing Radiation (BEIR). Since the IAEA was formed in 1957 it has made radiological protection one of its main areas of concern and, in consultation with ICRP, the World Health Organization (WHO) and other interested bodies, it prepares basic safety standards for radiation protection which serve as reference data for national legislation.

UNSCEAR was established in 1955 by the United Nations General Assembly as a result of international concern about the effects of fall-out from the testing of nuclear explosives. It was directed to assemble, study and disseminate information on observed levels of ionizing radiation and radioactivity (both natural and man-made) in the environment and on the effects of such radiation on man and his environment.

The BEIR Committee was established by the Division of Medical Sciences of the US National Research Council and includes eminent American scientists as well as those from other countries.

It is customary for each country to have its own national organization which specializes in all aspects of radiological protection and which acts as an advisory body for Government legislation. Such an organization would have its own research facilities and maintain close links with the ICRP and other international organizations. In the UK this role is undertaken by the National Radiological Protection Board (NRPB), an organization which was created by the Radiological Protection Act, 1970.

The Government's purpose in proposing the Act was to establish a national point of authoritative reference in radiological protection. The Act gave the Board the following functions:

by means of research and otherwise, to advance the acquisition of knowledge about the protection of mankind from radiation hazards and to provide information and advice to persons (including Government Departments) with responsibilities in the United Kingdom in relation to the protection from radiation hazards either of the community as a whole or particular sections of it.

The Act also empowered the Board to provide technical services to persons concerned with radiation hazards and to charge for these services. It also organizes training courses for radiological safety officers and others responsible for radiological protection.

NRPB has established its headquarters at Chilton, near the Harwell Laboratory and alongside the Radiobiology Unit of the Medical Research

Council (MRC), and carries out the bulk of its research and public health assessment there. Services are also provided from Chilton, Leeds and Glasgow where separate service centres have been established. NRPB is funded by the sale of its own services and by Parliament via the Secretary of State for Social Services.

Representative of the wide range of tasks undertaken at NRPB's Chilton laboratories are:

Studies of metabolism of plutonium and carbon-14
Chromosome aberration studies in human peripheral blood lymphocytes
High-energy neutron shielding studies associated with neutron radiotherapy
Development of thermoluminescent dosimetry system
Measurement of radon concentration in houses
Instrument calibration and testing
Inhalation studies on radioactive particles.

RADIATION HAZARDS

When radiation encounters living tissue it loses energy by a series of collisions with the molecular structure of the tissue and some or all of its kinetic energy is converted to heat. The collisions usually involve the removal of an orbital electron from a tissue atom, leaving that hitherto neutral atom with a net positive charge and hence in an ionized state. The passage of radiation through living tissue thus creates positive ions. The orbital electrons ejected by the radiation may go on to cause ionization of other atoms or may themselves be captured by nearby positive ions.

Alpha and beta particles, being electrically charged, are strongly influenced by the negative charges of the orbital atomic electrons and hence lose energy by a series of electrical interactions as they pass through material. The interactions may result in complete removal of orbital electrons, and hence ionization of the parent atoms, or they may simply excite the electrons by momentarily displacing them from their normal orbits. *Alpha and beta particles are said to cause direct ionization as they pass through matter.*

X-rays and gamma rays, on the other hand, possess no electrical charges and have to rely upon direct collisions with atomic matter in order to lose energy. A head-on collision between a gamma ray and the nucleus of an atom, for example, causes that nucleus to recoil from the impact and to move at high speed through the surrounding material. As it does so it collides with other nearby atoms and brings about their ionization. *X- and gamma radiation are therefore said to cause indirect ionization.*

Neutron interaction with living tissue is through collisions with the hydrogen content of the body, which appears mainly in the form of water (H_2O) (the total volume of water in an average adult male is 45 litres and weighs 45 kg). Hydrogen, with a single proton in its nucleus, has a very high scattering cross-section and therefore recoils readily when struck by a neutron, giving rise to intense ionization along its track. Neutron radiation, like that of X-rays and gamma rays, is therefore a source of indirect ionization.

Ionization of the atoms which make up living tissue is a very serious matter because it affects the chemical properties of the complex molecules of the individual cells from which the tissue is formed. Such damage can lead to dangerous biological effects in the cells forming part of the brain, for example, or of the vital organs such as the liver and heart. It can also lead to delayed cancers, some of which may be fatal. Less serious effects, as far as personal health is concerned, include reddening of the skin (erythema), dermatitis, ulceration and loss of hair (epilation).

If a group of people is subjected to a short-duration burst of radiation which resulted in a received dose of a few hundred millisieverts then only minor damage would result to their blood cells and there would be no lasting damage. If, on the other hand, a dose of between 2.5 and 5 sieverts were to be received by the group within the same short period then the chances are that half of the people in the group would die within a few weeks. If, however, the same large dose were to be acquired over a period of many years then very little, if any, adverse effects would be expected.

The human cell is composed mainly of water (about 80 per cent) and a complex arrangement of interlocking compounds which define the total character, physical appearance and behavioural pattern of the person formed by the cells. One such compound is the DNA molecule (the name is derived from deoxyribonucleic acid) which is found mainly in the core region of the cell. DNA embodies the body's genetic code and is able to pass on copies of itself containing hereditary characteristics through the reproductive process. Although no firm evidence for hereditary defects attributable to ionizing radiation have been found in the human offspring, such defects *have* been found in animals – mostly mice – by exposing them to such radiation. For safety reasons, therefore, it is assumed that ionizing radiation *can* cause hereditary defects in humans and this is taken into account when radiation safe working levels are being considered.

Biological damage

Biological damage caused by ionizing radiation can, if the damage occurs at a rate greater than that of the body's natural repair mechanism, be permanent and lead to severe illness, possibly death. It is important, therefore, when recommending safe working levels for radiation workers to take into account not only the magnitude of the dose which may safely be received but also the rate at which it may be acquired. Other very important considerations are the nature of the radiation, that is, whether it is in the form of X- or gamma rays, alpha or beta particles or neutrons; also, which part of the body is being irradiated.

In cases where damage to human cells may be transmitted to an offspring, or a later descendant, the radiation damage is said to introduce *genetic effects*. The most likely causes of genetic effects are irradiation of the cells forming the gonads, that is the ovaries or testes. If, however, the irradiated cells are located in some other body organ or tissue, or in a developing foetus, any biological damage resulting from the irradiation will be confined to the person or foetus irradiated; this is said to be a *somatic effect*.

It is well known from the medical histories of 19th century X-ray and radium workers that large doses of radiation can initiate tumours in most body tissues. It is also known from epidemiological studies of people and animals who have been exposed to lower, but nevertheless above the normally-permitted, doses of radiation that statistically detectable increases are found in the frequency of occurrence of some types of cancer. An example of this is the measurable increase in the induction of thyroid tumours and leukaemia in survivors of the Japanese nuclear explosions at Hiroshima and Nagasaki. However, although genetic effects resulting from radiation exposure have been observed in animal experiments, mainly in mice, *no such effects have ever been observed in any human population which could be attributed to either naturally-occurring or artificially-created radiation*. Surprisingly, even children who were later conceived by the 100 000 or so survivors of the Japanese bombs show no detectable differences from children who were conceived by parents unaffected by the bombs. On the other hand, clear evidence of somatic effects has been obtained from studies of children born in the 1930s to women who, at the time of their pregnancies, were subjected to radiation and also from studies of children born to women who were subjected to high radiation levels from the Japanese bombs during their pregnancies. The effects observed include mental retardation and microcephaly (underdeveloped head).

Risk factors

The relationship between received dose and biological damage depends upon many complex factors and no two people are affected in the same way by the same radiation dose; a similar ill-defined relationship exists between cigarette smoking and lung cancer.

It is well known that massive radiation doses can be lethal, especially if acquired over a short period of time, but great uncertainty exists about the risks, if any, associated with very low radiation doses such as those acquired from natural background. Estimates of risks associated with radiation dose are based mainly on studies of the Japanese bomb survivors and of groups of patients who acquired large doses of radiation from X-rays and radioisotopes for medical purposes. For safety reasons the ICRP assumes that *all* radiation doses carry some degree of risk and that a linear relationship exists between radiation dose and risk. This philosophy is known as a no-threshold linear hypothesis, that is no dose carries no risk; any dose carries some risk and that doubling the dose doubles the risk, and vice versa. There is a large body of scientific opinion which feels that this is an unduly cautious assumption and that there must exist a dose threshold (albeit different for each individual) below which there is no risk. On the other hand there is a small body of highly respected scientists who feel that the risks from low radiation doses are *underestimated* by the linear hypothesis. Since there is no clear evidence to substantiate either opinion the ICRP has cautiously adopted the no-threshold linear hypothesis in estimating dose–risk relationships and will continue to do so until there exists clear scientific evidence for doing otherwise.

The first ICRP recommendation for maximum permissible occupational radiation dose appeared in 1934 and was set at 700 mSv per year. This was later reduced to 150 mSv in 1950 following the large amount of new data acquired during the Second World War. In 1956 the figure was further reduced to 50 mSv where it has remained ever since. In formulating its recommendations the ICRP includes three fundamental principles which should always be considered where radiological protection is concerned; these are:

1   No practice involving radiation exposure should be adopted unless its adoption can be demonstrated to bring about a positive net benefit.
2   All radiation doses should be kept as low as is reasonably achievable, economic and social factors being taken into account.
3   All radiation doses should be limited to those set by the ICRP.

The ICRP recommendation of 50 mSv per year for radiation workers assumes uniform irradiation of the whole body and does not differentiate between a dose concentrated at, say, the lens of the eye and one evenly distributed over the soles of the feet. However, ICRP also publishes a list of so-called weighting factors which take into account variations in sensitivity to radiation of different parts of the body. These factors, in effect, express the *fractional* risk resulting from irradiation of a particular part of

Table 18.1
ICRP recommended annual dose limits for various organs and tissues

Tissue or organ	Dose limit (mSv)	Weighting factor
Gonads	200	0.25
Breast	330	0.15
Red bone marrow	420	0.12
Lung	420	0.12
Thyroid	500	0.03
Bone surfaces	500	0.03
Remainder*	500	0.30
Whole body total	50	1.00

*lens of eye = 150 mSv.

the body to the *total* risk which exists when the whole body is uniformly irradiated and enable the effective whole body dose to be inferred from a measurement of the partial body dose. In this way the overall risk is limited to the same level, irrespective of the way in which the dose is distributed.

Table 18.1 summarizes some of the more important recommended dose limits for various parts of the body and lists the accompanying weighting factors. The figures apply to authorized radiation workers whose acquired radiation doses are routinely monitored and supervised. *For members of the public, ICRP recommended dose limits are equal to one-fiftieth of those listed in the table, that is 1 mSv, although 5 mSv is permissible in some years if the average annual exposure of a lifetime does not exceed 1 mSv.*

Derived limits

Derived limit (DL) is a term associated with personal exposure to radiation and defines the maximum permissible concentration of a particular radioactive material which may be present in the air to be breathed by a workforce or which may be present on the surfaces of materials and structures to be handled by them.

Derived limits for contaminated air are known as *derived air concentrations (DACs)* and are expressed in terms of becquerels per cubic metre (Bq/m^3) of air. DACs take into account not only the type of radiation emitted by a particular material but also its chemical toxicity. For example,

Table 18.2

Derived air concentrations for radiation workers

Radionuclide	DAC (Bq/m³)
Tritiated water	8×10^5
Cobalt-60	5×10^2
Strontium-90	6×10^1
Caesium-137	2×10^3
Radium-226	1×10^1
Uranium-238 (natural)	7×10^{-1}
Plutonium-239	8×10^{-2}

the DAC for plutonium-239 (a highly toxic radioactive material) is ten million times less than that specified for the much less dangerous tritium (hydrogen-3 isotope) in the form of tritiated water vapour. Table 18.2 lists a few of the many DAC values recommended by the ICRP for radiation workers.

Derived limits specified for contaminated surfaces are expressed in terms of bequerels per square centimetre (Bq/cm²) of surface area and depend upon the chemical toxicity and type of radiation emitted by the material to be handled. Also taken into account is the class of laboratory or working area, for example whether it is a specially equipped radiochemical laboratory or simply a storage area. The material whose surfaces are being considered here is not itself radioactive (it may be a pair of ordinary scissors) but it is assumed that its surfaces have become contaminated with a material which *is* radioactive.

In some laboratories radioactive materials are handled in what are called 'fume cupboards' where radioactive dusts and gases are continuously extracted from the working area, or in 'glove boxes' where radioactive materials are handled through long rubber gloves connected to two port-holes in the front of the box (Figure 18.1). In working areas where dangerous levels of beta and gamma radiation may be present, the materials are positioned inside a 'viewing cell' constructed from thick-walled concrete and are handled with remotely controlled manipulators viewed through thick glass windows which absorb radiation and protect the worker (Figure 18.2).

Derived limits for contaminated materials vary considerably for different

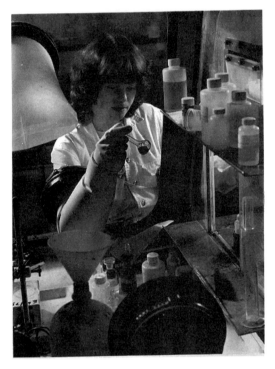

Figure 18.1   Plutonium nitrate being safely handled in a laboratory glove box

materials and different working conditions but typical figures for surface contamination of personal clothing range from 0.4 Bq/cm^2 for alpha-emitting materials to 4 Bq/cm^2 for beta-emitting materials.

Derived limits for whole body exposure are expressed in microsieverts per hour (μSv/h) and are derived from the assumption that radiation workers work a 40-hour week and a 50-week year and that the recommended annual dose should not exceed 50 mSv. This sets the maximum rate at which any radiation worker is permitted to acquire a radiation dose and is calculated as follows:

$$50 \text{ mSv/year} = \frac{50 \text{ mSv}}{50 \text{ weeks}} = 1 \text{ mSv per week} = 1000 \ \mu\text{Sv/week}$$

$$1000 \ \mu\text{Sv} \quad = \frac{1000 \ \mu\text{Sv}}{40 \text{ hours}} = 25 \ \mu\text{Sv per hour}$$

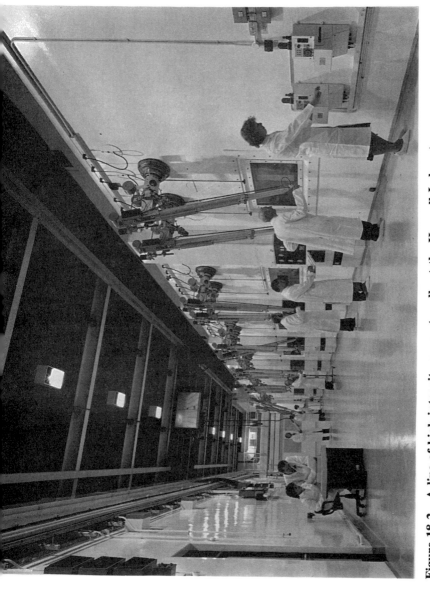

Figure 18.2   A line of high-integrity concrete cells at the Harwell Laboratory
The cells are used for radiochemistry experiments using remote handling techniques

It would appear, therefore, that any radiation worker could subject himself to a continuous radiation dose of 25 μSv per hour throughout every working day of his life and in doing so would not exceed the ICRP recommended dose rate of 50 mSv per annum. However, the most important criterion of the ICRP recommendations is not so much the recommended maximum dose but the stipulation *that all doses of radiation should be kept as low as reasonably achievable*. Any working environment in which workers received individual doses of 50 mSv per year on a regular basis over a number of years, for example, would be viewed by the ICRP as totally unacceptable. In fact, the Ionizing Regulations (1985) in the UK stipulate that any working environment which results in individual exposures in excess of 15 mSv in any year should be the subject of an internal managerial inquiry to determine ways in which the exposure could be reduced. Although there is no legal obligation to do so it is customary for copies of any report produced following such an inquiry to be made available to the Heatlh & Safety Executive (HSE).

Any individual exposure in excess of 30 mSv received in any calendar quarter is an event which must be reported to the HSE. Furthermore, all reports relating to the event must be sent to the HSE. These stipulations are mandatory and legally enforceable. It is, however, recognized that there are circumstances where such an exposure must be expected, although not on a regular basis. A radiation worker might, for example, be working on a contaminated piece of equipment and in doing so may accumulate 30 mSv in a three-month period. He may then be transferred to other non-active work for the rest of the year and so keep his exposure well below the 50 mSv permissible limit. This may be the only way of getting the job done in the time available by someone with specialized skills. Such a work practice would fall into the category of As Low As Reasonably Practicable (ALARP).

Background radiation

Table 18.3 lists typical values for annual radiation dose which would be acquired by someone living or working in high and low background regions of the UK, or in some other parts of the world. Also given is the annual radiation dose which would be acquired by someone residing at an altitude equivalent to that occupied by Concorde when crossing the Atlantic (about 10 miles). Assuming a transatlantic crossing time of 2.25 hours, and remembering that there are 8760 hours in a year, the annual value of 160 000 μSv results in each Concorde passenger acquiring a radiation dose of 40 μSv per flight, that is an acquired dose rate of 18.26 μSv per hour – a value

Table 18.3
Annual background radiation figures for selected localities

Place	Annual dose acquired from natural background radiation (μSv)
UK (mean)	1870
London (min)	800
Aberdeen (max)	1700
USA (mean)	1000
East coast (min)	900
Colorado (max)	2500
Sri Lanka	30 000–70 000
Brazil	17 000–120 000
10 miles altitude	160 000

which is close to the 25 μSv/h maximum recommended by the ICRP for radiation workers.

Collective dose

The effects of radiation from all sources on a large group of people is measured in terms of that group's collective dose and is expressed in man-sieverts (man-Sv), that is *the number of people within the group being considered multiplied by the average dose received by each person within that group* (this assumes uniform irradiation of the whole group).

For large populations the ICRP recommends an average received dose of 50 mSv per person over a period of one generation (30 years), which corresponds to an average annual received dose of 50 mSv ÷ 30 years = 1.7 mSv per person. For the UK, with a population of about 56 million, this figure corresponds to an annual collective dose of 56 million × 1.7 mSv = 95 200 man-Sv. For comparison the total dose received by each member of the UK population from natural sources of radiation has an average value of about 1870 μSv per annum, which corresponds to an average collective dose of 56 million × 1870 μSv = 104 720 man-Sv. Individual values vary considerably between different parts of the country, sometimes by as much as 1000 μSv per annum.

Natural radiation arises from cosmic rays from outer space, gamma rays from rocks, buildings and road surfacing materials, radon and thoron gases from the decay of uranium and thorium (see page 59) and radiation originating from within our own bodies, mostly from carbon-14 ($^{14}_{6}C$) and potassium-40 ($^{40}_{19}K$).

Artificial sources of radiation arise from the use of X-rays and radioisotopes in UK hospitals and clinics for medical treatment and diagnostic purposes and contributes about 250 μSv to the population's average annual dose, most of it due to X-rays. This figure corresponds to an average annual collective dose of about 14 000 man-Sv. Other sources of artificially-created radiation are the nuclear industry (waste discharges and occupational exposure), fallout from nuclear weapons testing, and miscellaneous sources such as luminous dials, gas mantles, air travel, television viewing screens and some garden fertilizers.

The collective dose to the UK population resulting from controlled discharges of radioactive materials by nuclear installations is about 80 man-Sv per year, a figure which corresponds to an average annual dose of about 1.5 μSv to each member of the population, assuming the dose is evenly distributed throughout the entire population. Of this amount, about 6 man-Sv are due to airborne discharges of gases and particulate matter from fuel fabrication and reprocessing plants and from nuclear reactors and spent fuel cooling ponds. A further 70 man-Sv arise from the discharge of low-level liquid wastes to the seas and rivers, mostly from the Sellafield reprocessing plant.

Of course, relatively few people are directly affected by discharges from nuclear installations and those living close to a particular installation will receive a much larger annual dose than the average annual value 2 μSv calculated for the population as a whole. Estimates indicate that the maximum annual dose values for such people range from about 5 μSv to 200 μSv for airborne discharges, and from about 50 μSv to 1300 μSv for low-level liquid discharges; mostly through eating fish containing radioactive caesium. A representative annual dose value for the most-exposed groups of people is 1000 μSv.

Although atmospheric testing of nuclear weapons ceased more than 20 years ago by those countries who were signatories to the 1963 test ban treaty, such tests are still occasionally carried out by a few non-signatory countries and it is these tests which 'top up' the otherwise steadily declining radioactive debris (fallout) which continues to spread itself around the surface of the earth. Since atmospheric testing of nuclear weapons first began, about 3 tonnes of plutonium have been deposited in this way, plus quantities of other radionuclides, the most important of which are strontium-90 ($^{90}_{38}Sr$) and caesium-137 ($^{137}_{55}Cs$).

The danger to living tissue from nuclear fallout comes about from the processes of inhalation and ingestion through diet, both of which cause internal radiation of the body. External irradiation comes about from gamma-emitting nuclides deposited on soil and buildings, etc. The average annual radiation dose received by people in the UK from nuclear fallout reached a peak of 80 μSv in the early 1960s but is now about 10 μSv; this figure corresponds to an annual collective dose of 560 man-Sv.

Occupational exposure of the 135 000 or so persons in the UK who are officially classified as radiation workers (also known as monitored workers) include such people as industrial and medical radiographers and people who work in universities, nuclear establishments and research laboratories in private industry. Few, if any, workers receive doses anywhere near the legal limit and most receive very much less. Radiation workers in the medical field, for example, receive average annual doses of about 0.7 mSv, whereas industrial radiographers working in factories receive about 1.7 mSv. The 2000 or so radiographers working in the field of pipeline and offshore installation inspection receive annual doses of about 27 mSv, an amount which is considerably higher than the average 2.5 mSv received by radiation workers in the UK nuclear industry.

The overall annual dose received by all categories of UK-registered radiation workers averages about 1.4 mSv, which corresponds to an annual collective dose for such workers of 135 000 $\times$ 1.4 mSv = 189 man-Sv, about 20 per cent of which is contributed by the nuclear industry. To this figure must be added the occupational dose received by the 200 000 or so non-monitored workers such as coal miners and other types of miner, dental workers and aircraft crew. The total annual occupation dose received by both categories of workers is about 450 man-Sv, which corresponds to an average personal occupational exposure for the UK population of 450 man-Sv $\div$ 56 million = 8 mSv.

The total annual radiation dose received by the UK population from all sources, both natural and artificial, has an average value of about 2150 mSv per person; a value which corresponds to a collective dose of 120 400 man-Sv. Table 18.4 and Figure 18.3 illustrate how these various sources contribute to the overall exposure. The figure shows that 87 per cent of the total exposure is due to radiation from natural sources, 11.5 per cent is from the medical industry and that the contribution from the discharge of nuclear materials to the environment is less than one tenth of one per cent.

EPIDEMIOLOGICAL SURVEYS

The ability to detect cancers which have been brought about by radiation is hampered by the tragically high rate of cancers of all types which affect the

Table 18.4
Average annual effective dose equivalents in the
UK from all sources of radiation

Source	Dose (mSv)
Cosmic radiation	300
Terrestrial gamma rays	400
Radon and thoron	800
Internal radiation	370
	1870
Medical	250
Weapons fallout	10
Nuclear power discharges	1.5
Occupational exposure	8
Miscellaneous sources	11
	280
Total	2150

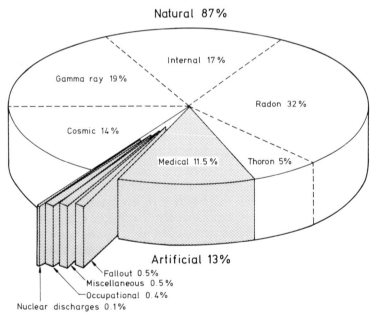

Figure 18.3 The composition of the total radiation exposure of the UK population (courtesy NRPB)

UK population as a whole; more than one person in five dies of cancer in the UK. Since cancers induced by radiation are indistinguishable from those induced by other means, it is impossible to detect with any degree of certainty the small number of cancers which *may* have been brought about by radiation.

The most obvious place to begin in studying radiation-induced cancers is, of course, the medical records of the 135 000 members of the UK population who are registered as radiation workers, plus others who work in or live near nuclear installations but who do not actually handle radioactive materials. Such records have been kept for many years by the UKAEA and BNFL and other organizations in the nuclear industry.

The Rose Report

A cancer mortality study was initiated in 1980 for the entire workforce of the UKAEA and the results were published in 1985: the authors were V. Beral et al. At the request of the UKAEA the survey was undertaken by an independent team of researchers selected by the Medical Research Council (MRC) and led by Professor G. Rose. The survey began in 1980 using data and medical records supplied by the UKAEA and was carried out at the London School of Hygiene and Tropical Medicine by the MRC's Epidemiological Monitoring Unit (EMU).

The aims of the survey were, firstly, to examine the mortality of all persons who had been employed since 1946 at the sites now administered by the UKAEA; secondly, to analyse the relationship, if any, between causes of death and exposure to radiation among employees; thirdly, to compare the conclusions with the risk estimates published by the International Commission for Radiological Protection.

The EMU supervised and checked the collection of personnel and radiation exposure information and carried out (independently of the UKAEA) the follow-up of ex-employees to identify deaths. The identification of deaths and assignment of cause of death were carried out without prior knowledge of any person's work history. A total of 39 546 men and women who had worked at Harwell, Culham, London Headquarters, Winfrith and Dounreay between 1946 and 1979 were studied. Personnel records of employees leaving Risley and Culcheth between 1965 had been destroyed before the survey began and could not therefore be included in the survey. The published results of the survey were as follows:

- The 39 546 employees were followed for an average period of 16 years after first undertaking employment with the UKAEA.

- Up to 31 December 1979 there were 3373 deaths among the employees and ex-employees, 937 of which were caused by cancers.
- Considered overall, the death rates from all causes were lower than those prevailing in the country as a whole but this is to be expected in any 'healthy workforce', where the chronically sick are not employed and where a potential employer such the UKAEA, British Rail, teaching profession, etc., insist on a thorough medical examination before offering employment to anyone.
- Leukaemia mortality rates were about 23 per cent above the national average but because of the relatively small number of deaths (35 overall), no statistical significance could be attributed to the unexpected increase. To highlight the dangers of inferring too much from poor statistics, one of the sub-groups studied was found to have an anomalously high leukaemia mortality rate and yet the members of this sub-group were *non-radiation workers and had been employed by the UKAEA for less than two years*.
- The incidence of cancer deaths from all causes was below the national average.
- About half the employees in the survey had been issued with film badges or other devices used to measure their external radiation exposure. Smaller numbers had also been monitored for possible internal contamination. The total dose recorded for these people was 660 Sv, an average of 32.4 mSv per employee with a radiation record.
- Mortality was not, in general, correlated with radiation exposure; the only exception to this was prostatic cancer (cancer of the prostate gland).
- There were, overall, 38 deaths from prostate cancer and the mortality rates for this cause of death were slightly above the national average for men who had a radiation record. However, for a small group of men who had been monitored for tritium, the death rates for this type of cancer were eight times the national average, even though their overall mortality rates, that is deaths from *all* causes, were well below the national average (about 60 per cent of that expected). *We therefore have the unusual situation of an abnormally healthy group of workers dying at an abnormally high rate from a certain type of cancer which in no other survey has been shown to be associated with exposure to tritium.* Tritium is not concentrated by the human metabolism in the prostate gland and no clear explanation can therefore be proffered for this phenomenon.
- The number of people included in the survey was not large enough to

enable accurate comparisons to be made with the ICRP risk estimates. In fact, the margin of error on any comparison is such that, at one extreme, the ICRP risk estimates could be considered as being *15 times too low* whereas at the other extreme it could be said that *low doses of radiation are beneficial!*

The UKAEA is seeking advice from the MRC on investigations which might be initiated in the light of the results of the survey and is updating its existing data base in anticipation of another survey to take place some time in the future.

Two scientific papers describing the UKAEA survey were published in the *British Medical Journal* on 17 August 1985; see references to Beral and Fraser at the end of this chapter.

The Smith–Douglas Report

This report describes the results of a study carried out on the mortality of workers at the Sellafield Plant operated by British Nuclear Fuels (BNFL). It was prepared by Dr P. G. Smith and Ms A. G. Douglas of the Department of Epidemiology at the London School of Hygiene and Tropical Medicine and was published in 1986.

The report was financially supported by the Imperial Cancer Research Fund and, through a grant to the London School of Hygiene and Tropical Medicine, by BNFL. Practical support was provided by the Imperial Cancer Research Fund, the National Radiological Protection Board (NRPB), the National Health Service Central Register, the Medical Statistics Division of the Office of Population Censuses and Surveys, and the National Insurance Records Branch of the Department of Health and Social Security (DHSS).

The report describes an in-depth mortality study of the medical and employment records and death certificates of 14 327 people who were known to have been employed at BNFL's Sellafield plant at any time between the opening of the site in 1947 and 31 December 1975 and included mortality data up to the end of 1983. Of all workers within this category, 96 per cent were successfully traced and 2277 of them were found to have died. Of those who died, 572 of them, that is 25 per cent, had contracted fatal cancer.

The report is extremely thorough and of considerable length and complexity. A brief summary of its findings are as follows:

1 The mortality of the Sellafield workers over the period studied, from all causes was, on average, 2 per cent less than that of the general

population of England and Wales and 9 per cent less than that of the population of Cumbria (previously known as Cumberland and the area in which the Sellafield plant is located). This finding is consistent with that expected from a healthy workforce.

2 The mortality of the Sellafield workers from cancers of all types was 5 per cent less than that of England and Wales and 3 per cent less than that of the county of Cumbria.

3 There was no significant difference in the number of cancer deaths occurring between radiation workers and non-radiation workers at the plant.

4 For cancers of specific sites, only one excess was detected among non-radiation workers and this was in the category of 'ill-defined and secondary cancers'. There was a significant deficit in the number of leukaemias detected among non-radiation workers and also significant deficits in the numbers detected for cancers of the lung, liver and gall bladder, and for Hodgkins Disease, among radiation workers.

5 When those radiation workers with higher levels of exposure were compared with those with lower levels, relationships between total exposure and cancer mortality were discovered for some cancers (bladder, multiple myeloma and leukaemia), although the associations were not statistically significant. However, when only the exposure received more than 15 years before death was considered, the associations with multiple myeloma and bladder cancer became significant. The authors of the report suggest that the bladder finding may be a chance occurrence as this has not been observed in other studies, whereas the multiple myeloma has been noted in American studies.

The Tiplady Report

One of the most important reports to have appeared in recent years concerning radiation and the general public was that published in 1981 by the Cumbria Area Medical Officer. The report describes a definitive survey carried out independently of the nuclear industry under the direction of Dr Tiplady, an epidemiologist and specialist in community medicine with the Cumbria Area Health Authority.

The purpose of the survey was to examine mortality data for leukaemia for those areas which now constitute the county of Cumbria over the period 1951 to 1978. Additionally, the survey analysed mortality data for *all* malignant diseases for the period 1974 to 1979 for the three Health

Districts of West Cumbria, East Cumbria and South West Cumbria. A brief summary of the report is as follows (the word 'significantly' implies statistical significance):

- The leukaemia mortality rate for the county of Cumbria was not significantly different from the national rate.
- The incidence of leukaemia in East Cumbria was not significantly different from the national rate, although the incidence of *all* malignancies in males was significantly lower.
- In West Cumbria (the district containing BNFL's Sellafield plant) the incidence of leukaemia in the 1967–1973 interval was not significantly higher than expected. However, for the 1967–1973 interval the incidence was significantly lower in males. Registration of *all* malignancies was significantly lower than expected throughout both time intervals.
- In South West Cumbria, although the incidence of leukaemia was not significantly higher than the expected value, the incidence of *all* malignancies was significantly higher in males and there was a significant rise in new cases of multiple myeloma in females.
- The incidence of myeloid leukaemia (the type known to be induced by radiation) was not significantly increased in any of the districts covered by the survey. There was, however, a significant increase in lymphatic leukaemia among West Cumbrian males.

The Black Report

In 1983 the UK Secretary of State for Health & Social Security requested Sir Douglas Black, Chairman of the British Medical Association, to enquire into allegations made in the Yorkshire Television programme 'Windscale – Nuclear Laundry' concerning localized clusters of childhood leukaemias in areas adjacent to the BNFL reprocessing plant at Sellafield (formerly known as Windscale). In essence, the programme claimed that deaths among young people from leukaemia were abnormally high in the Sellafield area and that this abnormality was caused by discharges of radioactive materials from BNFL's reprocessing plant.

The allegations were examined in depth by an independent advisory group set up and chaired by Sir Douglas Black and a report of their findings was published by HMSO in 1984; see the full reference on page 368.

The report is divided into four main areas covering firstly, examination of epidemiological evidence from 13 studies of different geological areas around Sellafield; secondly, environmental aspects of the Sellafield area, including drinking water, bacteria and viruses, the presence of other industry in the area, and details of radioactive discharges from the BNFL plant to the sea and air; thirdly, the radiation exposure of young people in the Sellafield area, the sources of radiation and the dose to the red bone marrow for each source (it is known that irradiation of the red bone marrow can induce leukaemia); fourthly, assessment of possible risks to young people living near Sellafield.

The report confirms the claim made in the Yorkshire Television programme that there exists a localized cluster of childhood leukaemias in the Sellafield area (four at Seascale and six at Millom) but points out that the incidence of this disease in West Cumbria as a whole is consistent with other parts of England and Wales and that although the incidence is unusual it is not unparalleled. The report also states that although the TV programme performed a public service in drawing attention to the Sellafield cluster there is no evidence to support the claim that the leukaemias were caused by the BNFL plant.

The report's recommendations cover three main areas: epidemiology, health implications of radioactive discharges, and regulatory mechanisms. In the first area it is recommended that a study should be carried out on the records of those cases of leukaemia and lymphoma (tumour consisting of lymphoid tissue) which have been diagnosed among young people up to the age of 25 and resident in West Cumbria, and that the Northern Children's Cancer Registry should be asked to re-analyse their existing data using data gleaned from the 1961, 1971 and 1981 population census returns, where appropriate. A particularly valuable recommendation in this area was that the MRC, or the Office of Population Censuses and Surveys (OPCS), should coordinate centrally the monitoring of small area statistics around major installations of all types which produce discharges that might present a carcinogenic or mutagenic hazard to the public. Such a monitoring service could give early warning of any untoward health effects.

In the area concerned with health implications of radioactive discharges the report recommends that more attention should be concentrated on measuring doses of radiation actually received by members of the public in West Cumbria, and in other relevant areas, and that more work should be done on improving the understanding of metabolic differences between adults and children.

Recommendations concerned with regulatory procedures state that the controls imposed upon BNFL by Government should be revised so that reviews of authorizations for the Sellafield plant take place more frequently. Greater emphasis should also be placed on the collection and consideration of relevant epidemiological data and any other human data relevant to the possible health consequences of discharges. The recommendations go on to state that the responsibility for monitoring (and interpreting of the results) the potentially serious environmental pollutants discharged by BNFL should be more clearly defined by Government. The results of any monitoring need to be considered in their entirety, and on a regular basis, by a designated body with significant health representation so that decisions taken about the control of authorized discharges from the Sellafield plant take fully into account all relevant factors.

When considering the findings of the Black Report it should be remembered that the national average incidence rate for childhood leukaemias and lymphomas is only 106 per million of the total population, that is one incident for every 9434 people, and that the highest incidence rate in the Northern region of England (see the reference to Craft, A.W. on page 368) occurs at Whittingham, Alnwick, Northumberland, and has a value of about 13.9 incidences per 1000 people (actually two children out of a local population of 144 children); this compares with an incidence rate of about 9.8 per 1000 for Seascale, Cumbria (four cases out of 411 children). Alnwick is close to the North East coast of England and on the opposite coastline to Sellafield.

For comparison, the annual national mortality rate for people who smoke 10 cigarettes per day is one in 200 of the UK population; for accidents on the road the rate is one in 5000.

Committee on Medical Aspects of Radiation in the Environment (COMARE)

COMARE was set up by the Department of Health and Social Security (DHSS) in November 1985 following a recommendation contained in the Black Report. Its terms of reference are:

> To assess and advise Government on the health effects of natural and man-made radiation in the environment and to assess the adequacy of available data and the need for further research.

COMARE's first chairman (elected to serve until 1989) is Professor Martin Bobrow, Prince Philip Professor of Paediatric Research at the United Medical and Dental Schools of Guy's and St Thomas' Hospitals. The other ten members of the Committee comprise professors and other specialists from ten different hospitals, universities and medical research establishments.

As its first task, the Committee was asked to consider the implications for the conclusions of the Black Report of new information provided by

BNFL and others. This it has done, and its first report appeared in July 1986. Some of the new data originated from Dr Derek Jakeman, a reactor physicist who once worked at the Sellafield plant (when it was known as Windscale) but who is now working at the UKAEA's research establishment at Winfrith in Dorset. Dr Jakeman claimed that several kilogrammes of uranium and associated fission products were discharged in oxide form directly to the environment in the mid-1950s and that this information was not made known to members of the Black Inquiry.

Although critical in its report of the manner in which the new data came to light, COMARE nevertheless accepts that the increased radiation doses received by young people living in the Sellafield area at the time of the unrecorded discharges are less than those received at the same time from natural background radiation and that there is therefore no reason to doubt the substance and essential conclusions of the original Black Report.

The new data supplied to COMARE was analysed by staff of the National Radiological Protection Board (NRPB) who calculated what effects the additional radiation doses may have had on the local population at the time. These calculations took into account new recommendations recently published by the International Commission for Radiological Protection (ICRP) concerning the rate of absorption of plutonium and americium in the human gut, and they also distinguished between the effects on bone marrow of radiation with high and low linear-energy transfer in living tissue.

The findings of the NRPB show that by taking into account the new data (and other considerations), the calculation of radiation risk to the Seascale population study published in the Black Report has increased the predicted number of fatal radiation-induced leukaemias resulting from the Sellafield discharges from the previously-calculated figure of 0.01 to 0.016. The risk to the average child in the study population from the Sellafield discharges is now calculated to be one in 75 000 with a maximum risk of one in 30 000 for children born in the mid-1950s. Bearing in mind the low number of leukaemias identified in the Black Inquiry (four children out of a population of 411) these revised figures are of no statistical significance.

MAFF Terrestrial Radioactivity Monitoring Programme

The Ministry of Agriculture Fisheries and Food (MAFF) has, for many years, carried out a comprehensive national programme for monitoring the presence of radioacitivity in fish and other marine materials, the results of which are published annually. Following recommendations made in the 1984 Black Report, and taking into account the growing increase in public interest in matters relating to radioactive discharges from nuclear installations, MAFF has extended its monitoring programme to include the terrestrial environment around all major nuclear sites in England and Wales which are authorized to discharge low-level radioactive wastes. The establishment of this new programme was announced to the House of Commons in July 1985 and was implemented in January 1986.

The new monitoring programme will cover all forms of agricultural and horitcultural produce, including milk, meat, fruit, vegetables, cereals, etc. Results will be published annually in such a way as to identify 'critical group' exposures and to indicate any general trends in the light of information on exposures to natural radioactivity and weapon fall-out in the northern hemisphere.

National Registry for Radiation Workers (NRRW)

This is a follow-up study designed by the NRPB to provide direct evidence as to the reliability of the current risk estimates for low doses of ionizing radiation. The aim of the study is to consider adverse health effects in relation to known radiation doses; the target population for the study therefore consists of all persons who have ever been monitored in the course of their work and for whom records have been kept, that is radiation workers.

The study population includes radiation workers in the nuclear industry (UKAEA, BNFL, CEGB, SSEB, Amersham International), Ministry of Defence, industrial radiographers and numerous research organizations. Data collection began in 1981 and has so far been restricted to radiation workers being monitored on or after 1 January 1976 because of the substantial amount of work involved in the collection and validation of data relating to ex-employees. However, arrangements are being made to move the starting date backwards in time wherever the necessary record exists. The data will be analysed at regular intervals as they accumulate; no analysis has been published at the time of writing.

REFERENCES AND FURTHER READING

The following list of references is given for those readers wishing to examine in greater detail the reports referred to in this chapter, and to study other reports relating to radiological protection in general.

Barton, C. J. et al., 'Childhood leukaemia in West Berkshire'. *Lancet*, vol. 2, no. 8466, p 1248, 30 November 1985

Beral, V. et al., 'Mortality of employees of the UKAEA 1946–1979', *Brit. Med. Journ.*, vol. 291, pp 440–447, 17 August 1985

Black, Sir Douglas. *Investigation of the possible increased incidence of cancer in West Cumbria*. Report of the Independent Advisory Group, London: HMSO, 1984

Cartwright, R. A., Miller, J. G., 'Lymphoid and haemopoietic malignancy case occurrence in the UK', *Lancet*, vol. 2, no. 8414, pp 1270–1271, 1 December 1984

Clough, E. A., 'The BNFL radiation mortality study', *Journ. Soc. Radiological Protection*, vol. 3, no. 1, pp 24–27, 1983

Clough, E. A., 'Further report on the BNFL radiation-mortality study', *Journ. Soc. Radiological Protection*, vol. 3, no. 3, pp 18–20, 1983

Craft, A. W. et al. 'Apparent clusters of childhood lymphoid malignancy in Northern England', (includes 8 good references), *Lancet*, vol. 2, no. 8394, pp 96–97, July 1984

Crow, J. F., 'Can we assess the genetic risks?', *Radiation Research*, Proceedings of 8th International Congress on Radiation Research, Tokyo, 1979

Darby, S. C., '*Protocol for the national register for radiation workers*', NRPB report R116, 1981

Ford, J., 'Australian studies dismiss bomb-test fears', *New Scientist*, vol. 100, p 868, 1983

Fraser, P. et al., 'Collection and validation of data in the UKAEA mortality study', (includes 12 good references), *Brit. Med. Journ.*, vol. 291, pp 435–439, 17 August 1985

Gloag, D., 'Risks of low-level radiation – the evidence of epidemiology', (includes 36 good references), *Brit. Med. Journ.*, vol. 281, pp 1479–1482, 29 November 1980

Goodwin, P., 'Register for UK radiation workers', *Nature*, vol. 255, p 517, 12 June 1975

Lloyd, O. L. et al., 'Mortality from lymphatic and haemopoietic cancer in Scottish coastal towns', *Lancet*, vol. 2, no. 8394, pp 95–96, 14 July 1984

Mancuso, T. F., Stewart, A., Kneale, G., 'Radiation exposures of Hanford workers dying from cancer and other causes', *Health Physics*, vol. 33, pp 369–384, November 1977

Neil, V. S. et al., 'Approach to the management of children with malignant disease in one district general hospital', *Brit. Med. Journ.*, vol. 283, pp 366–367, 1 August 1981

Oftedal, P., Searle, A. G., 'An overall genetic risk assessment for radiological protection purposes', *Journ. Medical Genetics*, vol. 17, p 999, 1980

Reisland, J. A., *Protocol for the study of the health of UK participants in the UK atmospheric nuclear weapons tests*, NRPB report R154 (1983)

Royal Commission on Environmental Pollution (1976), sixth report: *Nuclear power and the environment*, London: HMSO, Cmnd 6618

Shukla, P. T. et al., 'Is there a proportionality between the spontaneous and X-ray induction rates of mutations?', *Mutations Research*, vol. 61, no. 2, pp 229–248, July 1979

Smith, P. G., Douglas, A. G., 'Mortality of workers at the Sellafield plant of British Nuclear Fuels', *Brit. Med. Journ.*, vol. 293, p 845, 4 October 1986

Sources and effects of ionising radiation, UN Scientific Committee on the effects of atomic radiation, New York (1977)

Tiplady, P., *Leukaemia and other malignancies in Cumbria*, Cumbria Area Health Authority report, May 1981

Wade, B. O. et al. *Studies of environmental radioactivity in Cumbria, part 7; a summary of progress to December 1984*, AERE report R11743, London: HMSO, April 1985

19 Nuclear safeguards and public concern

Core melt-down represents the worst possible credible accident which could be envisaged for a nuclear reactor. It would come about through complete loss of coolant leading to an uncontrolled chain reaction and severe overheating of the fuel. Such an accident would be of the utmost severity and result in millions of pounds worth of damage and possible loss of life, but it *would not and could not result in a nuclear explosion*. This is because the fissile part of the fuel (U-235) amounts to only 3 per cent at most in thermal reactors, the rest being U-238; there is no physical configuration of that fissile part which could result in the formation of an initial mass suitable for initiating a nuclear explosion. The same applies to fast-reactor systems which use mixed oxide fuel (25 per cent PuO_2/75 per cent UO_2).

On the other hand, it cannot be denied that all reactor systems create Pu-239 from the U-238 content of their fuel inventory during normal operation (see Tables 9.7 and 9.8) and that any technically competent country possessing a fuel reprocessing plant could, if it so desired, chemically separate the Pu-239 content from spent nuclear fuel and use it for the production of nuclear weapons. In fact, this is how the world's five nuclear weapons states (USA, USSR, Britain, China and France) derive the Pu-239 for their own nuclear warheads using specially designed nuclear reactors.

Although U-235 is still widely used as a nuclear explosive it is a lot less efficient for such purposes than is Pu-239, which is also much easier to separate from spent fuel. Separating one element from another, for example Pu-239 from U-238, is chemically quite simple to do, whereas separating one isotope from another of the same element, for example U-235 from U-238 is very difficult to do and also very expensive. Hence the reason why Pu-239 is the world's most favoured nuclide for nuclear weapons production.

If a nuclear reactor is shut down at frequent intervals and its fuel removed for reprocessing, it is possible to extract the small amount of pure Pu-239 which will have been created during the reactor's short running time. This is an extremely inefficient way of operating a nuclear reactor but in doing so it does yield a very pure form of Pu-239 which is known as *weapons grade*. If, on the other hand, a nuclear reactor is operated for much longer periods, as is the case with nuclear power stations, the in-core residence time of the fuel can be as long as many years, during which time the Pu-239 created from the U-238 has had ample opportunity to capture core neutrons and be transformed into Pu-240 – a non-fissile nuclide. This is known as *reactor-grade* plutonium and would be considered by a nuclear weapons state as being unsuitable for use as a nuclear explosive. Nevertheless, it has been demonstrated that a nuclear explosion, albeit a relatively poor one, *can* be initiated by reactor-grade plutonium. It would be wrong, therefore, to assume that such plutonium would be of no use to anyone planning to make a nuclear device.

From what has been said so far, it is clear that the fissile nuclides U-235 and Pu-239 are both suitable for use in a nuclear weapon and that Pu-239 is created from the fertile nuclide U-238 by neutron capture in the core of a nuclear reactor. In order to prevent the spread of nuclear weapons, therefore, it is necessary to monitor worldwide stocks and movements of U-235 and Pu-239 and also that of U-238, without which Pu-239 would not exist. Furthermore, it is necessary to monitor the burn-up rating and movements of all spent fuel and the throughput of all reprocessing and enrichment plants. This is a formidable task, and yet one which international inspectors from the European Atomic Energy Commission (Euratom) and the International Atomic Energy Agency (IAEA) have been doing – as far as they are allowed to – since the Non-Proliferation Treaty (NPT) was signed and ratified on 5 March 1970.

NUCLEAR SAFEGUARDS

The work undertaken by IAEA and Euratom inspectors, and the means at their disposal, are aimed at preventing the diversion of nuclear materials from peaceful uses to nuclear weapons production. The whole concept is described by the term *Nuclear safeguards*, or simply 'safeguards'.

Although safeguards are backed-up by international treaties, European law, politics and diplomacy, the ground work is carried out by the inspection teams who rely heavily on on-the-spot measurements and personal inspection of documentation. The European Safeguards Research & Development Association (Esarda) was established to advance and harmonize R & D associated with safeguards within Europe and to ensure that safeguards inspectors were adequately equipped with instrumentation and technical back-up to do what is expected of them. The Association currently comprises nine partners, two of which are the UKAEA and BNFL.

At the end of 1986 there were 164 safeguards agreements in force with 96 states. Safeguards were applied in 51 non-nuclear weapon states and in four nuclear weapon states. More than 2050 inspections were carried out during 1986 at 595 installations in 53 non-nuclear weapon states and in four nuclear weapon states. Over 325 automatic photographs and TV surveillance systems were in operation in the field and 10 300 Agency seals which had previously been applied to nuclear materials were detached and subsequently verified at the IAEA headquarters in Vienna.

Over 1000 plutonium and uranium samples were analysed during 1986, and about 2840 analytical results were reported. Accounting and other safeguards data comprising 867 000 data entries were processed and stored in the Agency's computer. The sensitivity of the Agency's inspection and evaluation activities is illustrated by the fact that about 270 – mostly minor – discrepancies or anomalies were found in the 1986 records, most of which were satisfactorily explained upon subsequent appraisal or investigation.

At the end of 1986 the quantities of nuclear material under Agency safeguards, including that covered by voluntary-offer agreements with nuclear weapon states, amounted to 8.4 tonnes of separated plutonium, 13.2 tonnes of highly-enriched uranium, 194.5 tonnes of plutonium contained in irradiated (spent) fuel, 27 911 tonnes of low-enriched uranium and 47 402 tonnes of source material, that is fertile uranium-238.

The limitations of safeguards

No system of safeguards can ever be considered perfect, especially when there are still many countries throughout the world who are still not signatories to the NPT, or who refuse to allow Agency inspectors to examine their nuclear facilities. An example of safeguards limitations occurred on 18 May 1974 when the Indian Atomic Energy Commission announced to the world the detonation of a 'peaceful' nuclear explosion. Bearing in mind that India was not, at the time of the explosion, a signatory to the NPT (and is still not), and the fact that the NPT makes no distinction between a peaceful nuclear explosion and a nuclear weapon, the event marked a severe setback to the efforts of the NPT to prevent the spread of nuclear weapons.

Another disappointing setback to the principle of safeguards occurred on 7 June 1981 when the Israeli Air Force attacked Iraq's Nuclear Research Centre at Tuwaitha, near Bagdad, and severely damaged the Centre's materials-testing research reactor known as Tamuz-1. This was the second attack by Israel on the Tuwaitha Centre, the first having taken place in September 1980 but which resulted in very little damage. The excuse given by Israel for the attacks was that it believed Iraq was using the Tamuz reactor to embark upon a nuclear weapons programme. At the time of the attacks Iraq, but not Israel, were signatories to the NPT and the Tuwaitha Research Centre had been regularly inspected by the Agency's safeguards inspectors.

Following the 1981 attack, Agency inspectors again visited the Tuwaitha Centre and scrutinized fuel inventories and examined stocks of enriched uranium, depleted uranium and natural uranium, and the presence of all fuel assemblies, including those which were in the spent fuel storage pond. No evidence of any kind could be found to show that any nuclear materials had been diverted for the purpose of producing plutonium and all stocks tallied with those recorded during the previous inspection.

NUCLEAR ACCIDENTS

One of the things about nuclear power which seems to worry the general public is the likelihood of a major accident at a nuclear installation, leading to the release of large quantities of radioactive materials and massive loss of life. Such worries are nurtured by the nuclear industry's own legal obligation to report virtually all abnormal incidents to the Nuclear Installations Inspectorate (NII), even those which pose no significant threat to human or animal life or to the environment. The nuclear industry does not itself complain about such an obligation but it does, at times, give the impression to the public that the nuclear industry has a disproportionate share of the country's abnormal incidents, bearing in mind that every incident reported to the NII is automatically reported to representatives of the Media since to do otherwise would lead to even greater condemnation and accusations of a 'cover-up'. This is not to suggest that the nuclear industry does not make mistakes, because it most surely *does*. It sometimes does things which it should never have done and incidents described as accidents have, in many cases, been brought about, not by some unforeseen circumstance, but by negligence or human error. But the nuclear industry, like the aviation industry, the gas and petro-chemical industries, the medical profession and the coal mining industry is staffed by highly qualified but *ordinary* people who are as fallible in their behaviour as most other groups of people, and ordinary people occasionally make mistakes; they always have and they always will; it is inevitable. The acid test is surely 'how frequently have these mistakes occurred, how potentially dangerous have they been and how much loss of life or damage to the environment have they caused?'

The first nuclear reactor in Britain was built (at Harwell) in 1947 and since then not one member of the British public has ever come to any harm which could be attributed to radioactive materials; only a handful of the nuclear industry's own employees have lost their lives because of the nature of their work. So the nuclear industry can claim a safety record to be proud of when compared with other high-tech industries. To put things in perspective, the following tragic events have taken place since the birth of the nuclear power industry, and all were brought about through human error or because of inadequate instrumentation. It is earnestly hoped that the lessons learnt from each accident will lead to improvements in design or working practices which, in turn, will lead to a much safer world for everyone.

17 August 1947 More than 100 coal miners killed in the William pit explosion at Whitehaven, Cumberland.

21 October 1966 More than 100 schoolchildren killed at Aberfan, Glamorgan, when torrential rain caused a coal mining waste 'mountain' to slide into and engulf a school classroom.

1 June 1974 28 people killed, 90 injured and 4000 evacuated following an explosion at a Nypro plant at Flixborough, Humberside.

27 March 1977 Two 747 Jumbo jets, fully laden with fuel and passengers collided on the runway at Tenerife airport; 583 people died in the worst accident in aviation history.

25 April 1978 25 people killed in Waverly, Tennessee, USA, when two rail tankers carrying 288 000 gallons of butane gas exploded; 6000 people evacuated.

11 July 1978 More than 100 people killed at San Carlos de la Rapita, Costa Blanca, Spain, when a road tanker carrying liquid propylene crashed into a crowded camping site and exploded.

3 December 1984 At least 2000 people killed and 180 000 treated for ailments in Bhopal, India, when poisonous gas escaped from a tank containing 45 tonnes of methyl isocyanate.

1985 The worst year for accidents in the history of aviation; 1948 people killed worldwide.

Accidents in the nuclear industry have been relatively few and, in spite of the Chernobyl reactor accident, have resulted in little loss of life.

Windscale

The most serious nuclear accident to have occurred in Britain took place in 1957 at the Windscale (now Sellafield) reprocessing plant and involved a fire in one of two nuclear reactors known as the Windscale Piles; these were designed and operated as plutonium producers for military purposes but, since the fire, both have been de-fuelled and permanently taken out of service.

The Windscale fire began during what was expected to be a routine Wigner release (see Chapter 15, page 305). Unfortunately, due to inadequate instrumentation, the operation got out of hand and the core of the reactor burst into flames releasing large quantities of radioactive materials, mostly iodine, to the environment and effectively destroying the reactor. It was the lack of adequate instrumentation for monitoring and controlling the release of Wigner energy which led to thermal runaway and the fire. Fortunately, because of the much higher operating temperatures of the modern Magnox and AGR types of graphite-moderated reactors, the release of Wigner energy takes place continuously and automatically during normal operation and an accident of the type which destroyed the Windscale Pile could not happen in these types of reactors.

The Windscale Pile was a graphite-moderated reactor fuelled with natural uranium and cooled with air vented to atmosphere by way of a tall chimney fitted with a particulate filter close to its top. Thanks to the efficiency of the filter the amount of radioactive debris actually released to the environment during the Windscale fire was limited mainly to the radioisotope iodine-131 (8-day half-life). This was deposited on fields and buildings lying downwind of the Windscale site and reappeared in the milk of cows grazing in fields lying under the iodine plume. Since the drinking of milk represents a direct route to the human body, a restriction was quickly imposed on the use of milk produced over an area of about 200 square miles around the Windscale site for a period of six weeks. All milk produced within this area was emptied into drains and compensation paid to the farmers whose cows had produced it. Had the discarded milk been converted into cheese, instead of being thrown away, the radioactivity of the iodine it contained would have decayed to about 0.05 per cent of its initial value after about 80 days (10 half-lives) and after 5 months or so it would have virtually disappeared altogether, allowing the cheese to be sold and eaten without endangering anyone.

The fire in the Windscale Pile was a very serious nuclear accident, the worst ever to have occurred in Britain, but no clear evidence has been produced of any serious harm having been caused to anyone, or to the environment, and many valuable lessons have been learnt from it having happened. Improvements have been made, for example, in reactor design and operating procedures and also in the design and provision of electronic instrumentation used for site and area monitoring.

Three Mile Island

On 28 March 1979, a little after 4 am, an incident occurred at a nuclear power station located on an island in the Susquehanna river near the town of Harrisburg, Pennsylvania, USA; the station was known as Three Mile Island (TMI).

The incident at TMI culminated in a very serious accident and the reasons for it having happened did much to undermine the long-standing credibility of the nuclear industry worldwide. And yet, *not one individual came to any harm because of the accident* and the only damage which occurred was confined solely to the reactor core. The primary containment vessel of the PWR was unbreached and unaffected by the accident, as was the secondary containment building.

The TMI station is jointly owned by Metropolitan Edison (50 per cent), Jersey Central (25 per cent) and Pennsylvania Electric (25 per cent) and comprises two completely separate nuclear steam-generating plants known as Units 1 and 2. Each Unit consists of a single PWR and turbo-generating set licensed to generate 961 MW of electricity at a reactor core heat rating of 2772 MW; reactor coolant pressure is normally in the region of 2200 psi (about 150 atmospheres).

The TMI incident began with a breakdown in one of the feedwater pumps used to circulate water through the steam generators associated with the Unit-2 reactor and the turbine condenser. There was nothing unusual or worrying about this particular event; it was the sort of thing to be expected from time to time and the reactor's control system reacted to it exactly as it had been designed to do, that is it shut-down the turbine associated with that particular pump. It also automatically started up the emergency feedwater pumps (EFPs) so that heat from the steam generators which was previously being dissipated in the now shut-down turbine could be dissipated elsewhere. Unfortunately, and unknown to the operating staff in the control room, the 'block' valves on the two emergency feedwater lines had inadvertently been left closed so that water from the EFPs was unable to reach the steam generators. Up to this point the reactor was operating normally under automatic control and at about 97 per cent of full power.

The blockage of the emergency feedwater lines meant that no water was flowing through the secondary circuit of the steam generators and therefore that no heat was being removed from the primary coolant water which continued to circulate through the steam generators and through the reactor core in its own closed circuit driven by its own circulating pumps (see Figure 8.16). Naturally, with no heat being removed by the secondary circuit, the primary coolant began to overheat, causing the water in the reactor pressure vessel to expand. This, in turn, had the effect of raising the level of water in the pressurizer unit and consequently of increasing the pressure of the steam trapped above the water level. The coolant pressure of the reactor began to rise quickly above its normal value of 2200 psi and continued to do so until it caused the Pilot Operated Relief Valve (PORV) located at the top of the pressurizer to open at about 2205 psi. *All this took place within about 5 seconds following the automatic shut-down of the turbine.*

The PORV is an electromagnetically-operated device which functions in exactly the same way as does the weighted valve used on the lid of a domestic pressure cooker. Its purpose is to vent off unwanted steam and in doing so reduce internal pressure. The steam allowed to escape by the PORV is conveyed through pipework to a coolant drain tank where it is condensed to water. The tank has a storage capacity of 30 m^3 (about 1000 ft^3) and has its own cooling circuit. Relief valves on the tank allow surplus water to be discharged through floor drains into a sump in the floor of the containment building. Facilities exist for draining the sump and transferring its contents to an adjacent auxiliary building.

Unfortunately, the amount of steam released by opening of the PORV was insufficient to arrest the build-up of coolant pressure which continued to rise until it reached 2355 psi. At this point the reactor 'trip' mechanism was automatically actuated and control rods were rapidly driven into the core of the reactor to shut it down. *The time was now approximately eight seconds after the initial incident.*

With the PORV open and the reactor shut down, the temperature and pressure of the primary coolant water began to fall rapidly, as intended, and the only heat now being generated in the core was that due to radioactive decay of the fission products; the so-called decay heat (200 MW or so and falling rapidly).

At about 13 seconds after the initial incident the primary coolant pressure fell below the 2205 psi value which initially caused the PORV to open, whereupon the PORV should have re-closed. Unfortunately, it failed to do so, even though an indicator lamp in the control room was indicating that it had done so. In fact, the electromagnetic actuating mechanism for the PORV had operated as intended but the PORV itself remained stuck in the open position. *Up to this point the reactor had behaved exactly as the designers intended and what happened afterwards can all be traced to the failure of the PORV to re-close and the false indication given to the control room staff.*

At about two minutes after the initial incident the coolant pressure had fallen to 1600 psi, because of loss of steam through the PORV, and the High Pressure Injection System forming part of the Emergency Core Cooling System (ECCS) was automatically actuated. At this point emergency cooling water began pouring into the pressure vessel at a rate of about 4545 litres per minute (1000 gallons per minute), causing the level of water in the pressurizer to rise. Believing that the PORV had re-closed and fearing that the pressurizer might 'go solid' (fill completely with water) and damage the reactor, the reactor operating staff turned off one of the ECCS pumps and reduced the flow from the other. The reactor was thus starved of vital cooling water just when it needed it most.

The continued loss of steam from the still-open PORV eventually caused the pressure in the reactor coolant to fall below the saturation point, allowing it to start boiling – something which normally never happens in a PWR. The coolant water displaced by the resulting steam bubbles caused the water level in the pressurizer to rise still further, increasing the fears of the operating staff that the pressurizer was about to go 'solid'. To prevent this happening they began at once to drain off some of the coolant using the reactor's 'let-down' system. Approximately 8 minutes had now elapsed since the start of the incident.

It was at about this time that someone discovered that no emergency feedwater was being circulated through the secondary circuits of the steam generators and that this was because the block valves in the feedwater lines had wrongly been left in the closed positions. The valves were at once reset to the proper 'open' positions and coolant water once again began to flow through the secondary circuit of the steam generators. *No one, at this point was aware that the PORV was still stuck in the open position and that a mixture of steam and water droplets was still flowing out of the pressurizer into the drain tank in the containment building.*

At about 11 minutes into the incident an alarm in the control room indicated that the pressurizer drain tank was overflowing into the containment building sump and that slightly contaminated water from the sump was being pumped into a waste storage tank located in the adjacent auxiliary building. Pumping continued for about half an hour, during which time about 30 280 litres (8000 gallons) of water were transferred.

Approximately one hour after the initial incident, the four primary coolant pumps began vibrating loudly because of the steam–water mixture they were being forced to circulate. By this time the decay heat had fallen to about 50 MW but boiling of the coolant continued because of inadequate pressurization. The control room staff, worried that the pumps and their associated pipework would be damaged by the vibration, decided to shut-down all four pumps! Water and steam continued to pour out of the still-open PORV as the inadequately-pressurized and below-normal volume of primary coolant now boiled even more vigorously since none was being circulated through the steam generators where the secondary coolant had recently been restored. The outcome was swift and dramatic. The primary coolant continued to boil away until eventually the top of the core became exposed and without proper cooling it rapidly overheated. The overheating caused a chemical reaction to take place between the steam and the zirconium cladding of the fuel pins and resulted in the production of large quantities of hydrogen gas, some of which escaped into the containment building via the open PORV whilst the rest remained trapped in the reactor pressure vessel.

At about 2 hours 20 minutes into the incident the operating staff realized that the PORV must be stuck in the open position and so they immediately closed a back-up valve which stopped the loss of coolant from the pressurizer. Primary coolant pressure at once began to rise when this was done but with no circulating pumps operating the water continued to boil, causing even more of the core to become uncovered as the water turned to steam. It has been estimated that about two-thirds of the core eventually became uncovered in this way and that core temperatures had reached as much as 2200°C at one time, resulting in extensive damage to the fuel assemblies.

The ECCS pumps were re-started at about 8.25 am in an attempt to make up the loss of coolant but it was almost 10.30 am (6½ hours into the incident) before the core assembly was once again fully covered. Shortly after this a small explosion occurred in the containment building, sufficient to raise its internal pressure momentarily by 28 psi (about 2 atmospheres). This transient was insufficient to cause any damage to the containment building and was caused by the build-up of hydrogen gas due to the steam–zirconium reaction on the fuel cladding and the previously open PORV. The transient did, however, initiate the spray pumps in the containment building roof space (see page 152) and they operated for about 6 minutes.

After a number of other manoeuvres designed to rid the pressure vessel of steam and hydrogen the primary coolant pumps were re-started and measurements indicated that normal coolant flow through the steam generators had been re-established. The reactor, although badly damaged, was now once more under control and the incident had been arrested. The time was now nearly 8 pm, that is about 16 hours after the initial incident.

Incidental to the efforts of the control room staff to bring the reactor under control, a site emrgency was declared at about 7 am because of high radiation levels in the reactor building. At about 7.30 am the Civil Defence in nearby Harrisburg was notified of the incident, as was the Philadelphia Office of the Nuclear Regulatory Commission (NRC) and a national emergency was declared; NRC is the US equivalent of the British Nuclear Installations Inspectorate (NII). The NRC and the Pennsylvania Department of Environmental Resources at once dispatched teams of experts to the TMI site to assess the extent of the incident and began taking measurements of radioactivity. Also present were representatives from the site owners and Babcock & Wilcox (designers of the TMI installation), plus scientists, engineers and advisors from all over the USA and a huge contingent of reporters from the world's Media.

On Friday, 30 March, the State Governer, Richard Thornburgh, ordered the closure of several schools in the area of the plant and the evacuation of pre-school children and pregnant women living within five miles of the plant. Plans were also prepared for a mass evacuation for everyone living within 20 miles of the plant, should this be thought necessary. The possibility of mass evacuation was considered because of the potential danger from the large amount of hydrogen gas which had built up in the roof area of the containment building from the previously-open PORV. It was feared that an explosion could occur which might be powerful enough to breach the integrity of the containment building seal and release radioactive gases to the environment. In the event, the amount of hydrogen present in the containment building and the likelihood of an explosion had been greatly exaggerated by the NRC and there was never any need for mass evacuation. This was admitted by Roger Mattson, director of the NRC's division of system safety who, in the *Washington Post* of 2 May 1979, said 'there never was any danger of a hydrogen explosion of that much power; we just asked the staff the wrong questions'.

President Carter visited the site on 1 April to reassure the local population and later set up a commission of inquiry into the accident under the direction of Dr John G. Kemeny, president of Dartmouth College (USA). A second inquiry was set up by the NRC and headed by Washington lawyer Mitchel Rogavin, the purpose of which was to assess the performance of the NRC previous to, during and after the accident. A third inquiry, which included eight Pennsylvania cabinet ministers and various experts, was set up by the State Governor to look into the safety, economic and legal implications of the accident. President Carter later removed Dr Joseph

Henrie from his position as Chairman of the five-man team who make up the NRC, although he remained as a commissioner.

According to NRC estimates, the maximum increase in radiation measured at ground level just outside the TMI site boundaries at the time of the accident was less than 1 mSv and that anyone who has stayed just outside the site for 24 hours per day for the whole of the week following the accident would have received a radiation dose about equivalent to that which the annual background radiation in Denver or Manhattan exceeds that in Boston or Chicago. Similar reassurance about radiation levels were given on 3 May 1979 by Secretary Califino when testifying before a Senate Committee. He said that the two million people living within 50 miles of the TMI site had been exposed to a total of 35 man-Sv, which works out to an average exposure of 17 μSv per person in the Harrisburg area – about the same as the *additional* background radiation which a resident of Manhattan or Denver receives *every week* compared with someone living in Boston or Chicago.

As far as the TMI employees working at the site during the accident are concerned, 12 received a radiation dose of about 25 mSv and three received about 35 mSv. The maximum permissible dose under NRC regulations is 30 mSv per quarter, or an annual average of 50 mSv. According to measurements made by the NRC:

- No radio-iodine was detected in any of the 130 water samples taken by the NRC, the Department of Energy (DoE) and the Commonwealth of Pennsylvania.
- No radio-iodine was detected in any of the 147 soil samples taken by the NRC and the DoE, nor in any of the 171 vegetation samples taken within 3 km of the TMI site by the NRC, DoE and the State.
- No radio-iodine was found in any milk samples taken by the State but some was detected in 9 milk samples tested by the Federal Food & Drug Administration (FFDA); the amount detected was about 0.74 Bq of I-131 per litre. This is well below the threshold of 444 Bq above which the FFDA recommends that cattle should be fed on stored feed.

    To put things in perspective, the amount of I-131 detected in milk produced by cows grazing close to the Windscale site following the 1957 fire reached levels as high as 29 600 Bq per litre, that is *40 000 times greater than those detected in the vicinity of TMI*.

In 1985 the Pennsylvania Department of Health reported that a study it had made of cancer deaths of people living in the TMI area has so far failed to provide any evidence of increases which could be attributed to the TMI accident. The study covered cancer deaths recorded from January 1974 to December 1983 within a 20-mile radius of the TMI plant, as well as cancer deaths recorded in four specific areas downwind of the plant from 1 July 1982 to 30 June 1984. The Department's study involved a careful review and analysis of epidemiological data and statistical records – including death certificates – cancer registry files, TMI census data and registry files, school census data and physician survey returns.

At the request of the TMI Public Health Fund Advisory Board, and the Citizens Advisory Panel for the Decontamination of TMI, the Department's study team also examined the results of an independent survey carried out by a local couple, Norman and Marjorie Aamodt, which claimed to have found a dramatic increase in cancer deaths which was clearly linked to the TMI accident. The team found that some cancer deaths which the Aamodts had attributed to TMI included people who had been diagnosed as cancer patients *before* the accident, as well as heavy smokers who had died of *lung* cancer, as well as people who had moved into the area *after* the accident. In one instance, the Aamodt survey included a person who had died of causes other than cancer and who had been confused with a relative who had died of cancer *before* the accident. Needless to say, the Aamodt survey received widespread Media coverage and unjustifiably frightened many people who assumed the Aamodt's knew what they were talking about.

Lessons to be learned

The Kemeny report described in detail the extent of the damage suffered by the TMI reactor and estimated that probably 40 per cent of the reactor's 38 816 fuel elements had been severely damaged, plus many of the control rods. It also expressed the opinion that the main deficiencies in the safety aspects of the TMI reactor were not associated with the reactor's hardware but with the quality of its management. Similar misgivings about the quality of management were expressed by the Rogovin inquiry which went on to say *'the NRC itself is not focused, organized or managed to meet today's needs'*. It further said *'the most damning finding of all is that the TMI affair could easily have been avoided if lessons learned in one station had been passed on without delay to all others to which they might apply'*; this was a reference to a similar incident which took place at the Davis-Besse nuclear power station at Oak Harbour, Ohio, in September 1977 and which, like TMI, was also designed by Babcock & Wilcox. That plant, however, was operating at only 263 MW (thermal) when its PORV stuck open and within 21 minutes the operators had determined the cause of the trouble and operated a back-up valve which terminated the incident. No great harm was done to either plant or personnel but the potential for a more serious accident was always there and no warning about it was ever circulated to the operators of similar stations.

The recommendations made by the many inquiries into the TMI accident are too numerous to list here; suffice to say that the recommendations have been heeded and many practical steps have already been taken, for example:

- New operating procedures have been instituted at US nuclear plants to include an additional technical advisor to each shift.
- All 2500 licensed reactor operators have made a careful study of the sequence of events which took place at TMI in the hope that this will better prepare them for coping with similar problems happening in rapid succession.
- Operating staff are receiving more training on the theory and operation of nuclear reactors.
- A Nuclear Safety Analysis Centre has been established to receive reports of minor incidents and, if thought necessary, to initiate changes in equipment, design and operating practice.
- An industry-wide supervisory group known as the Institute of Nuclear Power Operators (INPO), which is supported by more than 50 utilities, has been set up to carry out annual inspections of nuclear plant and to set standards and accreditation schemes for training programmes and instructors.

Browns Ferry

Another potentially serious accident involving a nuclear power station occurred in March 1975 at a BWR station at Decatur, Alabama, USA, known as Browns Ferry. This particular accident is of great importance, not so much because of its potential seriousness, which is undeniable, but more because it illustrates so dramatically just how easy it is for something to happen which no one would have seriously considered when designing the reactor's safety systems. Who would have thought, in the year 1975, and in a country which had demonstrated its ability to send men to the moon and back, that one of its most technologically-advanced nuclear reactors would be put out of action by a man carrying a lighted candle? And yet this is exactly what happened.

One of the plant workers, searching for an air leak in a duct, was using the flicker of a lighted candle flame to indicate the presence of a draught of air. Unfortunately, the candle started a fire which spread to the insulation of electrical cables used to control many of the station's normal and emergency control systems for the Unit-1 reactor. The resulting damage to the normal and emergency cooling systems was so severe that the correct level of water in the reactor was maintained only by resorting to the use of control-rod drive pumps which were never intended to be used in this way. Needless to say, the accident quickly led to design changes and a tightening-up of maintenance procedures in the vicinity of electrical cable runs.

Chernobyl

Without any doubt, the most serious nuclear accident of all time was that which occurred at the Chernobyl nuclear power station in the USSR at 1.23 am on Saturday, 26 April 1986. The accident, like that which occurred at Three Mile Island in the USA in 1979, was primarily caused by human error, but tragically on a much greater scale. Ironically, the errors which led to the accident took place whilst experiments were being undertaken in an attempt to improve the overall safety of the station!

Thirty-one people (firemen and station workers) lost their lives fighting fires caused by the reactor going out of control, whilst many more suffered from radiation burns and have incurred an increased likelihood of contracting cancer; there were no immediate deaths among the local population. In all, 135 000 people were evacuated from within a 35-mile radius of the station following the accident and, according to the Soviet authorities, they were examined and tended by 1240 doctors, 3660 assistants, 920 nurses and 720 medical students.

The Chernobyl power station is located in the Ukraine region of the USSR, about 60 miles north of Kiev at a place called Pripyat – in effect a worker's settlement, the nearest town being Chernobyl, a few miles away. The station comprises four identical and independent reactor units of the so-called RBMK type. The reactor involved in the accident (Unit 4) was the youngest of the four units and had been in service for little more than a year since it first went critical in 1985.

The RBMK reactor is of a design unique to the USSR and is widely used in that country; there are 28 of them in service, seven more are under construction and eight more are planned. Together they produce about 10 per cent of the electricity used in the USSR.

The RBMK reactor is classified as a light-water cooled, graphite-moderated reactor (LWGR), producing 3200 MW of thermal power and 1000 MW of electrical power. It uses graphite as a moderator (like the

British Magnox and AGR), light water as a coolant (like the American BWR and PWR) and it has individual pressure tubes, instead of a single pressure vessel, in which the coolant is allowed to boil (like the Candian CANDU). The RBMK reactor is therefore, in effect, a combination of three very different types of reactors widely used throughout the rest of the world.

The RBMK reactor contains 204 tonnes of uranium fuel, enriched to about 2 per cent U-235, in the form of small UO_2 pellets stacked inside zirconium–niobium (Zr/Nb) alloy tubes (the fuel pins). The pins are located in 1693 pressure channels, through which the boiling water coolant flows, distributed throughout 2000 tonnes of graphite blocks which form the moderator and core of the reactor. The graphite assembly measures 12 m diameter and 7 m height and operates at a temperature of 700°C. The reactor is controlled by 80 manually-operated control rods, 21 SCRAM rods, 36 overcompensation rods, 12 automatic control rods and 21 shortened absorber rods. All are fabricated from a boron–steel alloy and clad in aluminium, and all are water-cooled. Refuelling of the reactor is carried out on-load by a top-mounted refuelling machine weighing more than 100 tonnes.

The accident

Engineers from the manufacturer of the turbo-generator sets installed at the Chernobyl power station wanted to carry out an experiment to see if the inertia spinning of a recently shut-down set could usefully be used to generate electrical power to operate vital safety systems, such as emergency coolant water pumps, in the event of an emergency shut-down of the reactor. To do this they needed to lower the power of the Unit-4 reactor to about one-third normal output; this they did, but because of the build-up of xenon poisoning in the core (see page 203) the thermal power dropped to about 30 MW which was much too low. In an attempt to re-establish power, the operators withdrew the manually-operated control rods, leaving only eight in the core. This violated the station operating rules which specfically states that there should never be less than 30 control rods in the core. As if this were not enough, they also violated five other safety regulations which included shutting off the automatic control system and the automatic shut-down mechanisms. Why they were allowed to do such irresponsible things is still shrouded in mystery.

When the 'stop valves' supplying steam to the turbo-generator set forming part of the experiment were deliberately closed (so that the turbine could 'free wheel') the reactor quickly began to overheat. The shift supervisor in charge of the reactor realized within half a minute or so that things were going terribly wrong and he ordered all control rods to be driven back into the core so that the reactor could be shut down. Sadly, it was all too late and whilst the rods were actually in the process of being driven into the core the reactor ran away.

With no means of controlling the reactor the fuel rapidly overheated, causing the water coolant in the pressure tubes to flash into steam; this, in turn, caused the fuel temperature to rise even further since steam is very much inferior to water as a coolant. Within a fraction of a second the pressure tubes ruptured because of the increasing steam pressure and the Zr/Nb cladding of the fuel split open allowing volatile fission products, gases and fragments of fuel to be released into the pressure tubes. The build-up of steam pressure eventually caused a steam explosion which blew the top plate (weighing more than 1000 tonnes) off the reactor allowing the massive fuel charging machine to fall into the reactor core along with the debris of the shattered top plate. The falling debris ruptured the remaining coolant pressure tubes, causing even more steam to be released.

The reaction of the steam on the zirconium cladding of the fuel resulted in the build-up of hydrogen gas which mixed with the air and exploded. This second explosion hurled burning debris far from the reactor building and started 30 separate fires, some of which threatened nearby oil and hydrogen supplies as well as the other three reactor units. By this time, the overheated graphite moderator was burning fiercely and the reactor core was open to the sky with all its fuel split open. A huge plume of radioactive gas and debris rushed high into the sky and drifted downwind to fall on nearby towns and farmland and, eventually, on to Sweden and other countries.

It has been estimated by the Soviet authorities that about 740 000 TBq of radioactivity were released into the atmosphere on the first day of the accident and that by the eleventh day the integrated total had risen to about 1.85 million TBq. All fires had been successfully extinguished by the fifth day following the accident. The burning graphite moderator was eventually extinguished by smothering it with 800 tonnes of limestone (which reacted with the fire to produce CO_2) dropped by repeated helicopter passes over the exposed core. About 40 tonnes of boron carbide were also dropped into the core to absorb neutrons and stop the reaction of the fuel, plus 2400 tonnes of lead to absorb the radiation being released, along with 1800 tonnes of sand and clay. Eventually the whole of the reactor core was entombed in thick concrete and the undamaged reactors at the Chernobyl station were re-started.

Environmental aspects

The first indication to people living outside the USSR that a nuclear accident had occurred was two days after the accident when abnormally high levels of radioactivity were detected in Sweden. Because of the direction of the wind at that time it was not until 2 May that radioactivity from Chernobyl was detected in Britain. Airborne levels of activity in the south of England reached a peak during the afternoon of 2 May but quickly fell back to nearly normal levels the following day. Similar increased levels of radioactivity were detected in Scotland on 3 and 4 May. The most important of the radioisotopes detected were iodine-131 and 132 (I-131, I-132), caesium-134 and 137 (Cs-134, Cs-137) and lanthanum-140 (La-140).

At the request of the Department of the Environment, the NRPB coordinated all monitoring data derived from British nuclear installations (BNFL, CEGB, SSEB and UKAEA) and from NRPB centres at Chilton in Oxfordshire, Leeds and Glasgow. These stations form a network of long-established monitoring points around the British coastline and at the centre of the country. The data included radionuclide concentrations measured in air (expressed in Bq/m^3), in rainwater (expressed in Bq/litre) and deposited on grass (expressed in Bq/m^2). The Ministry of Agriculture Fisheries and Food (MAFF), and the Scottish and Welsh Offices, were responsible for assessing the information derived from measurements in milk and food-stuffs. General gamma-ray measurements were also made at many points out of doors at the various stations.

Some of the airborne activity was deposited on the ground where, in the south of England, it gave rise to concentrations of a few hundred Bq/m^2 on grass. In Wales, Scotland and the North East of England, where there had been sporadic heavy rainfall, the activity levels were much higher at several thousand Bq/m^2. Radioactivity deposited on the ground became a direct source of gamma radiation affecting people and animals and, more importantly, it appeared in foodstuffs, such as green vegetables, and in milk produced by cows grazing on contaminated grass where it became concentrated in the human body through the pathways of eating and drinking.

During 3 and 4 May, most parts of Britain were still affected by the radioactive cloud and additional rainfall caused some high levels of measured radioactivity. However, the density of the cloud had by this time become dispersed and diluted and radioactivity levels dropped.

On 20 June 1986 the British Government imposed a temporary ban on the movement and slaughter of lambs in parts of Scotland, Wales and Cumbria because lambs in these particular areas had been grazing on land which had been heavily affected by Chernobyl fallout due to heavy rainfall following the accident at a time when their growth rate was high, making their meat more radioactive than is normal. Fully-grown sheep and lambs from other areas were not affected by the ban. The ban was progressively lifted from 21 August onwards, depending upon the areas affected, because lambs gain bodyweight very quickly and the total radioactivity per kilogramme of bodyweight falls accordingly, thereby reducing its concentration and danger to the public. However, the ban was reimposed in August 1987 in selected areas because the caesium had not been washed into the soil as deeply as was originally expected and was reappearing in the growing grass and hence in the grazing sheep.

General gamma-ray measurements ranged from about 0.1 μSv/hour to 0.6 μSv/hour for the low and high rainfall areas, respectively, and compare with normal values of 0.1 μSv/hour to 0.2 μSv/hour due to natural background radiation.

The National Radiological Protection Board (NRPB) has estimated that the Chernobyl reactor accident has added about 70 μSv to the average annual effective dose equivalent over the period May 1986 to April 1987 for people living in the UK and that this might eventually reach 100 μSv per annum over subsequent years. The figure of 70 μSv represents about 3 per cent of the 2150 μSv annual dose calculated for 1986 prior to the accident. If it is assumed that *all* levels of radiation (including that from natural sources) are dangerous then it must be assumed that the extra radiation exposure to people living in the UK due to the Chernobyl accident will one day give rise to some additional cancer deaths in these people. In fact, it has been estimated – using the most pessimistic assumptions – that about 125 extra cancers (not all fatal) will occur over the next fifty years or so in the UK because of Chernobyl. This figure should be compared with the 7 million expected cancer deaths from all causes over the same period, 60 000 of which *must* be attributed to natural background radiation in the UK if *all* levels of radiation are assumed dangerous!

Approximately 40 000 people die in Britain every year from lung cancer – most of which are unrelated to radiation – and about 1300 per year from the effects of natural background radiation.

Lessons to be learnt

An astonishingly frank and detailed account of what happened at Chernobyl on 26 April 1986 was presented to an assembly of 500 nuclear experts from 45 nations when they met at the IAEA's HQ in Vienna during the week 25–29 August 1986. The presentation was made by a 23-man Soviet delegation led by Academician Valery Legosov, a nuclear chemist and deputy director of the prestigious Kurchatov Institute of Atomic Energy, near Moscow, and accompanied by a 382-page report which was submitted to member countries of the IAEA.

Professor Legosov offered explanations but no excuses for what happened at Chernobyl and said that the experiments being carried out on the reactor that night were analogous to the captain of a jetliner suddenly opening and closing exit doors and switching off vital safety systems for experimental purposes whilst the aircraft was in-flight and fully laden with passengers. In essence the Soviet delegates admitted that the RBMK type of reactor had certain design weaknesses and implied that they had much to learn from the design features of Western reactors. They also stated that all existing RBMK reactors, plus those under construction, would be modified to prevent a similar accident occurring again and that all of the 135 000 people who were evacuated from the Chernobyl area would be medically monitored for the rest of their lives.

Professor Rudolph Rometsch, the Swiss nuclear scientist who acted as Chairman during the Soviet presentation, concluded that the Chernobyl post-mortem had generated 13 further proposals for improving safety in the nuclear industry, these are:

1   An international research programme on accident sequences;
2   International collaboration on man-reactor relations;
3   A conference to explore the balance of automation versus human action so as to minimize operating errors;
4   International training and the setting up of an accreditation scheme for reactor operators;
5   A review of present training standards to take into account lessons learnt from Chernobyl;
6   Upgrading of international standards related to fire protection;
7   The formulation of international standards for acceptable levels of radioactive contamination of foodstuff;
8   Greater international collaboration on decontamination methods;
9   International collaboration on environmental modelling of radioactive pathways;
10 International collaboration to support Soviet studies of radiation exposure of people;
11 Setting up of an international workshop to agree ways of studying long-term radiation effects;
12 International co-operation on ways of treating highly-irradiated victims;

13   A study should be made of any long-term effects brought about by
      dosing populations with non-radioactive iodine.

The Oklo reactor

The story which follows is presented to end this somewhat gloomy chapter
on a lighter note and to demonstrate to the reader that the concept of the
nuclear reactor is not as novel as we have been led to believe.

In 1972 the French Commissariat à l'Energie Atomique (CEA)
announced that it had discovered the remains of a 2000-million year old
light-water reactor at Oklo, now part of the Gabon Republic in West
Africa. The discovery came about after CEA geologists had measured an
abnormally low concentration of U-235 in rock samples containing natural
uranium, that is 0.44 per cent U-235 instead of the expected 0.72 per cent.
As can be imagined, the announcement was greeted with undisguised
scepticism but following a 2-year intensive research programme the results
of 74 scientists from 20 countries were shared at an IAEA-sponsored
conference in 1975 which was convened on the floor of an open-cut mine in
Gabon alongside the remains of which everyone now acknowledges to be a
prehistoric natural reactor.

The generally-accepted view of what happened at Oklo is that about
2000 million years ago, uranium leached out of rocks by flowing water
eventually concentrated at a single waterlogged point deep underground
and built up to an amount sufficient to form a critical mass. The surrounding water, functioning as a neutron moderator, was able to sustain a
neutron chain reaction once it had been initiated with a stray neutron or
two, and the whole mass then spontaneously began operating as a nuclear
reactor, albeit at the very low thermal power of 20 kW or so.

The phenomenon which took place at Oklo could never take place in
present times because of the relatively short half-life of U-235 (700 million
years) compared with that of U-238 (4 700 million years). The concentration of U-235 in natural uranium today is only 0.72 per cent whereas at the
time of the Oklo reactor the concentration was nearer 3.5 per cent, that is
comparable with that found in the enriched fuel used in present-day
light-water reactors. As stated in Chapter 8, page 144, it is physically
impossible to sustain a nuclear chain reaction in natural uranium using light
water as a moderator because of the very high capture cross-section of such
a moderator.

The power level of the Oklo reactor was controlled automatically by the
density of the light-water moderator in which the uranium was immersed.

As the power built up, so the moderator heated up and expanded, causing it to become less dense and therefore less effective as a moderator. If the moderator had boiled away, the chain reaction would have stopped immediately because the U-235 would have lost is moderator. The natural configuration of the reactor was such that it became self-regulating.

It has been estimated that the Oklo reactor operated for betwen 200 000 and 600 000 years before the U-235 concentration fell below that which was required to sustain the chain reaction. The passage of time has since reduced the U-235 concentration of the semi-depleted Oklo uranium to the present-day level of 0.44 per cent. Had the Oklo uranium not been subjected to nuclear fission its U-235 concentration would have been the same today as it is in all other parts of the world, that is 0.72 per cent.

Appendix: The British nuclear power industry

Figure A.1 illustrates how the British nuclear industry and affiliated bodies are organized and administered. The following is a brief description of each organization and a summary of the role played by that organization in matters affecting the nuclear industry.

Department of Energy, Thames House South, Millbank, London SW1P 4QJ

The Department of Energy is responsible for all aspects of nuclear power generation and utilization of energy policy. Its Atomic Energy Division advises on nuclear energy research and development and is the focus for interdepartmental and international energy matters.

United Kingdom Atomic Energy Authority, 11 Charles II Street, London SW1 Y4QP

The UKAEA is a government-owned organization which is responsible to the Department of Energy which also appoints its Chairman. Its main role is research into all aspects of atomic energy and nuclear power, although some of its laboratories derive income from commercial undertakings outside the nuclear field; the Harwell Laboratory, for example, offsets more than half its running costs from income derived from diversification projects. The UKAEA comprises the following laboratories:

The Harwell Laboratory, Oxfordshire (the main nuclear research laboratory).

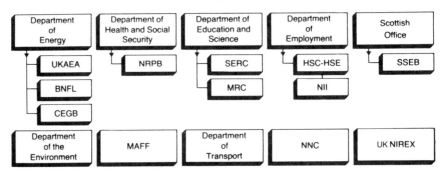

Figure A.1   Organization of the British nuclear power industry and its affiliated bodies

Risley Nuclear Power Development Establishment, Cheshire (nuclear engineering research).

Dounreay Nuclear Power Development Establishment, Caithness (fast reactor research).

Springfields Nuclear Power Development Laboratories, Lancs (nuclear fuel research).

Windscale Nuclear Power Development Laboratories, Cumbria (AGR development and fast reactor fuel manufacture).

Winfrith Atomic Energy Establishment, Dorset (thermal reactor research; also site of the 100 MWe Steam Generating Heavy Water Reactor).

Safety and Reliability Directorate, Culcheth, Cheshire (concerned with all aspects of safety and reliability associated with nuclear power).

Culham Laboratory, Oxon (fusion research; also site of the Joint European Torus (JET) project).

British Nuclear Fuels Plc (BNFL), Risley, Warrington, Cheshire WA3 6AS
BNFL was formed in 1971, having previously been part of the UKAEA. Although wholly owned by the Department of Energy, it operates as a private commercial company and provides a complete nuclear fuel cycle service to the electricity generating boards of the UK and to operators of nuclear generating stations in other parts of the world. Its principal activities are the extraction of uranium from ore concentrates, the enrichment of uranium, the manufacture of fuel for all types of reactors (British and foreign) and the reprocessing of nuclear fuel after it has been used in a reactor. BNFL operates nuclear power stations at Calder Hall (Cumbria) and Chapelcross (Dumfriesshire) and nuclear fuel plants at Sellafield (Cumbria), Capenhurst (Cheshire) and Springfields (Lancs).

Central Electricity Generating Board, Sudbury House, 15 Newgate Street, London EC1A 7AU

The CEGB is a nationalized industry responsible to the Department of Energy. It is responsible for power generation in England and Wales and is concerned with all aspects of power generation, particularly reactor technology; it operates twelve nuclear power stations with a combined net effective generating capacity of about 7600 MWe. A thirteenth nuclear power station is at present being built for the CEGB at Heysham, near Liverpool, and has a design output of about 1040 MWe.

South of Scotland Electricity Board, Cathcart House, Spean Street, Glasgow G44 4BE

The SSEB is a nationalized industry responsible to the Department of Energy. It is responsible for power generation in Southern Scotland and operates two nuclear power stations at Hunterston with a combined net effective generating capacity of 1300 MWe; a third nuclear power station is at present being built at Torness, near Edinburgh, and has a design output of about 1040 MWe.

National Radiological Protection Board, Chilton, Didcot, Oxon OX11 0RQ

NRPB is a government organization formed in 1970 and administered by the Department of Health and Social Security. Its aims are to promote information and advice on all aspects of radiological protection, including radiation hazards associated with nuclear power. NRPB has close ties with the International Commission for Radiological Protection (ICRP).

Science and Engineering Research Council, Polaris House, North Star Avenue, Swindon, Wilts SN21 1ET

SERC is administered by the Department of Education and Science. It is responsible for the support and encouragement of scientific research and education in universities and similar institutions in its own establishment and in collaboration with international organizations. SERC operates the Rutherford Appleton nuclear and astro-physics laboratory at Chilton (Oxon) and the Daresbury nuclear research laboratory (Cheshire).

Medical Research Council, 20 Park Crescent, London W1N 4AL

MRC was originally set up in 1913 and adopted its present title in 1920. Though not a government department it receives an annual grant-in-aid from parliament via the Department of Education and Science and receives funds from other government departments, industry and the Health and Safety Executive. It also receives funds from international agencies such as the World Health Organization (WHO). The function of the MRC is to promote the balanced development of medical and related biological research in Britain and to advise Government and other organizations on matters relating to medical research. MRC provides a nuclear radiation advisory service to NRPB.

Health and Safety Commission, Regina House, 259–269 Old Marylebone Road, London NW1 5RR

HSC is a government-funded organization which was set up in 1975 to administer the Health and Safety Act of 1974. It also reviews health and safety legislation and submits proposals for new and revised legislation.

Health and Safety Executive (address as for HSC)

HSE is the operating arm of the HSC. Its three-man executive is appointed by HSC to implement the Health and Safety at Work Act of 1974 and to report to the Commission.

UK Nirex Ltd, Harwell, Didcot, Oxon OX11 0RA

The United Kingdom Nuclear Industry Radioactive Waste Executive Limited (UK Nirex Ltd) was established in 1982. It is jointly owned by the CEGB ($42\frac{1}{2}$ per cent), BNFL ($42\frac{1}{2}$ per cent), SSEB ($7\frac{1}{2}$ per cent) and the UKAEA ($7\frac{1}{2}$ per cent); the Secretary of State for Energy holds a special single share which enables the Government to guarantee long-term continuity. UK Nirex is responsible for the management and disposal of most solid low-level and intermediate-level radioactive wastes.

Department of the Environment, 2 Marsham Street, London SW1

This government department has responsibility for monitoring and protecting the physical environment of the UK land mass and its approval must be obtained for the construction of any nuclear installation.

National Nuclear Corporation, Booths Hall, Chelford Road, Knutsford, Cheshire WA16 8QZ

NNC is the monopoly supplier and contractor for nuclear power stations in the UK. Formed in 1973 its aims are to promote and develop the British nuclear power industry. Shareholders of the NNC are: UKAEA (35 per cent), GEC (30 per cent), British Nuclear Associates (a consortium of 7 British companies) (25 per cent). The operating company of NNC is the Nuclear Power Company (NPC).

Nuclear Power Company Ltd, Cambridge Road, Whetstone, Leicester LE8 3HL

NPC is the operating arm of the National Nuclear Corporation. Formed in 1975, NPC is responsible for reactor design and construction and, at the time of writing, is responsible for construction of the AGR power stations being built at Torness for the SSEB and at Heysham for the CEGB.

HM Nuclear Installations Inspectorate, Thames House North, Millbank, London SW1 P4QL

NII is a branch of the HSE. Its duty in relation to all nuclear power stations under the Nuclear Installations Act of 1965 is to see that the appropriate standards are developed, achieved and maintained by the licensee, to see that the necessary safety precautions are taken, and to monitor and regulate the safety of the plant by means of its powers granted under the licence.

Ministry of Agriculture Fisheries and Food, Whitehall Place, London SW1

MAFF is responsible, among many other things, for monitoring and protecting UK rivers and the surrounding seas. Its approval must be obtained before any radioactive materials may be discharged to the sea, or to waterways which connect with the sea.

Department of Transport, 2 Marsham Street, London SW1

This government department is responsible for UK inland surface transport industries, including British Rail, the bus industry, freight and ports, the national motorway and trunk road network. Its approval must be obtained for the transportation of radioactive materials which exceed a notifiable level and for the type of container used.

Amersham International, White Lion Road, Amersham, Bucks HP7 9LL

Although not directly involved with the nuclear power industry, most of its radioisotopes are produced in nuclear reactors at the Harwell Laboratory. Its products are transported to hospitals and many other establishments throughout the UK and are exported to all parts of the world. Amersham International is subjected to the same rules and regulations which exist for all other nuclear plants in the UK.

Royal Commission on Environmental Pollution, Church House, Great Smith Street, London SW1 P3BL

This is a government-appointed body which advises on matters, both national and international, concerning the pollution of the environment and on the possibilities of danger to the environment. Its sixth report, published in 1976, was concerned with *Nuclear Power and the Environment* (Cmnd. No. 6618).

British Nuclear Energy Society, 1–7 Great George Street, London SW1 P3AA

BNES is a learned society formed from eleven constituent bodies covering all aspects of nuclear energy. It holds meetings, international symposia and conferences and is affiliated to the European Nuclear Society.

Further reading

The following list is presented for those readers who may wish to pursue the subject of nuclear power at a technical depth greater than that presented here.

Burn, D. *Nuclear Power and the Energy Crisis. Politics and the Atomic Industry*. Macmillan, 1978.
ISBN 0-333-15511-4.

Cameron, I. R. *Nuclear Fission Reactors*. Plenum Press, London. 1982.
ISBN 0-306-41073-7.

Cheshire, J. H. and Surrey, A. J. *Estimating the UK Energy Demand for the year 2000; a Sectoral Approach*. Sussex University's Science Policy Research Unit. 1978.

'Comments on the Green Paper on Energy Policy.' Energy Commission Paper No. 23. March 1979.
ISSN 0140-7996.

Cottrell, Sir Alan. *How Safe is Nuclear Energy?* Heinemann. London. 1981.
ISBN 0-435-54175-7.

The Development of Atomic Energy; Chronology of Events 1939–1984. UKAEA, London. 1984.

Digest of UK Energy Statistics. 1986. HMSO.
ISBN 0-11-412302-0
ISSN 0307-0603.

'Energy; Global Prospects 1985–2000.' Report of the Workshop on Alternative Energy Strategies (the WAES report). McGraw-Hill. 1977.

'Energy Policy: a Consultative Document.' Cmnd. 7101. HMSO. February 1978.

Environmental Impact of Nuclear Power. Proceedings of a conference held in London, 1981. British Nuclear Energy Society.

'Evidence Submitted 1974/75 to the Royal Commission on Environmental Pollution.' HMSO.

'First Report from the Select Committee on Energy. Session 1980–1981. The Government's Statement on the New Nuclear Power Programme. Volume 1: Report and Minutes of Proceedings.' February 1981. HMSO.

Flowers, Sir Brian (Chairman). 'Royal Commission on Environmental Pollution. Nuclear Power and the Environment. Sixth Report.' Cmnd 6618. HMSO. September 1976.

Glasstone, S. and Sesonske, A. *Nuclear Reactor Engineering*. Van Nostrand. New York. Third edition. 1981.
ISBN 0-442-20057-9.

Gowing, M. *Britain and Atomic Energy 1939–1945*. Macmillan. 1964.

Greenhalgh, G. *Power Tomorrow; Sizewell-B. The CEGB's Case*. Kogan Page. London. 1986.
ISBN 1-85091-173-8.

Iliffe, C. E. *An Introduction to Nuclear Reactor Theory*. Manchester University Press. 1984.
ISBN 0-7190-0953-7.

Kaye, G. W. C. and Laby, T. H. *Tables of Physical and Chemical Constants*. Longman. London. 15th edition. 1986.
ISBN 0-582-46354-8.

Leach, G. *A Low Energy Strategy for the United Kingdom*. International Institute for the Environment and Development. London. 1979.

Libowitz, G. G. and Whittingham, M. S. *Materials Science in Energy Technology*. Academic Press Inc. London. 1979.
ISBN 0-12-447550-7.

Longstaff, M. *Unlocking the Atom*. Frederick Muller. London. 1980. ISBN 0-584-10457-X.

McKay, A. *The Origins of the Atomic Age*. Oxford University Press. 1984. ISBN 0-19-219193-4.

Marshal, W. *Nuclear Power Technology*. Oxford University Press. (three volumes). 1983. ISBN 0-19-851948-6.

Nero, A. V. *A Guidebook to Nuclear Reactors*. University of California Press. London. 1979. ISBN 0-520-03482-1.

Nuclear Power and the Energy Future. Royal Institution Forum. Symposium Press. London. October 1977.

Patterson, W. *Nuclear Power*. Penguin Books. 1976. ISBN 0-14-02-1930-7.

Power Reactors in the World. International Atomic Energy Agency. Vienna. 1986 edition. ISBN 92-01-59-1861.

Roberts, L. E. J. *Nuclear Power and Public Responsibility*. Cambridge University Press. 1984. ISBN 0-521-24718-7.

Rose, J. W. and Cooper, J. R. *Technical Data on Fuel*. British National Committee, World Energy Conference. Scottish Academic Press. Edinburgh. Seventh Edition. 1977.

Semat, H. and Allbright, J. R. *Introduction to Atomic and Nuclear Physics*. Chapman and Hall. London. 1973.

Smith, C. O. *Nuclear Reactor Materials*. Addison-Wesley. London. 1967.

Taylor, R. H. *Alternative Energy Sources*. Adam Hilger. Bristol. 1983. ISBN 0-85274-476-5.

Todd, R. W. *An Alternative Energy Strategy for the United Kingdom*. Centre for Alternative Technology. Machynlleth, Powys, Wales. 1977.

Valentine, J. *Atomic Crossroads; Before and After Sizewell*. Merlin Press. London. 1985. ISBN 0-85036-336-5.

van Buren, A. *Nuclear or Not; Choices for our Energy Future*. Heinemann. 1978.
ISBN 0-435-54770-4.

White, H. E. *Introduction to Atomic and Nuclear Physics*. Van Nostrand. London. 1964.

Williams, R. *The Nuclear Power Decisions*. Croom Helm. London. 1980. ISBN 0-7099-0265-4.

'Working Document on Energy Policy for the Energy Commission.' Energy Commission Paper No. 1. Department of Energy. October 1977.

'World Energy Resources 1985–2020.' Executive Summary of Reports on Resources, Conservation and Demand to the Commission of the World Energy Conference. IPC Science & Technology Press. Guildford. 1978.

Index